D1087717

THE FRONTIERS COLLECTION

THE FRONTIERS COLLECTION

Series Editors:
D. Dragoman M. Dragoman A.C. Elitzur M.P. Silverman J. Tuszynski H.D. Zeh

The books in this collection are devoted to challenging and open problems at the forefront of modern science, including related philosophical debates. In contrast to typical research monographs, however, they strive to present their topics in a manner accessible also to scientifically literate non-specialists wishing to gain insight into the deeper implications and fascinating questions involved. Taken as a whole, the series reflects the need for a fundamental and interdisciplinary approach to modern science. Furthermore, it is intended to encourage active scientists in all areas to ponder over important and perhaps controversial issues beyond their own speciality. Extending from quantum physics and relativity to entropy, consciousness and complex systems – the Frontiers Collection will inspire readers to push back the frontiers of their own knowledge.

Information and Its Role in Nature
By J. G. Roederer

Relativity and the Nature of Spacetime
By V. Petkov

Quo Vadis Quantum Mechanics?
Edited by A. C. Elitzur, S. Dolev,
N. Kolenda

Life – As a Matter of Fat
The Emerging Science of Lipidomics
By O. G. Mouritsen

Quantum–Classical Analogies
By D. Dragoman and M. Dragoman

Knowledge and the World
Challenges Beyond the Science Wars
Edited by M. Carrier, J. Roggenhofer,
G. Küppers, P. Blanchard

Quantum–Classical Correspondence
By A. O. Bolivar

Mind, Matter and Quantum Mechanics
By H. Stapp

Quantum Mechanics and Gravity
By M. Sachs

Extreme Events in Nature and Society
Edited by S. Albeverio, V. Jentsch,
H. Kantz

**The Thermodynamic
Machinery of Life**
By M. Kurzynski

**The Emerging Physics
of Consciousness**
Edited by J. A. Tuszynski

Weak Links
Stabilizers of Complex Systems
from Proteins to Social Networks
By P. Csermely

Michal Kurzynski

THE THERMODYNAMIC MACHINERY OF LIFE

With 193 Figures and 8 Tables

 Springer

Prof. Michal Kurzynski
Adam Mickiewicz University
Faculty of Physics
Umultowska 85
61-614 Poznan, Poland
e-mail:
kurzphys@main.amu.edu.pl

Series Editors:

Prof. Daniela Dragoman
University of Bucharest, Physics Faculty, Solid State Chair, PO Box MG-11,
76900 Bucharest, Romania   email: danieladragoman@yahoo.com

Prof. Mircea Dragoman
National Research and Development Institute in Microtechnology, PO Box 38-160,
023573 Bucharest, Romania   email: mircead@imt.ro

Prof. Avshalom C. Elitzur
Bar-Ilan University, Unit of Interdisciplinary Studies,
52900 Ramat-Gan, Israel   email: avshalom.elitzur@weizmann.ac.il

Prof. Mark P. Silverman
Department of Physics, Trinity College,
Hartford, CT 06106, USA   email: mark.silverman@trincoll.edu

Prof. Jack Tuszynski
University of Alberta, Department of Physics, Edmonton, AB,
T6G 2J1, Canada   email: jtus@phys.ualberta.ca

Prof. H. Dieter Zeh
University of Heidelberg, Institute of Theoretical Physics, Philosophenweg 19,
69120 Heidelberg, Germany   email: zeh@urz.uni-heidelberg.de

Cover figure: The cover image shows a detail of a visualization of metastable molecular conformations. Courtesy of D. Baum, J. Schmidt-Ehrenberg and H.-C. Hege (Zuse Institute Berlin, www.zib.de/visual).

Library of Congress Control Number: 2006921157

ISSN 1612-3018
ISBN-10 3-540-23888-3 Springer Berlin Heidelberg New York
ISBN-13 978-3-540-23888-1 Springer Berlin Heidelberg New York

Springer is a part of Springer Science+Business Media
springer.com

© Springer-Verlag 2006
Printed in Germany

The use of general descriptive names, registered names, trademarks, etc. in this publication does not imply, even in the absence of a specific statement, that such names are exempt from the relevant protective laws and regulations and therefore free for general use.

Typesetting by Stephen Lyle using a Springer TeX macro package
Production and final processing by LE-TeX Jelonek, Schmidt & Vöckler GbR, Leipzig
Cover design by KünkelLopka, Werbeagentur GmbH, Heidelberg

Printed on acid-free paper   57/3100/YL - 5 4 3 2 1 0

To Krystyna, Ania, and Paweł

Preface

Thermodynamics was created in the first half of the 19th century as a theory designed to explain the functioning of heat engines converting heat into mechanical work. In the course of time, while the scope of research in this field was being extended to a wider and wider class of energy transformations, thermodynamics came to be considered as a general theory of machines identified with energy transducers. Important progress in biochemistry in the first half of the 20th century, and in molecular biology in the second half, made it possible to think of treating even living organisms as machines, at least on the subcellular level. However, success in applying thermodynamics to elucidate the phenomenon of life has been rather mitigated.

Two reasons seem to be responsible for this unsatisfactory situation. Nineteenth century thermodynamics dealt only with simple (homogeneous) systems in complete equilibrium. Although during the 20th century a nonequilibrium thermodynamics was developed, starting with the Onsager theory of linear response and ending with the Prigogine nonlinear theory of dissipative structures, these theories still concern the originally homogeneous systems. Because living organisms are complex systems with a historically frozen spatial and functional structure, a thermodynamics of both nonequilibrium and complex systems is needed for their description. The first goal of the present book is to formulate the foundations of such a thermodynamics.

The great advances in molecular biology in the 20th century concerned the structure but not the dynamics of biomolecules. The latter was assumed to be as fast as in small non-biological molecules, so that the following statement rooted in the conventional theory of chemical reaction rates still remains the dogma of modern biochemistry: *enzymes accelerate reactions by reducing the free energy of activation.* Only in the last two decades has more and more experimental evidence been gathered to show that the internal dynamics of biomolecules is as slow as, if not even slower than, biochemical reactions. The second goal of the book is to consider some possible consequences of this fact.

This book can be considered as an introduction to a new branch of science, a monograph and a textbook alike. It is the fruit of over a dozen years of the author's research into the statistical physics of biomolecules. No less important has been his experience obtained in lectures delivered for graduate students of biophysics and medical physics at the Faculty of Physics of the Adam Mickiewicz University, Poznań. The book is mainly concerned with theory and has been written by a theoretician, although it is addressed to all physicists and physicochemists, from graduate students to experienced researchers. The author hopes that some biochemists, molecular biologists and physicians will also take the trouble to read this book. It is assumed that the reader is acquainted with the notions of derivative, integral, ordinary differential equations, and probability theory on the level of a one-year academic course in the foundations of mathematics. Maybe some mathematicians and computer scientists interested in biological applications will also find these topics of interest.

Many theoreticians reading the book may find the formalism presented here somewhat oversimplified, whereas experimentalists will note the almost complete lack of description of experimental tools. Biochemists will criticize the selection of particular problems, while molecular biologists may find the presentation of recent crucial research insufficiently detailed. The descriptive presentation in biological terms is as a rule in disharmony with the explanatory and generalising presentation in physical terms. It is neither straightforward nor comfortable to work at the meet of several branches of science. The reader must forgive the author.

Besides Chap. 1, concerning the relationship between theory and experiment in biophysics, four problems compose the content of the book. In Chaps. 2 and 3, the nonequilibrium thermodynamics of complex systems is constructed from the very beginning, the nonequilibrium state being considered as a partial equilibrium state. In Chaps. 4 and 5, the organization of the biological cell and its main macromolecular components are reviewed and presented as a structure frozen in a historical process of life evolution. In Chaps. 6, 7 and 8, the biological processes on the subcellular level are considered as coupled chemical reactions proceeding within individual compartments of various organelles as well as transport processes across membranes. All these processes are catalyzed by specific enzymes so that particular attention is paid to the kinetics of enzymatic reactions and its control. Coupling of several reactions through a common enzyme is considered in the context of free energy transduction, the process of essential biological importance. All biological molecular machines, including pumps and motors, can be effectively considered as chemochemical machines. Chapter 9 discusses evidence for and consequences of the lack of partial thermodynamic equilibrium in the internal dynamics of biological macromolecules operating under steady-state conditions.

The book is written in such a way that it can in principle be read independently of the Appendixes. The latter are addressed mainly to physicists. Appendixes A and B require knowledge of more advanced mathematics. Appendix C, rather trivial for chemists and molecular biologists, is devoted to beginners on their first meeting with molecular biochemistry. Appendix D, the closest to the author's recent interests, presents a branch of science that has only just started to develop.

The author has appreciated discussions with many specialist scientists in various fields. His thanks go to each and every one of them. A lot of the discussions were possible through the support of the Alexander von Humboldt Foundation and several grants from the Polish State Committee for Scientific Research. Special thanks go to Jack Tuszynski for discussing the main theses and providing encouragement for the writing of this book. I am grateful to Genowefa Slósarek for a critical reading of some chapters and identification of certain elements that the reader might find difficult to understand, and Maria Spychalska for adjusting the English in several chapters.

Poznań, Poland   *Michał Kurzyński*
October 2005

Contents

1 Biophysics: An Experimental Tool of Biology or the Physics of Animate Matter?

For most biologists and physicians, biophysics tends to conjure up images of more or less complicated experimental instrumentation constructed by physicists. Nobody can deny the utility of physical equipment in biology and medicine. Progress in cytology would not be possible without optical and electron microscopy. The unraveling of metabolic pathways is to a large degree facilitated by using radioactive isotopes as tracers. The time course of particular biochemical processes can be observed by means of infrared spectroscopy. Optical spectroscopy is applicable to biomolecules that contain fluorophores, either intrinsic or externally attached. X-ray crystallography, neutron and electron diffraction techniques, not to mention high resolution nuclear magnetic resonance, have contributed to dramatic advances in molecular biology (Darnell et al., 1999; Morange, 2000). In the last decade, various techniques of single biomolecule imaging and manipulation have been developed (Ishijima and Yanagida, 2001).

However, physics is much more than a powerful experimental tool. Physical theory is equally important, because it provides the conceptual tools and language necessary for appropriate description of the relevant phenomena. It seems to the author, who has long been engaged in solid state physics research, that just this branch of physics gives an exemplary proportional representation of theory and experiment. Elementary particle physics is involved more with theory than experimental effort. Unfortunately, biophysics, especially in the area servicing the needs of molecular biology, is placed at the other extreme, where theoretical efforts have so far been very poor (Blumenfeld, 1974). The basis for molecular biology at the beginning of the 21st century is still the physical chemistry of the first half of the 20th century.

Finding new areas of application for its powerful experimental tools, molecular biology adopted almost without change the original theoretical interpretations of the techniques used, notwithstanding their different initial targets of investigation, which were not nearly as complex as the newly studied biological systems. In particular, this led to an almost universally accepted picture of biomolecules devoid of any complex internal dynamics beyond rapid vibrations around well defined 'tertiary structures'. Their impressive images are well suited to the pages of popular science magazines and introductory textbooks but are not a realistic representation of the underlying phenomena. This view was suggested, with accidental consistency, by biochemistry *in vitro* considering biological macromolecules as ordinary small molecules with vanishingly fast vibrational internal dynamics, and structural X-ray studies which assume that the difference between crystals of biomolecules and ordinary harmonic crystals can be brought down simply to the numbers of atoms in the unit cell. This simplistic picture has been particularly seductive to enzymologists, preoccupied with optimal orientations of several catalytic molecular groups in reaction transition states. In this way, they could conveniently avoid considering the dynamical properties of enzymes, simply assuming that the appropriate states present themselves as equilibrium thermal fluctuations (Fersht, 1999).

It is easy to understand the reason why elementary particle physics has oversubscribed theory. This is obviously dictated by the enormous costs associated with the construction of ever larger particle accelerators. Theoretical physics has always been cheaper than experiment. Hence it is much less clear why the theoretical basis of biophysics is so underdeveloped. The blame can probably be equally shared by biologists and theoretical physicists. The former undoubtedly displayed internal resistance to learning mathematical concepts more advanced than elementary algebra and rudimentary probability theory. The latter may have somewhat arrogantly and nonchalantly formulated various 'universal' theories of biological processes using an abstract language comprehensible only to a narrow group of experts (e.g., Fröhlich and Kremer, 1983; Davydov, 1982; Del Guidice et al., 1988). Consequently, we face an element of mutual distrust that will be difficult to overcome.

In such a situation, it will perhaps be helpful to refer to the names of a few theoretical physicists whose concepts have genuinely influenced contemporary biology. Sixty years ago, in 1944, Erwin Schrödinger, one of the co-originators of quantum mechanics, in his small book *What is life?* (Schrödinger, 1967) asked directly what was the chemical nature and the physical structure of molecules undergoing replication in the process of chromosome division. The answer to Schrödinger's question was the double helix of DNA, proposed ten years later by Crick and Watson.

In 1948 John von Neuman, another co-founder of quantum mechanics, compared a biological organism to a mechanical computing automaton, a computer, on whose theory of operation he happened to be working. This analogy (von Neuman, 1963), which identifies metabolism with hardware and the genome with software, is very deep and has been used right up to the present time. However, it ought not to be treated too literally. Recent experiments with the so-called refuse RNA are a good example. For a long time it has seemed that the structure of genomes of the eukaryotic organisms, with divided genes and intron sequences that do not code proteins, resembles the structure of information recorded on a computer disk with fragments of current program and data files divided by older, out-of-date fragments. Now, the latter analogy appears to be much oversimplified as it becomes more and more clear that some parts of RNA transcribed from introns play important regulatory functions and can in no way be treated as historical refuse (Dennis, 2002; Mattick, 2004).

The third physicist and theoretician Ilya Prigogine, a cofounder of the contemporary thermodynamics of open systems, claimed that the openness of biological organisms was the main property allowing a resistance to the inexorable destructive effects of the second law of thermodynamics (Prigogine and Stengers, 1984; 1997). Although his concept of the analogy between living organisms and dissipative structures (Nicolis and Prigogine, 1977; Prigogine 1980) cannot be treated as universal, the idea that biological processes are nonlinear and proceed in far from thermodynamic equilibrium conditions, has gained increasing approval.

The dissipative structures manifest themselves as either a temporal or a spatial self-organization, in simple cases reduced to periodic oscillations. Though the mechanism of stable temporal oscillations (openness and positive feedback) seems to be common to both animate and inanimate systems, the spatial and functional organization of living organisms on the subcellular level is considered as a frozen (historical) structure rather than a dissipative structure (see Chaps. 4 and 5). Examples of von Neumann's and Prigogine's ideas clearly indicate that the most general concepts cannot flow one way from theoretical physics to biology, but that a reverse flow is necessary. At present, both theoretical physicists and mathematicians are diligently studying the ideas of Charles Darwin.

I am tempted to mention the name of a fourth outstanding scientist, Manfred Eigen, although he would be bewildered at being called a theoretical physicist. He considers himself a chemist, or more exactly a biophysical chemist, and has inspired many crucial experiments. Nevertheless, his methodology is very close to that used in the most advanced areas of physics. He uses an advanced mathematical apparatus and co-works with theoreticians. This has so far brought about a theory of self-organization and macromolecular evolution that has given, for example, the notion of quasi-species so important in contemporary efforts to combat AIDS (Eigen et al., 1989; Eigen, 1993).

Unfortunately, the works of Eigen constitute a glaring exception against a much less bright background. There is still much to be done in the field of physical chemistry of biological macromolecules and nonequlibrium thermodynamics of biological processes. However, it seems very likely that in the near future, having constructed the essential elements of a theoretical apparatus, we will see biophysics not as an experimental tool of biology but simply as the physics of animate matter. This would be analogous to the way we view astrophysics as the physics of stellar matter. One of the tasks I set for myself when writing this book was to prepare the reader for such a change in the perception of what biophysics is becoming.

The subject of the book is biological processes occurring on the subcellular level. These are still macroscopic phenomena to be described in terms of thermodynamics, provided that special limitations are clearly stated and some assumptions made. Four points seem to be especially important and will be considered in more detail:

- The structure of the cell and metabolic pathways are kinetically frozen and constant in time. Therefore, the thermodynamics proposed to describe subcellular biological processes must be that of complex, spatially inhomogeneous systems.
- Thermodynamic nonequilibrium is treated as a partial equilibrium in which the vast majority of dynamical variables, characterizing a system on the microscopic level, have reached their equilibrium values determined by instantaneous values of some thermodynamic variables. Formally, the state of partial equilibrium differs from that of complete equilibrium by the number of thermodynamic variables. These can be either slow or constant. The slow thermodynamic variables reach their constant, complete equilibrium values in the process described deterministically by time-irreversible equations of nonequilibrium thermodynamics.
- The biological systems are thermodynamical and open, subjected to thermodynamic forces that can usually be assumed constant. The relations between the fluxes (time derivatives of thermodynamic variables) and the thermodynamic forces are, as a rule, nonlinear.

Thus the linear thermodynamics close to the complete equilibrium state, commonly used in biology, is an incorrect approximation. Besides the examples of biological clocks, the states of biological systems are stationary and do not make spatial dissipative structure as claimed by Prigogine (1980).

- The approximation of the partial equilibrium state holds only for time periods longer than the times of intramolecular relaxation. Biological systems are characterized by a distinct hierarchical organization with at least one intermediate (mesoscopic) level made by biological macromolecules with sizes from a few to a few tens of nanometers. Progress in understanding the dynamics of such macromolecules, in particular enzymatic proteins, compels one to change the traditional approach to biological processes. The intramolecular dynamics proves to be as slow as, if not even slower than, the biochemical reactions. If this is so, the dynamics should have a much greater effect on these reactions than predicted by the conventional theory of chemical reaction rates, assuming partial equilibrium. Rejection of the partial equilibrium assumption requires one to replace the conventional reaction rate constants by more sophisticated quantities, the mean first-passage times, and allows one to treat molecular biological machines transducing free energy as specially biased Maxwell demons.

Some of the topics in the present book have already been described more briefly in an earlier book (Tuszynski and Kurzyński, 2003), where biological processes occurring on the subcellular level were considered jointly with those occurring on the supracellular level of tissues and organs.

2 Statistical Description of Matter

2.1 Molecular Structure of Matter

Thermodynamics is a theory of physical phenomena proceeding on the macroscopic scale, while properties of the objects it deals with are determined by their microscopic structures. Almost ten generations of researchers have worked out the formulation of three basic statements which are presently no longer questioned. We list them, recalling the most important steps on the way toward their justification (Ochoa and Corey, 1995; Darnell et al., 1999; Morange, 2000; Ishijima and Yanagida, 2001).

Statement 1. *Matter on a macroscopic scale is composed of a huge number of molecules.*

- D. Bernoulli (1738) – combining Newtonian mechanics with probability theory, the origin of the kinetic theory of gases.
- A. Avogadro (1811) – formulation of a hypothesis that the same volumes of chemically distinct gases contain the same number of molecules.
- A.K. Krönig (1856), R.E. Clausius (1857) – complete statistical explanation of the gas laws.
- J. Loschmidt (1865) – determination of Avogadro's number to be of the order of 10^{23} molecules per mole.
- J.C. Maxwell (1866), L. Boltzmann (1872) — statistical explanation of transport phenomena.
- A. Einstein (1905), M. Smoluchowski (1906) – fluctuation theory of Brownian motion.
- T. Svedberg, J.B. Perrin (1909) – confirmation of the thermodynamic fluctuation theory from a study of diffusion in colloid solutions and emulsions.
- G. Binning, H. Rohrer and coworkers (1982, 1986) – construction of the scanning tunnel microscope and then the atomic force microscope, allowing observation and manipulation of single molecules.

Statement 2. *Molecules are composed of atoms, while atoms are composed of negatively charged electrons and positively charged nuclei.*

- J. Dalton (1805) – explanation of the chemical laws of constant and multiple ratios.
- M. Faraday (1833) – laws of electrolysis.
- D. Mendelejew, L. Meyer (1869) – the periodic system of the elements.
- J.H. van't Hoff (1874) – foundations of stereochemistry.
- S. Arrhenius (1884) – ionic theory of electrolytes.
- J.J. Thomson and E. Wiechert (1897), R.A. Millikan (1911) – determination of the mass and charge of the electron.
- E. Rutherford (1911) – discovery of the atomic nucleus.
- M. von Laue (1912), W.H and W.L. Bragg (1913) – X-ray diffraction on crystals.
- N. Bohr (1913), W. Heisenberg and E. Schrödinger (1926) – electrons and nuclei interact through the electromagnetic forces according to the laws of quantum mechanics.
- F. Hund, W. Heitler and F. London (1927) – foundations of the quantum-mechanical theory of the chemical bond.
- M. Born, J.R. Oppenheimer and successors (after 1927) – the motion of the nuclei in a given electronic state is determined by an effective 'adiabatic' potential that describes both the strong chemical bonds and much weaker non-chemical inter- and intramolecular interactions,
- several independent groups (1986) – observation of quantum processes proceeding in single atoms confined to electromagnetic or optical traps.

Statement 3. *Living organisms are also composed only of molecules.*

- F. Wöhler (1828) – synthesis of an organic compound from inorganic matter outside the living organism.
- H. and E. Büchner (1897) – protein enzymes can catalyze metabolic processes outside the organism.
- C. Embeden, O. Mayerhof, H. Krebs (1930–40) – study of metabolic pathways of the glycolysis and the citric acid cycle.
- F. Lipmann and H. Kalckar (1941) – ATP as the universal carrier of biological free energy.
- O. Avery, C. Macleod and M. McCarty (1944) – DNA as the carrier of genetic information.
- F. Sanger (1953) – sequence of amino acids in proteins is exactly determined.

- L. Pauling and R.B. Corey (1951), J. Kendrew (1957), M. Perutz (1959) – the spatial structure of proteins.
- J. Watson and F. Crick (1953) – model of DNA structure is able to explain replication and transcription.
- E. Racker (1960), P. Mitchel (1961), P.W. Boyer (1965) – mechanism of the membrane synthesis of ATP.
- F. Crick and others (1960–70) – genetic code.
- M. Eigen (1970–80) – RNA is subject to Darwinian evolution.
- P. Berg, H. Boyer, S. Cohen (1970–80) – recombinational technology of DNA (genetic engineering).
- H. Temin, D. Baltimore (1970) – discovery of the reverse transcriptase in retroviruses (priority of RNA versus DNA).
- E. Naher and B. Sakmann (1976) – patch-clamp technique enabling observation of ionic current through single protein channels.
- T. Cech, S. Altman (1983) – discovery of the autocatalytic intron splicing from RNA (priority of RNA versus proteins).
- Many independent groups (1990–2000) – various techniques of observation and manipulation of single biological macromolecules.
- Global cooperation within the Human Genome Project and, independently, the Celara Genomics Company (2001) – a 'sketch' (over 95% base pairs) of the full nucleotide sequence in human DNA.

The piece of history presented here is certainly incomplete and probably not quite representative. However, it unambiguously implies that, only at the end of the 1930s, when statements (1) and (2) were already well founded, did molecular biology finally spring up. Until recently, the limits of applicability of thermodynamics have not been clear when describing the processes proceeding in biological systems.

2.2 The Principle of Mechanical Determinism

The entire rational structure of physics is based on the *principle of mechanical determinism*, formulated over 300 years ago by Isaac Newton (Prigogine and Stengers, 1984; 1997; Newton, 1993). Using contemporary language, we can express it in the statement:

The *law of motion* and the *state* of the physical system at a given moment of time unambiguously and uniquely determine the state of this system at all other moments of time, both in the future and in the past.

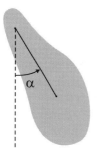

Fig. 2.1. The pendulum. A rigid body moving in a given plane under the influence of gravity. The distance between the mass centre and the rotation axis is fixed

States are characterized by appropriately long sequences of numbers. The principle of determinism endows a sense to the notion of *time* as a parameter that orders states linearly along *trajectories of motion* which are the curves in a generally higher-dimensional *space of states*.

To avoid being too abstract, let us consider a simple example. The mechanical state of a pendulum (Fig. 2.1) is determined by values of the deflection angle α and the angular velocity ω. Here, the space of states is reduced to a two-dimensional plane or, more precisely, an infinitely long strip with boundaries $\alpha = -\pi$ and $\alpha = \pi$ identified with one another. (Such a set can be imagined as the surface of a cylinder made from the strip by gluing its edges.)

Figure 2.2a shows several trajectories of the pendulum motion in the absence of friction. The uniqueness of the time evolution means that one and only one trajectory crosses each point in the state space. Locally, at each point, the law of motion determines a vector tangent to the trajectory, with length equal to the rate (i.e., the time derivative) of the state change along this trajectory (see Fig. 2.2b). The tangent vector field to the trajectories on the whole space of states is described by a certain differential equation referred to as the *equation of motion*. For the pendulum it is the system of two equations

$$\dot{\alpha} = \omega , \qquad \dot{\omega} = -I \sin \alpha , \tag{2.1}$$

where I is a constant and the dot denotes differentiation with respect to time. $\dot{\alpha}$ represents the horizontal component and $\dot{\omega}$ the vertical component of the tangent vector (see Fig. 2.2b). The course of the trajectories is found by solving this system of equations.

The two first-order differential equations (2.1) are equivalent to one second-order differential equation, involving a second time derivative. On multiplying the deflection angle α by the constant distance r between the mass center and the rotation axis, one gets the *position* $q = \alpha r$ of the mass center on a circle described by r. Twofold

(a)   (b)

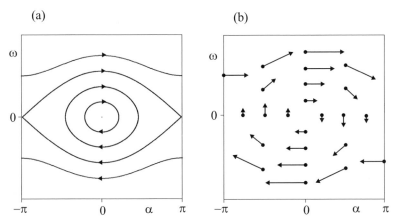

Fig. 2.2. (a) Trajectories of the pendulum in the absence of friction. For small values of the total energy, the pendulum *oscillates* around the equilibrium state $(\alpha, \omega) = (0,0)$ (trajectories with shape close to an ellipse) whereas for large values of the total energy the pendulum *rotates* with the angular velocity never falling to zero (closed trajectories crossing the line $\alpha = \pm\pi$). The stable equilibrium state $(\alpha, \omega) = (0,0)$ and also the unstable equilibrium state, the upside down state $(\alpha, \omega) = (\pm\pi, 0)$, make one-element trajectories. (b) Field of vectors tangent to the space of states shown in (a)

differentiation of this relation with respect to time, after taking into account (2.1) and multiplying both sides by the mass m, results in the expression

$$m\ddot{q} = -mrI \sin\frac{q}{r} \ . \tag{2.2}$$

It has the form of *Newton's second law*: the mass times the *acceleration* (the second derivative of position with respect to time) equals the *force* acting on the pendulum treated as an effective material point.

Generally, in classical mechanics, the motion of a system with n *degrees of freedom* characterized by n *position* coordinates $q_1, q_2, \ldots, q_n$, is determined by a system of n *Newton equations*

$$m_i\ddot{q}_i = F_i(q_1, q_2, \ldots, q_n) \ , \tag{2.3}$$

where m_i are masses related to the positions q_i, $i = 1, 2, \ldots, n$. On introducing n *momenta* p_i conjugate to the positions q_i, one can replace the system of n second-order differential equations (2.3) by the system of $2n$ first-order *Hamilton equations* (Penrose, 1979; Newton, 1993):

$$\dot{q}_i = \frac{p_i}{m_i} = \frac{\partial \mathcal{H}}{\partial p_i} \ , \qquad \dot{p}_i = F_i = -\frac{\partial \mathcal{H}}{\partial q_i} \ . \tag{2.4}$$

Thus the state of a mechanical system with n degrees of freedom is described jointly by a vector of $2n$ components $(q_1, \ldots, q_n, p_1, \ldots, p_n)$. The space of states of classical mechanics spanned by n positions and n momenta is referred to as the *phase space*, and trajectories in this space as the *phase trajectories*. A real-valued function on the phase space, the *Hamiltonian* $\mathcal{H} = \mathcal{H}(q_1, \ldots, q_n, p_1, \ldots, p_n)$ is interpreted as representing the system's total energy. The pendulum is a classical mechanical system with one degree of freedom. Its Hamiltonian is the sum of the kinetic and potential energies:

$$\mathcal{H}(q, p) = \frac{p^2}{2m} + mr^2 I \left(1 - \cos\frac{q}{r}\right) , \tag{2.5}$$

and the corresponding Hamiltonian equations (2.4) are of the form

$$\dot{q} = \frac{p}{m} , \qquad \dot{p} = -mrI \sin\frac{q}{r} . \tag{2.6}$$

In quantum mechanics the space of states is an infinite-dimensional space of *wave functions* and the equation of motion is the partial differential equation known as the *Schrödinger equation*, also involving the first derivative with respect to time t, like (2.1) and (2.4) (Newton, 1993). In the case of a system of many indistinguishable particles, the proper language is rather that of a *quantum field*. Let us stress clearly that the evolution of the wave function or, more generally, the quantum field is fully deterministic, only the process of measurement that couples the microscopic object to a macroscopic observer indicates some elements of indeterminism.

2.3 Irreversibility of Macroscopic Processes

Thermodynamic processes are also deterministic and the time evolution of physical quantities characterizing them is determined by the appropriate differential equations of motion. As an example let us consider the simple process of enzymatic catalysis *in vitro* (in a test tube) described by a system of two coupled chemical reactions (Cantor and Schimmel, 1980, Chap. 16):

$$E + R \underset{k'_-}{\overset{k'_+}{\rightleftarrows}} M \underset{k''_-}{\overset{k''_+}{\rightleftarrows}} E + P . \tag{2.7}$$

The symbol R denotes the reagent molecule, P the product molecule, E the free enzyme and M the intermediate enzyme–substrate complex.

Without the enzyme E, the reaction R $\rightleftharpoons$ P hardly proceeds at all. The enzyme is a biological *catalyst* – it takes part in the reaction and speeds it up considerably but it is not consumed, in the sense that the molecule E used in the first reaction (2.7) is recovered in the second.

For chemists, the scheme (2.7) with the *forward* and the *reverse reaction rate constants* k'_+, k''_+ and k'_-, k''_-, respectively, denotes the fulfillment of two *kinetic equations*:

$$\frac{\mathrm{d}}{\mathrm{d}t}[\mathrm{P}] = -k''_-[\mathrm{E}][\mathrm{P}] + k''_+[\mathrm{M}] \ ,$$
$$\frac{\mathrm{d}}{\mathrm{d}t}[\mathrm{M}] = -k'_-[\mathrm{M}] + k'_+[\mathrm{E}][\mathrm{R}] + k''_-[\mathrm{E}][\mathrm{P}] - k''_+[\mathrm{M}] \ . \tag{2.8}$$

The symbol whereby a compound is given in square brackets denotes its concentration in moles per dm^3. The course of each reaction in a given direction increases the concentration of one compound and decreases that of another. In the closed reactor, two conservation laws are satisfied:

$$[\mathrm{R}] + [\mathrm{P}] + [\mathrm{M}] = [\mathrm{R}]_0 = \text{const.} \ , \quad [\mathrm{E}] + [\mathrm{M}] = [\mathrm{E}]_0 = \text{const.} \ , \tag{2.9}$$

so that only two concentrations are independent variables.

Using the relations (2.9) and introducing dimensionless variables, the *molar ratios*

$$x \equiv [\mathrm{P}]/[\mathrm{R}]_0 \ , \quad y \equiv [\mathrm{M}]/[\mathrm{E}]_0 \ , \tag{2.10}$$

the dimensionless time expressed in units of the reciprocal reaction rate constant k''_+, $k''_+ t \to t$, as well as the parameters

$$a \equiv [\mathrm{R}]_0 k''_-/k''_+ \ , \quad b \equiv [\mathrm{R}]_0 k'_+/k'_- \ , \quad c \equiv [\mathrm{E}]_0/[\mathrm{R}]_0 \ , \quad d \equiv k'_-/k''_+ \ , \tag{2.11}$$

we can rewrite (2.8) in the form

$$\dot{x} = -c[a(1-y)x - y] \ ,$$
$$\dot{y} = d[b\,(1-y)(1-x-cy) - y] + [a\,(1-y)x - y] \ . \tag{2.12}$$

As in (2.1), the dot denotes differentiation with respect to time.

The time-constant equilibrium solutions to (2.12), denoted by x^{eq} and y^{eq}, satisfy the conditions $\dot{x} = \dot{y} = 0$, whence

$$y^{\mathrm{eq}} = a\,(1-y^{\mathrm{eq}})\,x^{\mathrm{eq}} \ , \quad y^{\mathrm{eq}} = b\,(1-y^{\mathrm{eq}})(1-x^{\mathrm{eq}} - cy^{\mathrm{eq}}) \ . \tag{2.13}$$

These equations relate values of the parameters a and b to values of the more directly interpretable parameters x^{eq} and y^{eq}.

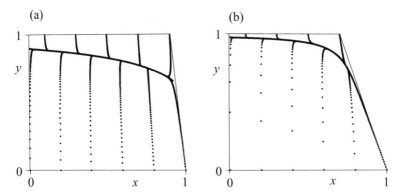

Fig. 2.3. Time course of the enzymatic reaction described by (2.12). The variable x determines the value of the molar ratio of the reaction product P, and the variable y the value of the molar ratio of the enzyme–substrate complex M. For both diagrams, it is assumed that $x^{\mathrm{eq}} = y^{\mathrm{eq}} = 0.75$ and $d = 1$. For (a) $c = 0.1$, and for (b) $c = 0.3$. Successive points are drawn for $\Delta t = 0.005$, from which the rates of particular process stages can be evaluated. (Drawings were made using the program DiGraph written by Tomasz Jarus)

The system of two differential equations (2.12) for the variables x and y is a counterpart to the system of two differential equations (2.1) for the variables α and ω. Following the conservation laws (2.9), the set of all possible states is the trapezium $0 \le y \le 1$, $0 \le x \le 1 - cy$. However, the trajectories do not circle around the equilibrium point as in the case of the pendulum, but tend to this point. Figure 2.3 shows numerical solutions to (2.12) for two different sets of values of the parameters. Note that the trajectories tend to the state of thermodynamic equilibrium in two stages. First, they reach a common trajectory, covering almost exactly the solution of the equation $\dot{y} = 0$, and then following this trajectory, although not quite along it, which is not allowed by the principle of determinism, they tend to the final equilibrium. The distinguished trajectory determines the *steady state* stage of the kinetics. We shall consider this in more detail in Chap. 7.

When compared to the determinism of mechanics, the determinism of chemical kinetics described by (2.12) is somewhat defective. The initial values of the molar ratios $x(0)$ and $y(0)$ do indeed determine the values of these ratios unambiguously and uniquely in all the *future* moments of time. However, if we try to find the values of x and y in the *past*, it may happen that what we obtain is not physically meaningful: either larger than unity or negative. The trajectories ex-

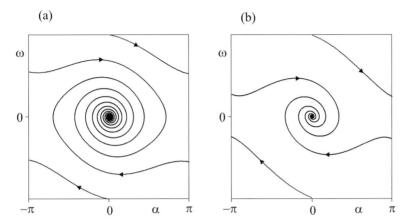

Fig. 2.4. Trajectories of a real macroscopic pendulum also tend to a unique equilibrium state. Solutions to (2.14) for $k = 0.2$ (**a**) and $k = 0.4$ (**b**). Only two trajectories are shown, starting from rotational motions in the clockwise and anticlockwise directions, respectively. (Drawings made using the program DiGraph written by Tomasz Jarus)

trapolated backward leave the space of admissible states. In contrast to (2.1), equations (2.12) are *irreversible* in time.

In general, the laws of motion on the macroscopic level are irreversible, while those on the microscopic level are reversible in time. Trajectories of a real, macroscopic pendulum are described by a system of equations

$$\dot{\alpha} = \omega , \qquad \dot{\omega} = -I \sin \alpha - k\omega , \qquad (2.14)$$

rather than (2.1), with an additional term $-k\omega$ describing *friction*. The course of the trajectories is shown in Fig. 2.4. Friction causes a continual decrease in the deflection amplitude as time goes by and a tendency toward the equilibrium state $(\alpha^{eq}, \omega^{eq}) = (0,0)$. Conversely, going back in time leads to a continual increase in the deflection amplitude until a value that can become physically meaningless.

Equations (2.1) of pure mechanics without friction and the trajectories in Fig. 2.2a are meaningful only on the microscopic level, of course, if quantum effects can be neglected. They can be applied, e.g., to describe torsional vibrations and internal rotations of the ethane molecule C_2H_6 (see Fig. 2.5) (Pauling and Pauling, 1975, Chap. 7). (As the rotational potential energy has *three* minima within the full rotation about the C–C axis, the angle α has to be replaced by 3α.) One of the tasks of statistical physics as far as we are concerned in

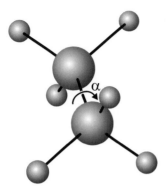

Fig. 2.5. Torsional pendulum as a model of the ethane molecule CH_3–CH_3. The bond length and angles are assumed constant. The potential energy has three minima for values of the angle α, corresponding to the position of the three hydrogen atoms bound to one carbon atom in between the three hydrogen atoms bound to another carbon atom

this chapter is to explain how macroscopic irreversibility is possible on the grounds of the reversible microscopic laws of motion of atoms and particles.

2.4 Instability of Motion as the Origin of Irreversibility

In contrast to (2.12), the equations of mechanics (2.4) not only seem to have physically sensible solutions both for $t \geq 0$ and $t < 0$, but they are in fact invariant under the transformation (Penrose, 1979; Prigogine and Stengers, 1984; 1997; Newton, 1993)

$$t, p_i \longmapsto -t, -p_i \; , \tag{2.15}$$

i.e., time reversal combined with the simultaneous reversal of all momenta. The symmetry (2.15) means that if $(q(t), p(t))$ is a solution to the set of equations (2.4), then $(q(-t), -p(-t))$ must also be its solution. It is shown schematically in Fig. 2.6a, where q and p represent the (in general) n-dimensional vectors $q = (q_1, q_2, \ldots, q_n)$ and $p = (p_1, p_2, \ldots, p_n)$. In fact, one could not distinguish whether a movie of the motion of the frictionless pendulum were projected from the beginning or from the end. In the mechanical system, not only the *reconstruction* of the past, but also *reversing* towards the past seems possible. Figure 2.6a suggests a 'journey in time' by two applications of the momentum reversal operation.

Although the property of reversibility of motion does not depend on the state space dimension, in the case of macroscopic systems composed of a huge number of molecules, this no longer seems to be the case. Let us imagine that the plane in Fig. 2.6 represents the practically infinite-dimensional space of states of some water in a pool and let state 1 represent throwing a stone into the water, whilst state 2

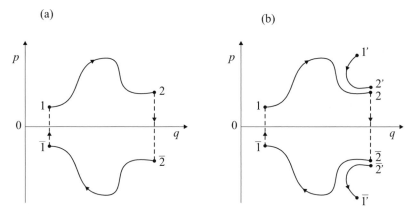

Fig. 2.6. (a) From the symmetry of the equations of mechanics it follows that, together with the trajectory segment $1\,2$, the space of states must also include the trajectory segment $\bar{2}\,\bar{1}$ which is the symmetrical reflection of the segment $1\,2$ in the plane $p = 0$. The segment $\bar{2}\,\bar{1}$ of the trajectory describes the *reverse* motion with respect to that described by the segment $1\,2$. A system which, starting from state 1, reaches state 2 after time t, will return to state 1 after momentum reversal (transition from 2 to $\bar{2}$), the free evolution to $\bar{1}$ through the next period of time t, and the renewed momentum reversal. (b) In the case of unstable motion, an infinitesimally small inaccuracy in the momentum reversal can lead the system onto a quite unexpected trajectory $\bar{2}'\,\bar{1}'$. Symmetrical to the trajectory segment $\bar{2}'\,\bar{1}'$ is the trajectory segment $1'\,2'$. This can be considered an alternative to the trajectory segment $1\,2$ when, knowing with a finite accuracy that the present state of the system is 2, one tries to reconstruct its past

corresponds to the resulting waves reaching the bank. Now the reverse process, admissible by mechanics, beginning with the waves moving back from the bank (state $\bar{2}$) and ending by ejecting the stone from the water (state $\bar{1}$), is never observed. If we see such a phenomenon in a movie we suppose immediately that it is being projected from the end to the beginning. A clear contradiction between the irreversibility of macroscopic phenomena and an expected reversibility of mechanics, referred to by Loschmidt in the 1870s as the 'irreversibility paradox', seemed to his contemporaries to be an argument against the newly-born statistical physics, whose aim was to derive the macroscopic laws of thermodynamics from the microscopic laws of mechanics.

The apparent contradiction between the irreversibility of macro-scopic phenomena and the reversibility of the laws of mechanics is explained by the asymmetric *instability* of trajectories (see Fig. 2.6b) (Penrose, 1979; Prigogine and Stengers, 1984; 1997). A stone can be thrown into water (state 1) in many different ways, getting a similar result each time (state 2). A stone ejected from the water (state $\bar{1}$) can be observed only if the initial state $\bar{2}$ is prepared with an accuracy that is practically unrealizable. A slight inaccuracy results in the system taking a completely different trajectory $\bar{2}'\,\bar{1}'$. Also an uncontrollably small inaccuracy in the reversal of momenta can lead to unpredictable results, which puts the traveler getting into the above time vehicle at serious risk.

The symmetry (2.15) implies that the practical impossibility of getting back to the past is equivalent to the practical impossibility of determining the state of this past. Two indistinguishable final states 2 and 2' can be a consequence of two different initial states 1 and 1' (see Fig. 2.6b).

Unstable mechanical systems prove to be practically irreversible, so there is no essential difference between equations (2.12) and (2.4). Just as not all values of x and y have physical meaning, not all states (q, p) in the phase space can be practically realized (with a sufficient accuracy) (Brillouin, 1964). Apart from the exact upside down state, all the other trajectories of a planar pendulum are stable only because it is a system of one degree of freedom, its phase space is two-dimensional, and continuous trajectories in such a space must be stable. Stable systems of a higher number of degrees of freedom are extremely rare. Recent investigations provide increasing evidence that mechanical systems of only a few degrees of freedom (including our planetary system, a traditional subject in the study of mechanics for over 300 years) occupy in the phase space only small regions filled with fully stable trajectories. Much more common is a partly stable motion (stable for chosen degrees of freedom) or completely unstable motion (Berry, 1978).

2.5 Statistical Ensembles.
Mixing and the Trend Toward Equilibrium

The great methodological role of the principle of mechanical determinism rests with the distinction between general *laws* (the form of the equations of motion) and individual *facts* (the data characterizing a state of a specific system at a specific time). Although we have rather well recognized laws of physics, a sufficiently accurate determination of the state of a given system poses practically insurmountable difficulties. Thus the *methodological* worth of mechanical determinism does not have to automatically imply its *prognostic* worth, that is, a practical prediction of states at any time in the future or reconstruction of states in the past.

The difficulty in determining the initial state arises for several reasons (Brillouin, 1964; Prigogine and Stengers, 1984; 1997). The first is the *complexity* of systems, which means that a complete characterization of a state requires the determination of the values of too many quantities. A cubic micrometer of liquid water or an object similar in size to the procaryotic cell contains some 3×10^{10} molecules, whose translational degrees of freedom are described by a state vector with 2×10^{11} components. The second reason is related to the already mentioned *instability* of motion. A state of an unstable system, even one of only a few degrees of freedom, must be determined with very high accuracy in order to predict its evolution over a reasonable time span. From the point of view of information processing, the description of a stable state of a complex system or an unstable system with a small number of degrees of freedom requires a practically infinite number of digits (bits).

The third reason is more fundamental in nature. In order to determine the state of a system, we need to perform a *measurement*, taking the risk of disturbing this state. The unavoidable disturbance when coupling a macroscopic measuring device with a microscopic quantum object makes it impossible to fully determine the wave function and thus to make use of the determinism implied in the Schrödinger equation. A disturbance of the system's state by the act of observation is not restricted to quantum physics. It is obvious, for example, that the more accurately a biological object is studied, the more it is affected by the measurement. The act of collecting very detailed data can eventually prove pointless because of the death of the object during their collection.

Statistical physics has found a simple way of circumventing the problem of the impossibility of knowing the initial state. Instead of a single system it considers a set (or 'ensemble') of many identical copies of the system differing only in the initial state. Thus, a single phase trajectory is replaced by a *phase flow* (Penrose, 1979). The trajectories included in the flow can be stable (when a small change in the initial state does not lead to significant differences in their course) or unstable (when a small change in the initial state brings about major consequences).

Particularly interesting for statistical physics are extremely unstable mechanical systems having the property of *mixing* the phase flow in an *exponentially fast* way (Penrose, 1979; Prigogine and Stengers,

1984; 1997). After a period of time called the *stochastization time*, the states of the set of systems showing this ability reach a practically uniform distribution over the entire domain of phase space available for motion (Fig. 2.7a). The uniform distribution of the final states is independent of that of the initial states, knowledge of which proves unnecessary. The effectiveness of the methods of statistical physics in various applications is a strong argument supporting the thesis that unstable behavior with exponential mixing is typical of systems composed of many identical particles.

In the limit of a continuum (infinite and uncountable number) of systems whose states initially fill up a certain regular region of nonzero volume in the phase space, we get a phase flow like the one shown schematically in Fig. 2.7b. Continuity of the evolution demands that the regions containing the states of the ensemble do not lose their simple connectivity, whereas the Liouville theorem, known from classical mechanics, demands that the initial volume of this region be preserved. Therefore, a result of the evolution with exponential mixing must be a region of fantastically diversified shape spreading over the entire available domain of the phase space, but without filling it (Prigogine and Stengers, 1984; 1997).

Assuming that the initial distribution of states in the statistical ensemble looks like the one shown in Fig. 2.7b (continuum of states filling a region of nonzero volume), we can make a coarse-grained description of the actual ensemble, which can at best be composed of a very high but always countable number of elements. Of course, the result of the evolution whereby the states spread uniformly over the entire available domain of the phase space for systems with exponential mixing is independent of the distribution of the initial states and whether there is a countable number of states or a continuum of them. The operation of making a coarse-grained description has a deep physical meaning, as it takes into account the uncertainty of measurements in the state determination, with each point being replaced by a cubicle or a sphere of the size corresponding to the measurement error. The operation should be made directly before each physical description of the system's state distribution. In this sense the description of the distribution of the final states of the system shown in Fig. 2.7b is definitely not physical. For mechanical systems with mixing, the operation of making the description coarse-grained does not commute with the operation of evolution. The process of making the description coarse-grained after the evolution is completed leads to a uniform distribution

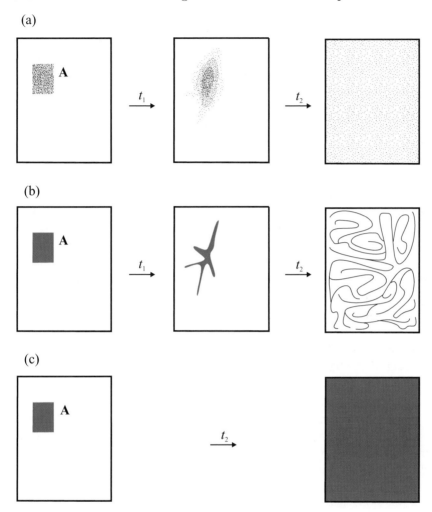

Fig. 2.7. Schematic picture of the time evolution of an initial set of states of a system having the property of exponential mixing. After a stochastization time, a set **A** spreads uniformly over the entire available domain of the phase space. Time t_1 is shorter and time t_2 longer than the stochastization time. (**a**) Set **A** is countable. (**b**) Set **A** is uncountable and has a nonzero volume. According to the Liouville theorem, the volume remains invariant during the evolution so that the spread of the set **A** over the phase space does not mean that it *covers* the entire available domain. (**c**) Coarse-graining of the set of states made both at the initial and final moment of time

of the system's states, continuously covering the entire region of the space of states available for motion (Fig. 2.7c).

The state of a statistical ensemble characterized by a uniform distribution of states of the systems making it up over the entire available domain of the space of states is called the state of *thermodynamic equilibrium* (Penrose, 1979; Prigogine and Stengers, 1984; 1997).

2.6 Probability and Entropy.
The Mechanism of Entropy Increase

The coarse-grained description of the statistical ensemble, presented schematically in Fig. 2.7, can be formally expressed using the concepts of probability theory (van Kampen, 2001, Chap. 1). Later on, we deal only with the statistical ensembles evolving according to the laws of classical mechanics. The states of the systems making up the ensemble are denoted by s and the whole space of states by $\mathbf{S}$. In general, each coarse-grained distribution of states in the ensemble is described by a certain *probability distribution density* function ρ determined on $\mathbf{S}$. The probability that a state of a system in the ensemble belongs to a given subset $\mathbf{A}$ of the phase space $\mathbf{S}$ is given by a volume integral, usually of a large number of dimensions:

$$P(\mathbf{A}) = \int_{\mathbf{A}} \mathrm{d}s\, \rho(s) \,. \qquad (2.16)$$

The probability distribution density has to be normalized to unity, i.e., the integral over the whole phase space is

$$\int_{\mathbf{S}} \mathrm{d}s\, \rho(s) = 1 \,, \qquad (2.17)$$

which means that each system is certainly in some state.

For a given probability distribution density ρ and a chosen *dynamical variable*, that is a real-valued function determined on $\mathbf{S}$,

$$\mathcal{X}(s) = x \,, \qquad (2.18)$$

and treating the function $\mathcal{X}$ as a *random variable* of the probability theory, we can define its *expectation value*

$$X = \langle \mathcal{X} \rangle \equiv \int_{\mathbf{S}} \mathrm{d}s\, \rho(s)\mathcal{X}(s) \,. \qquad (2.19)$$

The operation of relating the dynamical variables to their expectation values is linear, being an integration:

$$\langle a\mathcal{X} + b\mathcal{Y}\rangle = a\langle\mathcal{X}\rangle + b\langle\mathcal{Y}\rangle \, , \tag{2.20}$$

for any two dynamical variables $\mathcal{X}$ and $\mathcal{Y}$ and two arbitrary real numbers a and b. The difference between the dynamical variable and its expectation value

$$\mathcal{X} \equiv \mathcal{X} - \langle\mathcal{X}\rangle \tag{2.21}$$

is referred to as a *fluctuation*. From the property (2.20), it follows that the expected value of a fluctuation equals zero. A non-zero measure of a fluctuation is its *standard deviation*, i.e., the square root of the expectation value of the square of the fluctuation:

$$\sigma = \sqrt{\langle(\mathcal{X})^2\rangle} \, . \tag{2.22}$$

Let us assume that we know the probability $P(\mathbf{A})$ of a state of a given system belonging to a subset $\mathbf{A}$ of the state space $\mathbf{S}$ and let the system be subject to observation. Having found (measured) that the system is actually in a state belonging to $\mathbf{A}$ (in the probability theory the set $\mathbf{A}$ is interpreted as an *event*), the observer gains a certain *amount of information* (Brillouin, 1964) about the system investigated and the original probability $P(\mathbf{A})$ changes into certainty. The amount of information is larger as the probability $P(\mathbf{A})$ is smaller. Conversely, if $P(\mathbf{A})$ is large, the amount of information gained is small. In other words, the amount of information I is a decreasing function of P. The amount of information corresponding to the occurrence of two independent events (the joint probability is equal to the product of the corresponding probabilities) should be the sum of the amounts of information corresponding to the occurrence of the separate events. The only function of probability having all these properties is the logarithm:

$$I(\mathbf{A}) = -k \log_b P(\mathbf{A}) \, , \tag{2.23}$$

where k and b are some constants. The logarithm of unity equals zero: if the event $\mathbf{A}$ was sure to happen before the observation, $P(\mathbf{A}) = 1$, and we do not gain any new information, i.e., $I(\mathbf{A}) = 0$. The formula (2.23) was introduced by Shannon in his famous paper on information theory in 1948.

Choice of the constants k and b determines a unit of the amount of information. Assuming $k = 1$ and $b = 2$, we get one *bit*, the amount of information gained after realization of one of the two equally possible ($P = 1/2$) events represented by a *binary digit* 0 or 1:

$$I = -\log_2 \frac{1}{2} = 1 \, . \tag{2.24}$$

The amount of information contained in a number written with the help of the three decimal digits equals

$$I = -\log_2 P_1 P_2 P_3 = -\log_2 \frac{1}{9} - \log_2 \frac{1}{10} - \log_2 \frac{1}{10} = 9.28 \text{ bits} .$$

(The probabilities of the choice of a value of each digit are independent of each other, so the amounts of information are to be added; the first digit cannot be zero and assumes only 9 values whereas the remaining assume 10 values.) The amount of information contained in an arbitrary sequence of 8 binary digits is somewhat smaller:

$$I = -\log_2 \left(\frac{1}{2}\right)^8 = \log_2 2^8 = 8 \text{ bits} ,$$

i.e., 1 byte.

If one knows not only the probability $P(\mathbf{A})$ of belonging to a certain region $\mathbf{A}$ of the state space $\mathbf{S}$, but also the whole probability distribution density $\rho(s)$ on $\mathbf{S}$, then following (2.19), one can determine the *expected amount of information* to be gained by an observer after establishing the state s of the system (Brillouin, 1964):

$$S = -k_B \int ds\, \rho(s) \ln \rho(s) . \tag{2.25}$$

Equation (2.25) is identical to the expression first proposed by Boltzmann in 1872, and later, for the general case of an arbitrary statistical ensemble, by Gibbs in 1902, in order to give a statistical interpretation of *entropy*. This is the name given by Clausius in 1865 to the quotient of the bound energy (the part of the internal energy that cannot be used to perform work) and temperature. In this formula, ln stands for the natural logarithm of base $e = 2.718\ldots$, while $k_B = 1.38 \times 10^{-23}$ J/K is the value of the constant occurring in the ideal gas equation divided by the number of molecules, known as the *Boltzmann constant*. Entropy is expressed in the units of energy divided by the units of temperature.

Equation (2.25) needs some explanation. Of course, only a logarithm of a dimensionless quantity makes mathematical sense. However, the probability density is expressed in the units of the inverse volume in the phase space. As follows from the form of the Hamilton equations (2.4), the product of any generalized position and the conjugate momentum is always expressed in the units of energy multiplied by the units of time, i.e., in the units of action. The unit of volume in the phase space is the unit of action raised to a power n, where n is the number of degrees of freedom of the system. Having determined the unit of action, we can move to the dimensionless function of probability density and make mathematical sense of (2.25). The determined unit of action corresponds to the accuracy to which the observer can specify the microscopic state s of the system. This is, of course, somewhat arbitrary, but there is a natural physical limit to this accuracy given by the quantum of action, the Planck constant $h = 6.6 \times 10^{-34}$ J s.

Let $\mathbf{A}$ be a given subset of the phase space. We calculate the entropy (2.25) for the probability density to be constant in the region $\mathbf{A}$ and zero elsewhere (see Fig. 2.7b):

$$\rho(s) = \begin{cases} \Omega^{-1} & \text{if } s \text{ belongs to } \mathbf{A} , \\ 0 & \text{if } s \text{ does not belong to } \mathbf{A} . \end{cases} \qquad (2.26)$$

For the probability to be normalized to unity as in (2.17), the constant Ω must represent the volume of the subset $\mathbf{A}$:

$$\Omega = \int_{\mathbf{A}} ds . \qquad (2.27)$$

Hence, we have

$$S = -k_B \int_S ds\, \rho(s) \ln \rho(s) = k_B \int_{\mathbf{A}} ds\, \Omega^{-1} \ln \Omega = k_B \ln \Omega , \qquad (2.28)$$

so the entropy is proportional to the logarithm of the volume of region $\mathbf{A}$, or more precisely, to the logarithm of the number of units of volume contained in $\mathbf{A}$, determined by the accuracy of the state specification. In its original formulation, Boltzmann's entropy is proportional to the logarithm of the number of microstates (states of individual systems comprising the ensemble) in a given macrostate (state of the statistical ensemble).

Since according to the Liouville theorem the volume in the phase space of classical mechanics is preserved during evolution, the entropy of a statistical ensemble calculated for the probability distribution $\rho(s)$, evolving strictly according to the equations of motion, also remains unchanged (Fig. 2.7b). However, as stated earlier, only the probability distribution coarse-grained each time we wish to describe the ensemble makes physical sense. In systems with mixing, the coarse-grained probability distribution extends on a certain time scale over the entire available subset of states (Fig. 2.7c). Consequently, the entropy of the coarse-grained distribution in the state of thermodynamic equilibrium is *greater* than in the initial state. The instability of motion, hence the phenomenon of mixing, leads to the law of entropy increase in time (Penrose, 1979; Prigogine and Stengers, 1984; 1997).

2.7 The Law of Large Numbers.
Physical Realizations of Statistical Samples

It might seem that the probability, and thus also the entropy, being a measure of the information the observer had before the measurement, characterizes the observer and not the system. But the probability can be experimentally determined (Jaynes, 1978). Let us consider for example the tossing of a coin. According to the principle of insufficient reason (formulated by Daniel Bernoulli, whose father and uncle spent much time playing this game), the probabilities of getting heads or tails are the same, and each equals $P = 0.5$. However, after a series of tosses we may find that tails has appeared on average 0.6 times per toss. If the number of tosses is high enough, we conclude that the coin is non-symmetric. In further experiments we expect to toss tails with probability $P = 0.6$, so the value of the probability does indeed characterize the system studied (its asymmetry) and not the observer.

The value of probability is determined in a *statistical experiment* carried out for a sufficiently large *statistical sample*, assuming that the components of the sample are *statistically independent*, which is understood in the sense that there is no correlation between the random variables describing particular elements of the sample. Two random variables (that we identify with dynamical variables) $\mathcal{X}$ and $\mathcal{Y}$ are *non-correlated* if the expected value of their product is equal to the product of their expected values:

$$\langle \mathcal{X}\mathcal{Y} \rangle = \langle \mathcal{X} \rangle \langle \mathcal{Y} \rangle \, . \tag{2.29}$$

From the definition of fluctuation (2.21) and linearity (2.20), it follows that

$$\langle \mathcal{X}\mathcal{Y} \rangle - \langle \mathcal{X} \rangle \langle \mathcal{Y} \rangle = \langle \ \mathcal{X} \ \mathcal{Y} \rangle \, , \tag{2.30}$$

so the lack of correlation of the variables $\mathcal{X}$ and $\mathcal{Y}$ implies that the corresponding *fluctuation correlation function* equals zero:

$$\langle \ \mathcal{X} \ \mathcal{Y} \rangle = 0 \, . \tag{2.31}$$

A consequence of the condition of statistical independence is the extremely important theorem known as the *law of large numbers*. Consider a statistical sample of N identical random variables (that is, dynamical variables) $\mathcal{X}^{(l)}$ with the same expected value X and standard deviation σ:

$$\left\langle \mathcal{X}^{(l)} \right\rangle = X \, , \qquad \left\langle (\ \mathcal{X}^{(l)})^2 \right\rangle = \sigma^2 \, . \tag{2.32}$$

We assume that the variables $\mathcal{X}^{(l)}$ are non-correlated:

$$\left\langle \, \mathcal{X}^{(l)} \, \mathcal{X}^{(l')} \right\rangle = 0 \,, \tag{2.33}$$

for $l \neq l'$. The *statistical mean* or *average* of the series of variables $(\mathcal{X}^{(l)})$ is a random variable equal to their arithmetical mean:

$$\overline{\mathcal{X}} \equiv \frac{1}{N} \sum_{l=1}^{N} \mathcal{X}^{(l)} \,. \tag{2.34}$$

Linearity (2.20) implies that the expected value of the statistical mean $\overline{\mathcal{X}}$ equals the expected value X:

$$\langle \overline{\mathcal{X}} \rangle = \left\langle \frac{1}{N} \sum_{l=1}^{N} \mathcal{X}^{(l)} \right\rangle = \frac{1}{N} \sum_{l=1}^{N} \langle \mathcal{X}^{(l)} \rangle = X \,, \tag{2.35}$$

while the statistical independence (2.33) of the variables $\mathcal{X}^{(l)}$ gives

$$\left\langle (\,\triangle\overline{\mathcal{X}})^2 \right\rangle = \left\langle (\overline{\mathcal{X}} - X)^2 \right\rangle = \left\langle \left[\frac{1}{N} \sum_{l} (\mathcal{X}^{(l)} - X) \right]^2 \right\rangle \tag{2.36}$$

$$= \frac{1}{N^2} \left\langle \left[\sum_{l} \triangle\mathcal{X}^{(l)} \right]^2 \right\rangle = \frac{1}{N^2} \sum_{ll'} \langle \, \triangle\mathcal{X}^{(l)} \, \triangle\mathcal{X}^{(l')} \rangle = \frac{\sigma^2}{N} \,.$$

(In the last but one expression, in the sum of N^2 components, only N components are different from zero.) As a consequence, the standard deviation of the statistical mean $\overline{\mathcal{X}}$ from its expected value X is

$$\sqrt{\left\langle (\,\triangle\overline{\mathcal{X}})^2 \right\rangle} = \frac{\sigma}{\sqrt{N}} \,. \tag{2.37}$$

In the limit $N \to \infty$ this deviation tends to zero, so the value of the statistical mean $\overline{\mathcal{X}}$ exactly equals the expected value X:

$$\overline{\mathcal{X}} = X \,. \tag{2.38}$$

This is the thesis of the law of large numbers. The statistical mean $\overline{\mathcal{X}}$ from a sufficiently large sample is a random variable which, in contrast to the variables $\mathcal{X}^{(l)}$, shows *no fluctuations at all*.

The probability P of tossing tails in a series of Bernoulli experiments can be considered as the expected value of a dichotomous (i.e., taking only two values) random variable $\mathcal{P}$ determined by the formula

$$\mathcal{P} = \begin{cases} 1 & \text{if tails}\,, \\ 0 & \text{if heads}\,. \end{cases} \tag{2.39}$$

Indeed, equations (2.19) and (2.16) imply that $\langle \mathcal{P} \rangle$ determines the probability of the set of states **A** leading to tails. By virtue of the law of large numbers, P is equal to the mean value of $\overline{\mathcal{P}}$, hence the fraction describing the ratio of the number of favorable tosses to the number of all tosses in a large enough sample. A possible difference between the mean value obtained for an insufficiently large sample and the real value P is interpreted as an error in measurement in a statistical experiment (Jaynes, 1978).

The law of large numbers allows one to identify the expected values of the probability theory with the mean values from a sufficiently large statistical sample. In statistical physics the quantity $X = \langle \mathcal{X} \rangle$ is the *mean* or *average value* of the dynamical variable $\mathcal{X}$ and the name *expected value* is not used (van Kampen, 2001, Chap. 1). A statistical sample can be a set of *many* identical experiments performed at different times on the same *single* system or *one* experiment (measurement) performed simultaneously on *many* identical systems making up a statistical ensemble. Practically speaking, there is no problem in realizing the two types of sample to determine the probability of tossing heads or tails. (We can toss the same coin many times or toss a sack of coins once.) However, use of the two methods in the statistical description of macroscopic physical systems calls for caution.

Ludwig Boltzmann, one of the founders of statistical physics, emphasized that the main feature of a macroscopic measurement is to take the time average of a given dynamical variable measured in consecutive microscopic states of the observed system. This feature follows from the finite time of measurement and from the inertia of the macroscopic measuring device. A measurement by such a device automatically realizes a statistical experiment on a sample of the first kind. The problem is that we cannot in general expect the dynamical variables $\mathcal{X}(t)$ describing a given property of the same system at different moments of time to be statistically independent.[1] Another founder of statistical physics, Josiah Willard Gibbs, treated the macroscopic quantities as mean values obtained for a statistical ensemble, namely the one introduced in Sect. 2.5. This time the problem is that a statistical ensemble made of many copies of the same system was to its author a *concept*, an abstract idea, and not a genuinely existing object. We always deal with only one copy of a macroscopic system, so it is difficult to consider it as a subject of statistical experiments.

---

[1] The temporal evolution of the system's state defines the temporal evolution of the dynamical variable as a real function of this state.

(a)   (b)   (c)

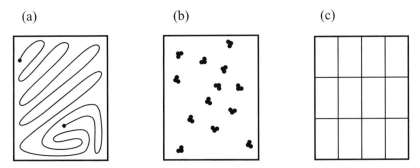

Fig. 2.8. Three physical realizations of a statistical sample. (**a**) Single copy of an ergodic system which, during the period of measurement, visits a sufficiently uniformly distributed set of states in the entire domain of the phase space available for the motion. (**b**) Ideal gas of many identical un-correlated molecules. (**c**) System of strongly interacting molecules that can be divided into many, approximately identical and practically independent many-molecule subsystems

If statistical physics is entitled to use the actual realizations of the two types of statistical sample, it is only because it deals with macro-scopic systems assumed to show the property of exponential mixing discussed in Sect. 2.5.[2] The property of mixing is mathematically very strong. First of all, it implies *ergodicity*, the equality of the time aver-age of a dynamical variable $\mathcal{X}$,

$$\overline{\mathcal{X}} \equiv \frac{1}{\tau} \int_0^\tau \mathrm{d}t\, \mathcal{X}(t) \,, \tag{2.40}$$

for a period of time τ of the order of the stochastization time, and the expected value $\langle \mathcal{X} \rangle$ defined by the formula (2.19):

$$\overline{\mathcal{X}} = \langle \mathcal{X} \rangle \,. \tag{2.41}$$

In this way the thesis (2.38) of the law of large numbers becomes valid without the need to assume a lack of correlation between elements of the sample $\mathcal{X}(t)$. Ergodicity means that, during the averaging time, the system visits a sufficiently uniformly distributed set of states in the entire available domain of the phase space (see Fig. 2.8a) (Penrose, 1979).

For over half a century, the condition of ergodicity seemed of key importance for statistical justification of the laws of thermodynam-ics, but today it is obvious that it is important only in the context of

---

[2]We should emphasize that mixing is a property of a system and not of a statistical ensemble.

the mixing condition, as only this condition ensures a trend toward thermodynamic equilibrium. Ergodicity is sufficient when considering the properties of a system that has already reached thermodynamic equilibrium, but the processes of reaching the equilibrium must be described in terms of another property of unstable systems with mixing, that is, a *time decay of correlation*: for any two dynamical variables $\mathcal{X}$ and $\mathcal{Y}$,

$$\langle \mathcal{X}(\tau)\mathcal{Y} \rangle \longrightarrow \langle \mathcal{X} \rangle \langle \mathcal{Y} \rangle , \qquad (2.42)$$

for a period of time τ of the order of the stochastization time. [In the state of thermodynamic equilibrium, $\langle \mathcal{X}(\tau) \rangle = \langle \mathcal{X} \rangle$.] The decay of correlations is a formal expression of the stochastization process (Penrose, 1979).

It should be emphasized that ergodicity or mixing can already characterize very small mechanical systems such as two rigid spheres on a billiard table, as was proved by Sinai in 1966 (Berry, 1978). Hence, the property of mixing alone is not sufficient to justify the foundations of statistical physics. Only for a *macroscopic* system composed of many identical molecules of one or several kinds can the process of stochastization end with a division of the system into many *identical* uncorrelated subsystems (Penrose, 1979). An individual macroscopic system in such a state makes a real statistical ensemble and its state can be sensibly treated as a state of thermodynamic equilibrium. (Recall that the notion of thermodynamic equilibrium refers only to a statistical ensemble or, more generally, a statistical sample.)

Each measurement of a dynamical variable in the form of a sum

$$\mathcal{X} = \sum_{l=1}^{N} \mathcal{X}_{\text{sub}}^{(l)} , \qquad (2.43)$$

where the index l numbers individual subsystems, is a statistical experiment to which the law of large numbers applies. Here, the sum on the right-hand side is not divided by the number of subsystems N, as in (2.34). The mean value of the variable $\mathcal{X}$ is nonzero and proportional to N, while its standard deviation, according to the law of large numbers, is proportional to $\sqrt{N}$. The *relative* fluctuation, defined as the ratio of the standard deviation to the mean value of a given quantity, is thus inversely proportional to $\sqrt{N}$, as in the original formulation of the law of large numbers, and tends to zero in the limit as $N \to \infty$. The density of the probability distribution of a system composed of uncorrelated identical subsystems is a product of identical functions of the probability distribution density ρ_{sub} defined for the states s_l of individual subsystems:

$$\rho(s) \equiv \rho(s_1, \ldots, s_N) = \rho_{\text{sub}}(s_1) \cdots \rho_{\text{sub}}(s_N) . \qquad (2.44)$$

The density of the probability distribution ρ in the form (2.44) is unambiguously determined by the density ρ_{sub} of the probability distribution of the subsystems, and it is all the same whether we speak about one or the other.

The type of subsystem into which a given macroscopic physical system is decomposed depends on the character of the system, its internal structure, and the strength of internal interactions. The simplest system of statistically independent subsystems is the ideal gas, that is, an ensemble of many identical particles whose interactions are negligibly small (Fig. 2.8b).[3] Despite its simplicity, the statistical physics of both classical and quantum ideal gases explains many nontrivial phenomena. Unfortunately, this model is not able to describe certain properties of the condensed phase, e.g., short-wavelength collective excitations or phase transitions, related to a spontaneous ordering of the system in which the interactions between particles and their mutual correlations are essential (Chandler, 1987). In such circumstances, a commonly adopted approach is to divide a given system into many, approximately identical parts, as shown in Fig. 2.8c. (Considering its arbitrary character, the division should be made in mind rather than in reality.) If the parts are sufficiently large, the correlations and interactions on their boundaries can be neglected, as in the ideal gas. An exception is the critical state of the system in conditions close to a continuous phase transition, when the correlations become infinitely long-range and the standard deviations infinitely high (see Appendix A.5).

All three physical realizations of a statistical sample can be used in the thermodynamic description of processes taking place in a living biological cell. However, because of the lack of spatial uniformity, the problem is not so simple as in the thermodynamics of abiotic systems, and we shall discuss it again several times.

---

[3]If the particles are indistinguishable, the elementary volume of the phase space, determining the accuracy of determination of the state of a macroscopic system by an observer [see the comment concerning the definition of entropy (2.25)] must comprise all $N!$ permutations of the elementary volumes with a fixed numbering of particles (Chandler, 1987).

2.8 The Relativity of Thermodynamic Equilibrium

The process whereby a macroscopic system reaches thermodynamic equilibrium does not have to be monotonic in time. It usually takes place in stages and is characterized by several stochastization times that can be significantly different. Thus the concept of thermodynamic equilibrium is not absolute and depends on the time scale on which a given phenomenon is considered (Haken, 1990; Prigogine and Stengers, 1984; 1997).

For example, let us consider the process in which a cup with hot coffee to which a small amount of cream has been added moves towards thermodynamic equilibrium (Palmer, 1982). On a time scale of a few *minutes*, the process of spontaneous mixing (called diffusion by physicists) will take place as a result of which, to begin with, the concentration of the cream in the coffee reaches equilibrium. Over a longer time – of the order of one *hour* the temperature of the coffee and the surrounding air will reach the same value. However, a cold coffee in a cup is not yet a system at full thermodynamic equilibrium. If the surrounding air is dry enough, water from the coffee must evaporate on a time scale of a few *days* (equilibrium is reached between the liquid and the gas phase). But complete equilibrium has still not been reached, as the porcelain is not everlasting and will eventually undergo sublimation or decomposition into dust.

The above example illustrates the *hierarchism* in the process of reaching thermodynamic equilibrium. The origin of the hierarchy of stochastization times are bottlenecks in the phase space (Fig. 2.9). A division of the space of states into subsets separated by bottlenecks follows from a specific organization of the system on the macroscopic or microscopic scale, which we shall discuss in the next chapter.

On a practical level, we can distinguish *fast*, *slow* and *very slow* processes for each time scale. By definition, *nonequilibrium thermodynamics* deals with the phenomena involved in reaching thermodynamic equilibrium by physical quantities determined by slow processes. The stochastization times corresponding to these processes are known as *relaxation times*. On the time scale of slow processes, the physical quantities determined by fast processes are already in a thermodynamic equilibrium, called *partial equilibrium*, while the physical quantities determined by very slow processes are *frozen* (Palmer, 1982).

Structures in frozen nonequilibrium are commonplace. For example, a certain amount of carbon, hydrogen and oxygen atoms taken in the proportion 1:2:3 can occur in the form of a diamond covered

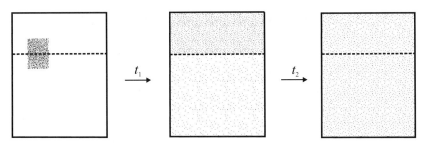

Fig. 2.9. As a consequence of the occurrence of a bottleneck (*broken line*) between two subsets in the space of states, the process of stochastization takes place in two stages: at first within the subsets (time scale t_1), and then between the subsets (time scale t_2)

with water in a pure oxygen atmosphere, in the form of cellulose (the main component of paper) also in a pure oxygen atmosphere, or as a mixture of gaseous carbon dioxide and steam. Only the latter form is in complete thermodynamic equilibrium, whereas the first two are, in normal conditions, frozen structures. In fact, the atomic composition of matter in the Universe, determined by the ratio of the total number of neutrons to protons, is also frozen. Neutrons and protons are the high- and low-energy states of the same particle known as a nucleon. At the mean temperature of the Universe, only 3 degrees above absolute zero, practically all nucleons should occur in the low-energy state, that is as protons. Thus, the Universe should be composed only of hydrogen, and the fact that there occur also carbon, nitrogen, oxygen and some other elements as well, proves the presence of a deep nonequilibrium. The frozen structures store a memory of the time of their formation. The ratio of neutrons to protons was established at an early stage in the evolution of the Universe, when the density of neutrinos became too small to effectively influence nucleon transformations. The diamond remembers the high pressure in the depth of the Earth's crust, and cellulose, the process of biosynthesis in a plant cell.

In certain states of matter, the stochastization times do not determine well-separable scales but form a more or less continuous spectrum. In such situations, only the time of observation allows a distinction as to whether a given structure is frozen or not. Two states of this type, the glassy state and the already mentioned critical state, have been for many years the subjects of intense study. It has been shown that biological matter on some time scales reveals many properties of the glassy state (Appendix D.3) (Frauenfelder et al, 1991; 1999).

In terms of nonequilibrium thermodynamics, the process of reaching a complete equilibrium is treated as passing through a series of states of partial equilibrium. It is not concerned with evolutionary processes on a time scale shorter than the time of stochastization to a state of partial equilibrium. By virtue of the ergodic theorem, valid for systems with mixing, the average over a statistical ensemble is equal to the average over the time of stochastization. Therefore, it should be emphasized that, at the stage of reaching a state of partial equilibrium, a given system does not have to be an actual ensemble of statistically independent subsystems.[4] Only after the stochastization time is the system, if macroscopic, divided into such subsystems, thanks to another property of systems with mixing, the decay of correlations. In the thermodynamic description of macroscopic systems, both temporal and spatial averages are taken, as only the additive dynamical variables in the form of a sum (2.43) over all component subsystems are considered. Only a few of the dynamical variables survive such a summation without vanishing. The mutually independent nonzero sums thus obtained have the physical meaning of *thermodynamic variables* and unambiguously characterize the *thermodynamic state* of a given system (Callen, 1985).

By assumption, the subject of nonequilibrium thermodynamics is only macroscopic systems. In order to note in them something more than a set of atoms or molecules, they must be viewed from a certain temporal and spatial perspective.

---

[4]Even in the ideal gas, the process of reaching equilibrium must be described taking into account not only the interaction but also correlations between the velocities of individual particles (see Appendix B.1).

3 Thermodynamic State

3.1 Global and Structural Thermodynamic Variables

Macroscopic physical processes are slow on the molecular time scale and extended on the molecular distance scale. The thermodynamic description carries out averaging both over time and over a set of statistically independent identical subsystems into which the system breaks up as a result of the disappearance of spatial correlations as time goes by. Only few dynamical variables survive such averaging without being canceled out. They have the form of the sum (2.43) over the component subsystems and are called *thermodynamic variables*.

According to the law of large numbers, means taken over a sufficiently large statistical ensemble do not undergo fluctuations. As a consequence, the thermodynamic variables can be identified with these means. Values of a complete set of thermodynamic variables uniquely characterize the *thermodynamic state* of a system. By definition, the thermodynamic state is, on a given time scale, a state of thermodynamic equilibrium. The values of the thermodynamic variables should thus remain constant on this time scale. Two kinds of such constants are known in physics (Newton, 1993).

Constants of the first kind result from a *continuous symmetry* of the microscopic equations of motion. *Energy* is constant due to invariance of the equations of mechanics with respect to translations in time. *Momentum* is constant due to invariance of these equations with respect to translations in space, and *angular momentum* is constant due to their invariance with respect to rotations. Only the first quantity is important in thermodynamics as it is always possible to choose a reference system so that both the momentum and the angular momentum of the system are equal to zero. Constancy of *electric charge* results from a more abstract *gauge symmetry* of the equations of electrodynamics.

Constants of the second kind are related to a *spontaneous breaking* of continuous symmetry during, e.g., a phase transformation (see Ap-

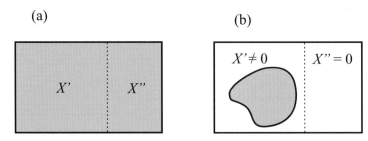

Fig. 3.1. Division of simple (**a**) and complex (**b**) thermodynamic systems into two subsystems. In the case of a simple system, the values of a thermodynamic variable characterizing a given subsystem (X' for the first and X'' for the second) are always proportional to the size of these subsystems. However, in the case of a complex system, it may be arbitrary. (**b**) presents a situation where the variable considered has a nonzero value only in the selected subsystem with internal constraints

pendix A.1). Spontaneous condensation fixes the value of the *volume*, whereas spontaneous solidification, forbidding individual molecules to leave their specific positions, determines the value of the *length* and other parameters that describe the *shape* (*form*) of the solid. Spontaneous orientational ordering of component entities determines the values of electric or magnetic moments which, when related to a unit volume, are called the *polarization* or *magnetization* of a system, respectively. Spontaneous breaking of some more abstract symmetries determines the values of certain discrete quantities that characterize the amount of matter, e.g., the number of atoms of a particular kind (constant in the absence of nuclear reactions) or the number of molecules of a particular kind (constant in the absence of chemical reactions).

The fact that the means over an ensemble are not canceled out is a consequence of the principle of *additivity* of thermodynamic variables. If a system is composed of two subsystems, primed and double primed (see Fig. 3.1), the variables characterizing the whole system are sums of the values of the variables in the two subsystems:

$$X = X' + X'' . \tag{3.1}$$

The division of the system into subsystems can be arbitrary; either real or devised in a physical space or in an abstract space of internal states. The only condition that must be met is that of the macroscopic character of the subsystems, as otherwise their description in terms of the thermodynamic variables would be meaningless.

(a)

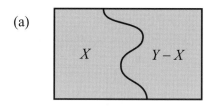

(b)

$$
\begin{array}{c}
\text{H} \quad \text{O} \\[-2pt]
\vert \quad \Vert \\[-2pt]
\text{R}-\overset{\displaystyle}{\text{C}}-\text{C}-\text{R}' \\[-2pt]
\vert \\[-2pt]
\text{H} \quad ketone
\end{array}
\quad \rightleftharpoons \quad
\left[
\begin{array}{c}
\text{H}\cdots\text{O} \\[-2pt]
\vdots \quad \vdots \\[-2pt]
\text{R}-\text{C}\!=\!\text{C}-\text{R}' \\[-2pt]
\vert \\[-2pt]
\text{H}
\end{array}
\right]
\quad \rightleftharpoons \quad
\begin{array}{c}
\text{H}-\text{O} \\[-2pt]
\vert \\[-2pt]
\text{R}-\text{C}-\text{C}-\text{R}' \\[-2pt]
\vert \\[-2pt]
\text{H} \quad alcohol
\end{array}
$$

Fig. 3.2. (a) Complex thermodynamic system in which internal constraints are illustrated symbolically as a wall dividing the system into two parts. The thermodynamic state of the system is specified by the values of two thermodynamic variables: global Y and structural X. The structural variable X specifies the state of one of the subsystems. The value of its counterpart for the other subsystem can be obtained using the additivity property (3.1). (b) Example of the chemical reaction of isomerization. A molecule of a compound designated R–C_2H_2O–R' (R and R' denote two arbitrary groups of atoms) may exist in two distinct chemical states (isomers) as a *ketone* or an *alcohol*. A transition of the molecule from one chemical state to another through a *transition state* (the structure in brackets) corresponds to the transition of a molecule through the semipermeable partition as in (a). In this case, the total number of molecules corresponds to the global variable Y, while the number of molecules of a given isomer corresponds to the structural variable X. During the reaction, the value of the variable X changes until it reaches its equilibrium value when the flux of molecules moving from left to right is balanced by the flux of molecules moving in the opposite direction

If the value of each variable characterizing a subsystem is proportional to the size of this subsystem (i.e., each variable is *extensive*) the whole system is said to be *simple* (Fig. 3.1a), otherwise it is referred to as *complex* (Fig. 3.1b). It is usually more convenient to speak about *constraints* on the motion rather than more or less abstract symmetries. Simple systems are uniform and restricted only by *external* constraints determining the values of *global* variables characterizing the system as a whole. Complex systems have an inner structure imposed by the presence of *internal* constraints determining the values of additional *structural* thermodynamic variables.

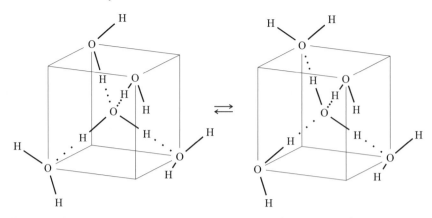

Fig. 3.3. Change in the electric dipolar moment (polarization) of a cluster of five molecules of water requires reorganization of the hydrogen bond system, just like a chemical reaction that requires reorganization of the covalent bond system (see Fig. 3.2b)

The internal constraints can be represented by a wall that divides the system into two parts (Fig. 3.2a). The wall is rigidly fixed and impermeable to molecules, and conducts neither energy nor charge. In order to characterize the system's thermodynamic state, one has to determine the values of the thermodynamic variables separately for each part or, according to the property of additivity (3.1), for the whole system and one chosen subsystem. A real wall can be purposefully constructed to work only to a certain degree as a constraint. A wall which is not rigidly fixed becomes a movable piston, and a wall incompletely impermeable to molecules becomes a semi-permeable partition (membrane). A diathermal wall conducts heat, while one made of a charge conductor conducts electrons.

An electrical conductor does not necessarily require the structure of a three-dimensional wall, and a semi-permeable partition does not have to be a macroscopic entity, but can appear on a microscopic scale, in the space of internal states of the molecules forming the system. Such a microscopic partition occurs for example in chemical reactions. Let us consider a simple unimolecular reaction of isomerization (Fig. 3.2b). The continuum of the states characterizing internal degrees of freedom of the molecule (lengths of particular bonds, planar valency angles and dihedral angles of rotation about the bonds, considered jointly with the conjugate momenta) corresponds to the inside of the whole box, while the two chemical states correspond to the two parts of the box. Passage through the partition corresponds to a reaction – conversion from one chemical state to another through the chemical transition state (structure in brackets). The partition symbolizes the bottleneck (see Fig. 2.8), a barrier of partly entropic nature (the lower number of effective degrees of freedom of the transition state following from strict orientation of particular molecular groups) and partly energetic nature (temporary cleavage of two covalent bonds).

Constraints that fix the value of the polarization or magnetization have a similar nature. In the condensed phase, molecules can usually occur in a small number of discrete orientational states of electric or magnetic dipolar moments. For instance, a water molecule linked by four hydrogen bonds to four neighboring molecules can assume six different orientations within a tetrahedron formed by these neighbors (Fig. 3.3). A change in the molecule orientation requires a transient breaking of hydrogen bonds, just as the chemical reaction requires a transient breaking of covalent bonds. Discrete states of magnetic moments result directly from their quantum nature.

On a short time scale, imperfect constraints behave as perfect ones and determine the values of appropriate thermodynamic variables, while on a longer time scale, the actual imperfections in the constraints determine the rates of change in the values of these variables. The imperfect structural constraints can be more or less fictitious. Even in simple systems in the absence of any structure, the processes of reaching complete equilibrium do not take place in the whole space occupied by the system simultaneously, but develop gradually, starting in small areas and then in larger ones, as if there were partitions dividing them. The reason for the slow rate of the changes is their spatially extended character. According to the principle of the continuity of the phenomena, at a constant rate of interaction propagation, the larger the magnitude, the lower its rate. This hierarchic description of nonequilibrium processes is offered by the thermodynamics of continuous media, and in particular hydrodynamics, which we shall not discuss in this book.[1]

3.2 The Clausius Entropy

It follows from the considerations in Sect. 2.7 that a macroscopic system can be treated as a real statistical ensemble only after it has reached a thermodynamic state. Only in such a state do the probability and thus the entropy make physical sense, i.e., only then are they

---

[1] In the thermodynamics of continuous media, the notion of *partial* equilibrium is replaced by that of *local* equilibrium.

measurable. Because the thermodynamic state is uniquely determined by the mean values of an appropriate set of thermodynamic variables, the corresponding probability distribution is also uniquely determined by this set (see the more detailed considerations in Appendix A.4). As a consequence, the entropy S in the thermodynamic state is not only a function of the probability density, but a direct function of the thermodynamic variables

$$S = S(E, X_1, \ldots, X_n) = S(X_0, X_1, \ldots, X_n) \,. \tag{3.2}$$

A distinguished variable is energy. The number of thermodynamic variables other than energy n is referred to as the number of *thermodynamic degrees of freedom* of the system. The function (3.2) was introduced into thermodynamics at the phenomenological level by Clausius in 1865, and we thus call it the *Clausius entropy*, in contrast to the statistical *Boltzmann–Gibbs entropy* (2.25), being in general only a function of the probability density. The Clausius entropy (3.2) contains all the information about the thermodynamic states of the system. The irreversible progression toward the complete equilibrium state is closely linked to the maximization of its value.

The Clausius entropy satisfies four universal postulates[2] that follow from experimental phenomenology and are well justified on the basis of the statistical interpretation of thermodynamics.

Zeroth Postulate. The entropy S exists for each thermodynamic state. It is a continuous and at least doubly differentiable function of the thermodynamic variables and is additive like them:

$$S(E' + E'', X_1' + X_1'', \ldots, X_n' + X_n'') \tag{3.3}$$
$$= S'(E', X_1', \ldots, X_n') + S''(E'', X_1'', \ldots, X_n'') \,,$$

where S' and S'' are the entropy functions of the subsystems [see (3.1) and Fig. 3.1].

First Postulate. The entropy S is a strictly increasing, i.e., monotonic function of energy E:

$$\left(\frac{\partial S}{\partial E} \right)_{X_1, \ldots, X_n} > 0 \,. \tag{3.4}$$

---

[2]The numbering of the postulates corresponds to the historical numbering of the 'laws of thermodynamics' (Kondepudi and Prigogine, 1998). The formulation of the postulates presented here is a modification of the more modern formulation due to Herbert B. Callen in 1960 (Callen, 1985).

(In this and further formulas, the lower subscript on brackets around a partial derivative indicates those parameters that are held constant on differentiation.)

Thanks to the monotonic character of the dependence of S on E, it can be inverted to give the dependence of the energy on the entropy and other thermodynamic variables:

$$E = E(S, X_1, \ldots, X_n) \,. \tag{3.5}$$

The entropy can be treated as a thermodynamic variable describing on the macroscopic scale the myriads [a picturesque expression used by Callen (1985)] of microscopic dynamical variables which disappear as a result of statistical averaging and have no direct influence on the distinguished variables $X_1, \ldots, X_n$.

Second Postulate. The entropy S is a convex function of all its arguments:

$$\left(\frac{\partial^2 S}{\partial E^2}\right)_{X_1, \ldots, X_n} < 0 \,, \qquad \left(\frac{\partial^2 S}{\partial X_i^2}\right)_{\ldots, X_{i-1}, X_{i+1}, \ldots} < 0 \,, \tag{3.6}$$

for $i = 1, 2, \ldots, n$, but in general this function is not monotonic. In conditions when $E = $ const. and in the absence or weakening of the constraints determining the values of variables X_i, a system initially in a state of partial equilibrium spontaneously evolves to the state of complete equilibrium determined by those values of the distinguished variables for which the entropy S reaches a maximum.

The convex character of the function S implies that, in conditions when $S = $ const. instead of $E = $ const. (e.g., for purely mechanical systems with $S = 0$), the energy E spontaneously takes a value that minimizes it as a function of the distinguished variables X_i (Fig. 3.4). It should be emphasized that the form of (3.2) does not indicate the path to the complete equilibrium state.

Third Postulate. There is a state for which

$$\left(\frac{\partial E}{\partial S}\right)_{X_1, \ldots, X_n} = 0 \,. \tag{3.7}$$

The value of the entropy in this state is zero $(S = 0)$.

(a)   (b)

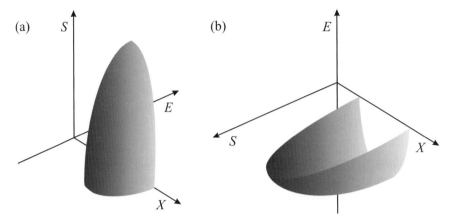

Fig. 3.4. Sketch of the dependence of entropy S on energy E and a thermodynamic variable X whose value is not fixed by constraints (**a**). The one-to-one entropy–energy dependence allows the energy E to be treated as a function of the entropy S and the variable X (**b**). In the plane E = const., the entropy S reaches a maximum as a function of X, whereas in the plane S = const., the energy E reaches a minimum as a function of X. For clarity only the range of negative energy values is considered

3.3 Temperature and Thermodynamic Forces. Equations of State

The zeroth and second postulates make it possible to define the important concepts of *thermal* and *dynamical* equilibrium (in particular mechanical, electrical and chemical) and the related physical quantities: *temperature* and *thermodynamic forces*. In order to do this we shall consider a system composed of two subsystems and for the sake of simplicity we shall distinguish only one global thermodynamic variable X other than energy E (Fig. 3.5). The same reasoning will hold for a system with an arbitrary number of degrees of freedom, either simple or complex.

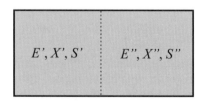

Fig. 3.5. Thermodynamic system composed of two subsystems. Walls represent internal and external constraints

Assume that the values of variables X' and X'' for the subsystems are determined by the constraints

$$X' = \text{const.} , \qquad X'' = \text{const.} , \tag{3.8}$$

but that the subsystems can exchange energy (they are separated by a *diathermal*, heat conducting wall) in such a way that

$$E' + E'' = E = \text{const.} , \tag{3.9}$$

with the whole system being isolated. According to the second postulate satisfied by the Clausius entropy, the energy E' spontaneously takes such a value that the total entropy of the system

$$S(E' + E'', X' + X'') = S'(E', X') + S''(E'', X'') \tag{3.10}$$

reaches a maximum, and its derivative with respect to E' therefore vanishes:

$$\frac{\partial S}{\partial E'} = \left(\frac{\partial S'}{\partial E'}\right)_{X'} + \left(\frac{\partial S''}{\partial E''}\right)_{X''} \frac{\mathrm{d}E''}{\mathrm{d}E'} = 0 . \tag{3.11}$$

From (3.9), we have

$$\frac{\mathrm{d}E''}{\mathrm{d}E'} = -1 , \tag{3.12}$$

and hence a necessary condition for the maximum entropy is

$$\left(\frac{\partial S'}{\partial E'}\right)_{X'} = \left(\frac{\partial S''}{\partial E''}\right)_{X''} , \tag{3.13}$$

or taking advantage of the one-to-one relationship between S and E,

$$\left(\frac{\partial E'}{\partial S'}\right)_{X'} = \left(\frac{\partial E''}{\partial S''}\right)_{X''} . \tag{3.14}$$

From the point of view of experimental phenomenology, the subsystems exchange energy until their *temperatures* become equal,

$$T' = T'' , \tag{3.15}$$

and energy is given by the subsystem with the higher initial temperature. The condition of *thermal equilibrium* (3.15) is equivalent either to (3.13) or (3.14). (More complex functions of derivatives occurring in these equations are irrelevant here since they lead to more complex equations.) The realization of a particular possibility is determined by the sufficient condition for a maximum entropy which demands that, immediately before reaching a maximum, the entropy increment

$$\Delta S = \frac{\partial S}{\partial E'} \Delta E' = \left[\left(\frac{\partial S'}{\partial E'} \right)_{X'} - \left(\frac{\partial S''}{\partial E''} \right)_{X''} \right] \Delta E' \qquad (3.16)$$

should be positive. Assuming that the primed system gives energy ($\Delta E' < 0$), the expression in brackets must be negative. If the temperature T is defined by

$$T \equiv \left(\frac{\partial E}{\partial S} \right)_X , \qquad (3.17)$$

this implies that $T' > T''$, so the primed system is indeed warmer than the doubly primed one. Thus, (3.15) corresponds to (3.14) and not to (3.13).

Let us recall that, in the traditional formulation of phenomenological thermodynamics, the definition of temperature for systems in thermal equilibrium was possible because of the *zeroth law of thermodynamics* (Kondepudi and Prigogine, 1998). In our formulation adapted from Callen (1985), it is possible thanks to the zeroth and second postulates. Since energy is an increasing function of entropy (Fig. 3.4), the absolute temperature (3.17) is non-negative. According to the third postulate, entropy takes the value zero at the lowest possible temperature $T = 0$, which is the content of the *third law of thermodynamics*.

On identifying the primed subsystem with the total system we are interested in and the double primed subsystem with the environment, the above considerations lead to the important practical conclusion that the thermodynamic state of an isolated system of energy E is the same as that of a system at thermal equilibrium with an environment at an appropriate temperature T (Fig. 3.6a).

Now assume that the partition separating the subsystems does not allow a change in the entropy components (we shall exhibit a realization of such conditions later on):

$$S' = \text{const.} , \qquad S'' = \text{const.} , \qquad (3.18)$$

but that the values of the thermodynamic variables X' and X'' can change subject to

$$X' + X'' = X = \text{const.} \qquad (3.19)$$

When X is interpreted as volume, for example, the partition can be a movable piston. When X is interpreted as a number of particles of a given kind, the partition is a semi-permeable membrane. When X is interpreted as charge, the partition is a conductor. In conditions of constant total entropy $S = S' + S'' = \text{const.}$, the value of X' (and also

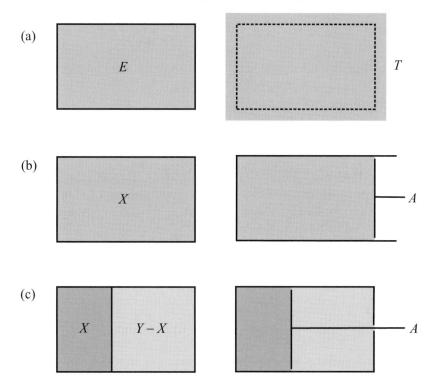

Fig. 3.6. (a) The thermodynamic state of an isolated system of energy E is the same as that of a system at thermal equilibrium with an environment at an appropriate temperature T. **(b)** The thermodynamic state of a simple system with a fixed value of the global thermodynamic variable X (e.g., volume) is the same as the state of this system in equilibrium with an environment that exerts an appropriate thermodynamic force A on it. This force is illustrated here as a piston applying some pressure to the system. **(c)** A thermodynamic state of a complex system specified by fixed values of the global variable Y and structural variable X, represented by the volume of the subsystem separated by a wall, is the same as the state of this system when a chosen subsystem is acted upon by an external force A, represented here by an internal piston, preventing it from reaching an equilibrium with the rest of the system

X'') will be established spontaneously in such a way that the total energy of the system

$$E = E'(S', X') + E''(S'', X'')$$ (3.20)

reaches a minimum, and its derivative with respect to X' therefore vanishes:

$$\frac{\partial E}{\partial X'} = \left(\frac{\partial E'}{\partial X'}\right)_{S'} + \left(\frac{\partial E''}{\partial X''}\right)_{S''} \frac{\mathrm{d}X''}{\mathrm{d}X'} = 0 \ . \tag{3.21}$$

Using (3.19), we have

$$\frac{\mathrm{d}X''}{\mathrm{d}X'} = -1 \ , \tag{3.22}$$

and hence a necessary condition for the minimum energy is

$$\left(\frac{\partial E'}{\partial X'}\right)_{S'} = \left(\frac{\partial E''}{\partial X''}\right)_{S''} \ . \tag{3.23}$$

The negative derivative of the energy with respect to the thermodynamic variable X, viz.,

$$A \equiv -\left(\frac{\partial E}{\partial X}\right)_S \ , \tag{3.24}$$

is called the *thermodynamic force* conjugate to X. In terms of the definition (3.24), the relation (3.23) can be interpreted as equality of the forces:

$$A' = A'' \ . \tag{3.25}$$

This is the condition of *dynamical equilibrium* between the two subsystems. The thermodynamic state of a closed system, whose global thermodynamic variable X assumes a fixed value, is the same as that of a system at mechanical equilibrium with an environment acting on it with the appropriate force A (Fig. 3.6b). The dynamical equilibrium means that the system responds to the external force A with an equal force acting on the environment. Physical interpretations of the forces conjugate to various thermodynamic variables are given in Table 3.1.

If internal constraints do not allow a mechanical equilibrium between the subsystems, the thermodynamic state of the system is characterized by two independent variables: the global one will henceforth be denoted by $Y = Y' + Y''$ and the structural (local) one by $X \equiv Y' = Y - Y''$, so that the system becomes complex. If B is the thermodynamic force conjugate to the global variable Y, the force conjugate to the structural variable X is

$$A \equiv -\left(\frac{\partial E}{\partial X}\right)_S = -\left(\frac{\partial E'}{\partial Y'}\right)_{S'} + \left(\frac{\partial E''}{\partial Y''}\right)_{S''} = B' - B'' \ . \tag{3.26}$$

The thermodynamic state of a complex system is characterized by the fixed values of the global variable Y and structural variable X, the same as the state of the system when the chosen subsystem is acted

Table 3.1. Some thermodynamic variables and the forces conjugate to them

Thermodynamic variable X	Conjugate force A
Volume V	Pressure P
Number of molecules N	Negative chemical potential $-\mu$
Charge Q	Negative electrical potential $-\phi$
Electric moment $\boldsymbol{P}V$	Negative electric field $-\boldsymbol{E}$
Magnetic moment $\boldsymbol{M}V$	Negative magnetic field $-\boldsymbol{H}$
Shape of a solid ϵV	Negative stress $-\boldsymbol{\tau}$

Note: In spatially homogeneous systems the notions of *polarization, magnetization* and *deformation* (*strain*) are used, defined as the *electric moment, magnetic moment* and *form* of a solid, respectively, related to the unit of volume. Polarization, magnetization and the conjugate electric and magnetic fields are vectors, i.e., they are characterized by three spatial components, whereas the strain of the solid and the stress acting on it are in general characterized by six numbers (three describe the stretching and three the torsion of the body).

upon by an external force A preventing it from reaching an equilibrium with the rest of the system (Fig. 3.6c).

Equation (3.26) may be generalized to the case of *thermal* nonequilibrium caused by a difference in temperatures. It suffices to identify the variable X with the entropy of one subsystem:

$$A = - \left(\frac{\partial E'}{\partial S'} \right)_{Y'} + \left(\frac{\partial E''}{\partial S''} \right)_{Y''} = -T' + T'' \, . \tag{3.27}$$

Definitions (3.17) and (3.24) [or (3.26)] determine $1+n$ relations between the temperature and thermodynamic forces and $1+n$ thermodynamic variables including the entropy:

$$T = T(S, X_1, \ldots, X_n) \, , \qquad A_i = A_i(S, X_1, \ldots, X_n) \, , \tag{3.28}$$

where $i = 1, \ldots, n$. The relations (3.28) are called the *equations of state*. They uniquely determine the fundamental equation (3.5) or the equivalent equation (3.2). The equations of state of complex systems can be found if the equations of state of the simple component subsystems are known.

As an example, let us consider the equations of state for an ideal gas composed of molecules of one kind. It is a system with two thermodynamic degrees of freedom ($X_1 = V$, $X_2 = N$). Besides the temperature T, the properties of the surroundings are characterized by two thermodynamic forces: pressure $A_1 = P$ and, in the case of the open system, chemical potential $A_2 = -\mu$. The fundamental equation (3.2) has the form (its derivation can be found in Appendix A.1):

$$\frac{S}{N} = \frac{S_0}{N_0} + k_B \ln \left[\left(\frac{E}{E_0} \right)^{3/2} \left(\frac{V}{V_0} \right) \left(\frac{N}{N_0} \right)^{-5/2} \right] , \qquad (3.29)$$

where ln denotes the natural logarithm, k_B is the Boltzmann constant, and the index 0 distinguishes values of S, E, V and N in a certain reference system.

From (3.29), the two well-known equations of state can be found, the first relating temperature to energy:

$$E = \frac{3}{2} N k_B T , \qquad (3.30)$$

and the second, pressure to volume:

$$PV = N k_B T . \qquad (3.31)$$

The relationship between the chemical potential and the number of molecules N (the third equation of state),

$$\mu = -k_B T \ln \left[\left(\frac{E}{N} \right)^{3/2} \left(\frac{V}{N} \right) \right] , \qquad (3.32)$$

is less well known but equally important. We shall use it in Chap. 6 to derive an expression for the thermodynamic force that drives chemical reactions.

3.4 Energy Transformations: Work, Heat and Dissipation

Using the definitions of temperature (3.17) and thermodynamic forces (3.24), a small change in the energy

$$E = E(S, X_1, \ldots, X_n) \qquad (3.33)$$

of a system with n degrees of freedom can be rewritten in the general form

$$\Delta E = \left(\frac{\partial E}{\partial S} \right)_{X_1,\ldots} \Delta S + \sum_i \left(\frac{\partial E}{\partial X_i} \right)_{S,\ldots} \Delta X_i = T \Delta S - \sum_i A_i \Delta X_i .$$
$$(3.34)$$

An energy change ΔE can be achieved in two ways: either by providing (or removing) *heat* Q, or by doing *work* on (or by) the system W:

$$\Delta E = Q + W \ . \tag{3.35}$$

Equation (3.35) expresses the content of the *first law of thermodynamics*, an empirical ascertainment that heat is not an independent physical quantity but a form of energy that emerges from work or changes into it (Kondepudi and Prigogine, 1999).

While the supply or removal of heat is linked to a change in the value of the entropy S, work involves a change in the values of the thermodynamic variables X_i of various kinds (mechanical, chemical, electrical, etc.). However, one can neither identify heat Q with the quantity $T\Delta S$ nor directly identify work W with the sum $-\sum_i A_i \Delta X_i$, since (3.35) can still be satisfied when something is added to and subtracted from it simultaneously. Rewriting it as

$$\Delta E = (Q + D) + (W - D) \ , \tag{3.36}$$

we obtain the general relationships:

$$Q + D = T\Delta S \tag{3.37}$$

and

$$W - D = -\sum_i A_i \Delta X_i \ . \tag{3.38}$$

The quantity D is called energy *dissipation*.[3]

As shown in the previous section, the thermodynamic state of a given system can be specified either by the values of certain thermodynamic variables fixed by appropriate constraints or by the values of the conjugate forces acting on the surroundings. If the environment acts on the system with identical forces, the system is in complete thermodynamic equilibrium. If, on the other hand, external forces do not counterbalance internal forces, the system is only in a state of partial equilibrium. The state of partial equilibrium can be treated formally as a state of complete equilibrium maintained by *fictitious* external forces that balance internal forces instead of incomplete constraints. Positive or negative work W is performed by or against the *actual* external forces A_i^{eq}:

$$W = -\sum_i A_i^{\mathrm{eq}} \Delta X_i \ , \tag{3.39}$$

---

[3]In Prigogine's terminology, the quantity D/T is referred to as *internal entropy production*.

whereas dissipation is related to differences, possible in the partial equilibrium state, between the values of these forces and those of the internal forces A_i determined by the equations of state (3.28):

$$D = \sum_i (A_i - A_i^{\text{eq}}) \Delta X_i .$$
(3.40)

The actual external forces A_i^{eq} are equal to the internal forces A_i in the state of complete equilibrium, thus explaining the notation.

A change in the system's thermodynamic state is called a *thermodynamic process*. Processes wherein no work is done on or by the system, $W = 0$, take place without any external constraints or manipulation by forces, and are thus referred to as *spontaneous*. Processes taking place without exchanging heat with the environment, $Q = 0$, are referred to as *adiabatic*. For processes that are simultaneously spontaneous and adiabatic, the relationships (3.35) and (3.37) take the form

$$\Delta E = 0 , \qquad T\Delta S = D .$$
(3.41)

Following the second postulate, the Clausius entropy in such processes tends to a maximum. Hence, due to the positivity of the temperature T, we always have

$$D \geq 0 .$$
(3.42)

From (3.37) and the inequality (3.42), we get the relation

$$T\Delta S \geq Q ,$$
(3.43)

which formally expresses the *second law of thermodynamics* (Kondepudi and Prigogine, 1999).

Processes occurring without dissipation, i.e., $D = 0$, are said to be *reversible* since under adiabatic conditions, $Q = 0$, they do not cause an increase in entropy and, as we know, such an increase is a result of the irreversibility of the process. It follows from (3.40) that reversible processes proceed only through states of complete equilibrium where internal forces exactly balance external forces. For reversible processes, the second law (3.43) takes the form of the equality

$$T\Delta S = Q .$$
(3.44)

Processes that are simultaneously adiabatic and reversible can be identified with *isentropic* processes, with $S = \text{const}$. We referred to such processes in the definition (3.26) of the thermodynamic forces. Later, we will see that a good approximation for the reversible processes are

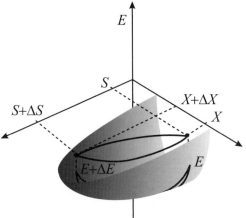

Fig. 3.7. Two exemplifying realizations of a process in the space of thermodynamic variables represented here by entropy S and a single variable X. The energy change ΔE depends only on the initial and final states, which are assumed to be the same for both processes. In contrast, work, heat and dissipation depend in general on the process paths

quasistatic processes proceeding sufficiently slowly. Only this kind of process is considered by traditional equilibrium thermodynamics. Let us stress that the general subject of our considerations are irreversible processes proceeding at a finite rate.

For arbitrary large changes in the state of the system, the equality (3.35) remains unchanged, whereas the expression (3.39) should be replaced by

$$
\begin{aligned}
W &= -\int \sum_i A_i^{\mathrm{eq}}\,\mathrm{d}X_i \\
&= -\int_{t_0}^{t} \sum_i A_i^{\mathrm{eq}}\Big(S(t), X_1(t), \ldots, X_n(t)\Big)\,\dot{X}_i(t)\,\mathrm{d}t .
\end{aligned} \qquad (3.45)
$$

The integral (3.45) is evaluated along the whole path of the process in the space of thermodynamic variables starting from the initial state $(S, X_1, \ldots, X_n)$ and ending in the final state $(S + \Delta S, X_1 + \Delta X_1, \ldots, X_n + \Delta X_n)$ (Fig. 3.7). The variable t is an arbitrary parameter that determines subsequent positions along this path. In particular, it can be time, but does not have to. The dot in the second line of (3.45) denotes differentiation with respect to the parameter t.

In contrast to energy change (3.34), the quantity W and also, according to (3.35) and (3.37), Q and D depend not only on the initial and final thermodynamic state of the system but, in general, on the whole path of the process as well as the velocity $\dot{X}_i$ along this path.

We say that energy E and entropy S are *functions of state*, whereas work W, heat Q and dissipation D are *functions of process*.

3.5 Free Energy and Bound Energy

Thermodynamics was formulated in the first half of the 19th century as a theory to explain the functioning of *heat engines* converting heat into work. Such engines had been in use for almost a century by that time (Newcomen in 1712, Watt in 1769). As we shall show in the following, it is not possible to convert heat into work in a stationary (cyclic) way when such an engine is coupled to only one source of heat at a constant temperature. The machine converting heat into work must also be coupled to a system of lower temperature (cooler). Accordingly, traditional thermodynamics dealt with processes taking place when coupling to an environment of variable temperature. For simplicity, these processes were assumed to be quasistatic, i.e., without dissipation.

This approach to thermodynamics has been adopted in many handbooks right up to the present time. Meanwhile, the processes studied in laboratories are for the main part those taking place at a *constant* temperature stabilized by high quality thermostats. Moreover, the phenomenon of dissipation is no longer neglected, as it provides much information on the processes taking place at the microscopic scale. Hence, the *irreversible isothermal processes* are of particular interest. In fact these are the working conditions of biological machines converting, not heat into work, but one type of work into another. The majority of processes considered in this book are just of this type.

Under isothermal conditions, $T = $ const., the temperature T can be brought inside the increment operation and the change (3.34) can be rewritten as

$$\Delta E = \Delta(TS) + \Delta F , \qquad (3.46)$$

where

$$\Delta F \equiv -\sum_i A_i \Delta X_i . \qquad (3.47)$$

The quantity F can be considered as a function of temperature T rather than entropy S. It then follows that

$$A_i = -\left(\frac{\partial F}{\partial X_i}\right)_T , \qquad (3.48)$$

which turns out to be much more practical than (3.24). Many other mathematical consequences of introducing the quantiy F are presented in Appendix A.2.

It follows from (3.46) that the energy E is the sum

$$E = F + TS .\qquad(3.49)$$

(We assume that $E = F$ for $T = 0$ when, due to the third postulate, the Clausius entropy is zero.) The system in the thermodynamic state under conditions $T = $ const. behaves as if it were composed of two subsystems. The first subsystem of the energy F can be called *dynamic* – its state is characterized by the thermodynamic variables X_i. These are, as we remember, selected macroscopic (*slow*) dynamic degrees of freedom of the system. The second, *thermal* subsystem of the energy TS is characterized by all the other (*fast*) microscopic degrees of freedom that determine the entropy S.

These two subsystems can interact with the environment and with each other and the measure of the interaction is the energy exchanged. On the basis of the relation (3.46), equation (3.37) can be rewritten as

$$Q + D = \Delta(TS) ,\qquad(3.50)$$

while the definition (3.47) allows us to write (3.38) as

$$W - D = \Delta F .\qquad(3.51)$$

It follows from the two relations and (3.46) that the work is the energy exchanged between the environment and a dynamic subsystem, heat is the energy exchanged between the environment and a thermal subsystem, and dissipation is the energy exchanged between the dynamic and thermal subsystems. According to the second law of thermodynamics, dissipation is always positive, which means that energy can only be transferred from the dynamic to the thermal subsystem, and never the reverse. The energy

$$F = E - TS\qquad(3.52)$$

of the dynamic subsystem is called the *free energy*, as only this energy can be used for doing work. Because of the unidirectional energy transfer from the dynamic to the thermal subsystems, the energy of the thermal subsystem TS cannot be used for doing work and that is why it is called the *bound energy*. A general scheme of the energy exchange relations in a system in a thermodynamic state under isothermal conditions $T = $ const. is shown in Fig. 3.8.

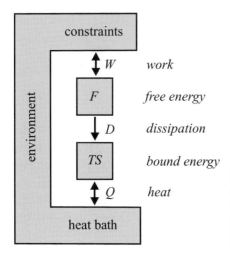

Fig. 3.8. Dynamic and thermal subsystems of the thermodynamic system under isothermal conditions $T =$ const. and their interaction with the environment and with each other

Equations (3.50) and (3.51) imply that, in isothermal processes, the quantities $Q + D$ and $W - D$ do not depend on the pathway of the process but only on the differences between the initial and final values of the entropy and free energy of the system. In contrast to (3.37) and (3.38), equations (3.50) and (3.51) hold for arbitrary large increments and there is no need to introduce a curvilinear integral of the type (3.45). Nevertheless, it should be emphasized that, for arbitrary large increments ΔF, the linear approximation (3.47) is no longer valid and then work W, dissipation D and heat Q will in general depend on the pathway of the process, hence on the curvilinear integral (3.45).

For the *isothermal reversible* processes, the dissipation equals zero, $D = 0$, and the direct relations hold:

$$\Delta F = W , \qquad T\Delta S = Q . \qquad (3.53)$$

In this case work and heat do not depend on the process pathway. For processes that are simultaneously *isothermal* and *adiabatic* (occurring without exchanging heat with the environment, $Q = 0$),

$$\Delta E = W , \qquad T\Delta S = D \geq 0 . \qquad (3.54)$$

Work done on or by the system equals its free energy increment and its entropy increases only due to the free energy dissipation. For processes that are simultaneously *isothermal* and *spontaneous* (occurring without work done on or by the system, $W = 0$),

$$\Delta E = Q , \qquad \Delta F = -D \leq 0 . \qquad (3.55)$$

(a)   (b)   (c)

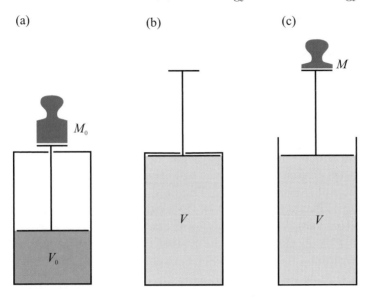

Fig. 3.9. Different modes of realization of isothermal gas decompression: (**a**) A certain amount of gas is enclosed in a container by a movable piston loaded at the beginning of the process by a load M_0, exerting such a pressure on the gas that it is restricted to the volume V_0. The movement of the piston is slow enough to ensure that the gas remains homogeneous all the time. For the sake of simplicity we shall assume that there is a vacuum in the part of the container outside the piston. (**b**) When the load is released the gas is decompressed to a maximum volume V limited by the container construction (the constraints). No work is done, so the system does not take heat from the thermostat, and all the free energy released is dissipated. (**c**) Work is done and some heat is taken from the thermostat when the load M_0 is reduced (in a jump or continuous way) to the load M, giving an effect equivalent to a restriction of the container construction keeping the gas volume at V

In this case the energy change of the system determines the quantity Q called the *process heat*, and dissipation is the only cause of the free energy decrease.

Let us illustrate the two ways to implement free energy change, work and dissipation, on a simple example of different realizations of the ideal gas isothermal decompression process (Fig. 3.9). It follows from the equation of state (3.30) that, in the process taking place under $T = $ const. conditions, the energy of the fixed number of molecules $N = $ const. does not change:

$$\Delta E = 0 \ . \tag{3.56}$$

Thus, a decrease in the system's free energy should exactly compensate an increase in the system's bound energy and it follows from (3.29) determining the entropy change that we have

$$-\Delta F = T\Delta S = Nk_{\mathrm{B}}T\ln\frac{V}{V_0} \, , \qquad (3.57)$$

where V_0 and V denote the initial and final volumes of the gas, respectively.

In the process of *spontaneous* gas decompression (the whole load is taken off at once, see Fig. 3.9b), the expression (3.57) describes dissipation of free energy only. The final volume V of the gas can also be obtained by taking off part of the load (Fig. 3.9c). Then some work (of negative sign) is being done over the rest of the load exerting the pressure P^{eq} on the gas, as the rest of the load will be elevated to a level determined by the difference between the initial and final volumes of the gas:

$$-W = P^{\mathrm{eq}}(V - V_0) = Nk_{\mathrm{B}}T\frac{V - V_0}{V} \, , \qquad (3.58)$$

where we have used the equation of state (3.31) for a fully equilibrated final state. The work (3.58) is smaller than the free energy released (3.57) and the difference is dissipated. The dissipated part can be reduced by taking the load off in small portions, which corresponds to replacing (3.58) by a sum of the appropriate increments. In the limit of infinitely small increments, when the external pressure P^{eq} becomes equal to the internal one P determined by the equation of state (3.31), the expression (3.58) is replaced by the integral

$$-W = Nk_{\mathrm{B}}T\int_{V_0}^{V}\frac{\mathrm{d}V}{V} = Nk_{\mathrm{B}}T\ln\frac{V}{V_0} \, , \qquad (3.59)$$

whose value equals the free energy released (3.57). Then, dissipation is reduced to zero and the process becomes reversible.

The distribution of the free energy released by a system into work and dissipation is presented in Fig. 3.10 for different modes of realization of the process of isothermal decompression of the ideal gas. It should be emphasized that, according to the law of conservation of the total energy of a system (3.56), the work is done at the expense of the heat collected from the thermostat. However, this does not contradict the second law of thermodynamics since, after the process, the free energy of the system has been reduced. According to the second law of thermodynamics in the Kelvin formulation, the exchange of heat into work is impossible only when, after the process, the system and its environment come back to the state they had before the process.

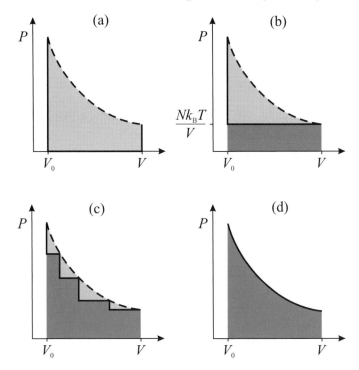

Fig. 3.10. Distribution of the free energy released by a system [the area under the isotherm $P = P(V)$] into work (*darkly shaded area*) and dissipation (*lightly shaded area*) in different realizations of the process of isothermal ideal gas decompression (see Fig. 3.9). The volume dependence of the pressure exerted by the external load is given by the *solid line*, and the volume dependence of the internal pressure determined by the equation of state is shown by the *broken line*. (**a**) The whole load at the piston is released in one step and the whole free energy released is dissipated. (**b**) Part of the load at the piston is released. The system does some work by elevating the load that is left to a level determined by the difference between the initial and the final volumes. Much of the free energy released is still dissipated. (**c**) More work can be done and less energy dissipated by taking the load off in small steps. (**d**) By decreasing the portions of the load subsequently taken off we get, in the limit, a realization of the reversible process without dissipation

3.6 Open Thermodynamic Systems: Steady State versus Dissipative Structures

Let X be a certain structural thermodynamic variable, $X \equiv Y' = Y - Y''$. The force conjugate to X [see the comments on the definition (3.26)] is, under isothermal conditions, determined by

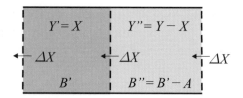

Fig. 3.11. Open thermodynamic system in a steady state

$$A = -\left(\frac{\partial F}{\partial X}\right)_T = -\left(\frac{\partial F'}{\partial Y'}\right)_T + \left(\frac{\partial F''}{\partial Y''}\right)_T = B' - B''. \qquad (3.60)$$

As discussed in Sect. 3.3, a state with a constant value of the structural variable X can be retained either through absolutely tight internal constraints or through an external force A^{eq} exactly equilibrating the internal force A. Otherwise the system is in a partial equilibrium state and evolves toward the state of complete equilibrium.

However, if both the external and internal constraints are not absolutely tight (the system as a whole is *open*), the change ΔX in the value of the thermodynamic structural variable X during its evolution, involving a transfer of some amount of X from one subsystem to another, can be compensated for by supplying the same amount of X *from outside* to the first subsystem and releasing the same amount of X *to the outside* from the second subsystem (Fig. 3.11). Because the value of the variable X in the whole system does not change, the values of the free energy and the bound energy (entropy) of the whole system remain unchanged:

$$F = \mathrm{const.}, \qquad S = \mathrm{const.}. \qquad (3.61)$$

In the absence of an external force, $A^{\mathrm{eq}} = 0$, a spontaneous change ΔX is associated with dissipation [see (3.40)]:

$$D = A\Delta X. \qquad (3.62)$$

To fulfill the conditions (3.61), D must be equal on the one hand to the work done on the system and, on the other, to the heat released to the environment:

$$D = W = -Q. \qquad (3.63)$$

The environment performs work on the system, not through an external force, but as a result of the flux of the quantity ΔX across the system (Fig. 3.11). The system's thermodynamic state does not change because the work performed balances the heat released.

In a nonequilibrium *steady state* maintained due to such a flux, the rate of dissipation, i.e., the amount of free energy transferred into a bound energy form over a unit of time Δt, remains constant:

$$\frac{D}{\Delta t} = \frac{W}{\Delta t} = -\frac{Q}{\Delta t} = \text{const.} \qquad (3.64)$$

In the limit of $\Delta t \to 0$, the ratio (3.64) is called the *dissipation function*[4] and denoted by Φ. In view of the second law of thermodynamics, it is always non-negative and, from (3.62), it takes the form

$$\Phi = A\dot{X} \geq 0 \,. \qquad (3.65)$$

The time derivative $\dot{X}$ has the meaning of the rate of change of the variable X and is referred to as the *flux* of the variable X.

Two examples of simple open thermodynamic systems in a steady state are shown in Figs. 3.12a and b. The direction of the flux $\dot{X}$ is determined by the condition that the product of this flux and the internal force A be positive. The flux of gas flows through a semi-permeable partition from the region of higher pressure P' to that of lower pressure P'', but the available volume grows in the opposite direction. The current in an electric conductor, understood as a movement of *positive* electric charge, flows through a resistor from a point of higher potential ϕ'' to a point of lower potential ϕ'.

The reasoning can be generalized to the steady state maintained by several simultaneously acting flows $\dot{X}_i$ related to forces A_i. Then the inequality (3.65) takes the form

$$\Phi = \sum_i A_i \dot{X}_i \geq 0 \,. \qquad (3.66)$$

It should be remembered that the second law of thermodynamics demands that the total sum (3.66) be non-negative, whereas particular components of the sum can be both positive and negative. According to (3.64), the components stand for the work done by particular fluxes per unit time, in other words, for the *power* of the fluxes. Hence, not only can the environment do work on a system, but the system can do work on the environment.

The porous partition (Fig. 3.12a) can be replaced by a turbine (Fig. 3.12c), while the resistor in the system of Fig. 3.12b can be replaced by an electric motor (Fig. 3.12d), and in this way some of the

---

[4]Prigogine refers to the product of the dissipation function and the temperature as the rate of *entropy production*.

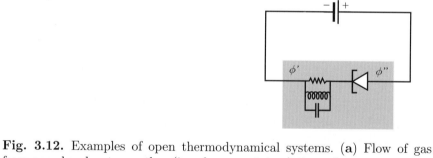

Fig. 3.12. Examples of open thermodynamical systems. (**a**) Flow of gas from one chamber to another (in a large container) through a porous partition. (**b**) Flow of electric current between the battery electrodes through a resistor. When the porous partition is replaced by a turbine (**c**) or the resistor is replaced by an electric motor (**d**), the systems can do some work. The systems (**a**–**d**) are in steady state conditions. Replacing the resistor by a redamped resonance circuit (**e**) leads to undamped oscillations that can also be used to do some work, although the steady state condition is no longer met

work done on a system can be recovered, preventing its total conversion into heat. The electric motor can be connected to the turbine and then gas can be pumped from the chamber with the gas under a lower pressure to that with the gas under a higher pressure. The possibilities of organization of open systems under conditions of incomplete thermodynamic equilibrium into various free energy transducers, i.e., machines, are unlimited. In this sense living organisms are such machines and the processes of energy transduction in living organisms at the subcellular level will be the subject of Chap. 8.

Here we should point out one more problem. In the above discussion, the thermodynamic system considered has been assumed to be in the steady state, i.e., the values of the fluxes $\dot{X}_i$ of the thermodynamical variables X_i and forces A_i, hence also the dissipation function (3.66), have all been assumed constant. However, is a realization of such a state admitted by the laws of thermodynamics in any conditions? In other words is the steady state always stable and is the non-negative dissipation function always constant or decreasing in time to a constant steady state value (for $A_i = 0$ the equilibrium value)?

To find an answer to these questions, we differentiate (3.66) with respect to time:

$$\dot{\Phi} = \sum_i \dot{A}_i \dot{X}_i + \sum_i A_i \ddot{X}_i . \tag{3.67}$$

The first component can be rewritten as

$$\sum_i \dot{A}_i \dot{X}_i = \sum_{ij} \left(\frac{\partial A_i}{\partial X_j} \right)_{X_i} \dot{X}_j \dot{X}_i . \tag{3.68}$$

It follows from the conditions of thermodynamic stability (Appendix A.3) that this expression for nonzero fluxes is always negative and thus leads to a decrease in the dissipation function. However, we cannot say this about the second term, which can be either positive or negative. For large forces A_i and far from thermodynamic equilibrium, the total sum (3.67) can assume positive values, although it does not have to, and this implies instability of the initial steady state. Then another state becomes stable. It can be a new steady state of 'broken symmetry' (one of a few alternatives) either periodically oscillating in time or with chaotic dynamics, hence in general the state referred to by Prigogine as a *dissipative structure* (Prigogine, 1980; Kondepudi and Prigogine, 1999) and by Haken (1990) as a *synergetic structure*.

The resistor in Fig. 3.12b can be replaced by a resonance circuit with a tunneling diode (Fig. 3.12e), rather than by an electric motor

(Fig. 3.12d). If the voltage at the source exceeds a certain threshold value, the negative resistance of the diode will compensate a small but finite and positive resistance of the resonance circuit and stable undamped oscillations will appear in the system. From the point of view of our discussion, Fig. 3.12b presents an exemplary thermodynamic system in a steady state, and Fig. 3.12e a thermodynamic system with a dissipative structure. In practice, the tunneling diode is supplied through a voltage divider shorted by appropriate capacities so that the current flowing through it is constant. In this way the difference between a steady state and a dissipative structure disappears and the two examples illustrate two alternative organizations of the inside of a black box representing a cyclic electric machine.

Machines and dissipative structures do not have to be organized on a macroscopic level. In the chemochemical machines which play a key role in biological processes (discussed in detail in Chap. 8), the organization is realized on a microscopic level or, to be more exact, on the mesoscopic level of the macromolecular enzymes coupling particular reactions. The conversion of a chemo-chemical machine into a dissipative structure is often equivalent, not only to a spontaneous temporal reorganization, but a spontaneous spatial self-organization as well.

3.7 Rate of Nonequilibrium Thermodynamic Processes

For systems under conditions close to complete thermodynamic equilibrium, we can assume that the fluxes $\dot{X}_i$ depend linearly on the thermodynamic forces A_i (a situation described as a *linear response to external stimulus*) (Kondepugi and Prigogine, 1999):

$$\dot{X}_i = \sum_j L_{ij} A_j \ . \tag{3.69}$$

The coefficients of proportionality L_{ij} are called Onsager's *kinetic coefficients*.

In the approximation (3.69), the dissipation function (3.66) is a quadratic function of thermodynamic forces:

$$\Phi = \sum_i L_{ij} A_i A_j \geq 0 \ . \tag{3.70}$$

The sign of Φ indicates that the diagonal kinetic coefficients are nonnegative:

$$L_{ii} \geq 0 \ . \tag{3.71}$$

This is not true in general for the off-diagonal coefficients. However, using statistical thermodynamics (see Appendix A.5), it can be demonstrated that, in the absence of magnetic fields, the matrix of kinetic coefficients is symmetrical:

$$L_{ij} = L_{ji} \,. \tag{3.72}$$

This expresses the so-called *fourth law of thermodynamics* due to Onsager.

Table 3.2 gives examples of several linear relations between fluxes and forces. Off-diagonal kinetic coefficients describe the so-called *cross-effects* and are related to *thermodynamic coupling* of various irreversible processes. This coupling is the basis for the *free energy transduction* phenomena which play significant roles in many processes taking place in biological cells. However, their description in terms of a linear approximation is by and large incorrect in such cases.

In the linear approximation, and with the symmetry (3.72), the second component of the sum (3.67) can be rewritten as

$$\sum_i A_i \ddot{X}_i = \sum_{ij} A_i L_{ij} \dot{A}_j = \sum_{ji} L_{ji} A_i \dot{A}_j = \sum_j \dot{X}_j \dot{A}_j \,. \tag{3.73}$$

Thus, in the linear approximation, the second component of (3.67) exactly equals the first component, and the whole dissipation function derivative (3.67) turns out to be an expression that is *always* negative for nonzero fluxes. A conclusion is *Prigogine's variational principle*: close to complete equilibrium, systems tend to the state of minimum dissipation.

Table 3.2. Examples of linear relations between fluxes and forces [see the definition (3.60) and Table 3.1]

Relation	Process	Law
$\dot{Q} = L_{QQ}(\phi'' - \phi')$	Electrical conduction	Ohm's law
$\dot{N} = L_{NN}(\mu'' - \mu')$	Diffusion, chemical reaction	Fick's law
$\dot{S} = L_{SS}(T'' - T')$	Heat conduction	Fourier's law
$\dot{Q} = L_{QS}(T'' - T')$	Thermoelectric effect	Seebeck's law
$\dot{S} = L_{SQ}(\phi'' - \phi')$	Electrothermal effect	Peltier's law
$\dot{N} = L_{NS}(T'' - T')$	Thermodiffusion effect	Soret's law
$\dot{S} = L_{SN}(\mu'' - \mu')$	Reverse effect to thermodiffusion	Dufour's law

The linear relationships (3.69) apply to both open and closed systems. They can be used, for example, for the simplest case of a closed system with only one nonequilibrium variable X that evolves toward the equilibrium value X^{eq}:

$$\dot{X} = LA. \tag{3.74}$$

The force A conjugate to X disappears in the state of complete equilibrium. Hence, in the linear approximation, the equation of state linking this force to X takes the form

$$X - X^{\mathrm{eq}} = -CA , \tag{3.75}$$

where C is the system's *capacity* (see Appendix A.3):

$$C \equiv - \left(\frac{\partial X}{\partial A} \right) . \tag{3.76}$$

Inserting one of these equations into the other gives a linear differential equation that describes the system's relaxation to the state of complete equilibrium:

$$\dot{X} = -\tau^{-1}(X - X^{\mathrm{eq}}) , \tag{3.77}$$

where

$$\tau = C/L . \tag{3.78}$$

The solution to (3.77), viz.,

$$X(t) - X^{\mathrm{eq}} = [X(0) - X^{\mathrm{eq}}] \, e^{-t/\tau} , \tag{3.79}$$

describes an exponential decay of the initial value $X(0)$ to the equilibrium value X^{eq} with a characteristic *relaxation time* τ. It follows from (3.78) that it equals the ratio of the capacity and the kinetic coefficient.

We shall often use linear equations of the type (3.77). Their range of applicability often greatly exceeds the range of validity of the linear approximations (3.69) and (3.75). The time course of chemical reactions, considered in more detail in Appendix B.3, represents one such case.

4 Origins and Evolution of Life

4.1 History in Physics

It has been customary to think about physics as a branch of science that is indifferent to history. Kepler's planets move along regular orbits and, after some period of time, return at the same velocity to the same position in space. Also in the case of unstable (chaotic) trajectories, as stated in Poincaré's famous theorem in the late 19th century, mechanical systems return arbitrarily close to the initial state. Equations of motion in classical mechanics and electrodynamics are time-reversal invariant. Similarly, by and large, in quantum mechanics and quantum field theory.

Though Boltzmann in his statistical mechanics introduced the 'arrow of time', this idea applied to the future and not to the past of a physical system. The trend toward equilibrium is irreversible and the system before reaching its thermodynamic equilibrium state may follow a variety of distinct pathways starting from various initial conditions. Consequently, knowledge of the final state is not sufficient to draw conclusions about the initial state of the system.

Each physical system has a *structure*, an *organization* or, in other words, *constraints* imposed on its motion. This feature contains history understood as a memory of an event in the past when this structure became 'frozen'. Indeed, it is frozen and not fixed as we fix initial conditions when solving equations of motion. Freezing is a kinetic process that does not contradict the second law of thermodynamics guaranteeing a trend toward the state of total thermodynamic equilibrium. However, the time it takes to reach the total thermodynamic equilibrium turns out to be much longer than the time courses of the events and processes we try to explain by assuming a specific structural organization of the system considered.

The process of the kinetic freezing of the structure is believed to follow the laws of physics. Nonetheless, it has so much randomness (Dawkins, 1986; Winkler-Oswatitsch and Eigen, 1992; Kauffman,

1993) that we are unable to deduce the structure from those very laws. Hence, it must be introduced into theory as an externally supplied piece of information. The randomness (in the sense of deterministic chaos) of practically every structure made spontaneously without the participation of any conscious constructor is a consequence of the time-reversal symmetry of the equations of microscopic physics (Prigogine and Stengers, 1984; 1997). According to this symmetry, the existence of any trajectory tending to the stable state of thermodynamic equilibrium (e.g., trajectory $1\,2$ in Fig. 2.6) implies the existence of a trajectory starting from an extremely unstable initial state in which the phenomenon of 'ergodicity breaking' (Palmer, 1982), i.e., a particular structural distinction, takes place (trajectory $\bar{2}\,\bar{1}$ in Fig. 2.6).

Our knowledge of present physical structure allows us to try to reconstruct the history of the Earth (van Andel, 1994), the Solar System (McSween, Jr., 1995), the Universe (Weinberg, 1980) or even time itself (Hawking, 1988). That these interests occupy many physicists is not shocking to anybody these days. There is a general consensus that the laws of physics are well understood today and it is time to apply them to systems and processes with high degrees of complexity.

Without a doubt, the greatest challenge for physicists today is the phenomenon of life. Living systems are extremely complex and organized hierarchically. This is clearly a result of the process of evolution that is almost as old as the Earth itself. Not surprisingly, biology was the first branch of science that attempted to reconstruct past events from today's knowledge of the biosphere. This quest started with the discovery of fossils of long-extinct species. A healthy dose of creativity and imagination applied to sets of more or less complete skeletal remains led to depictions of various extinct animal and plant species (Briggs and Crowther, 1990). The history of life on Earth, viewed from this perspective, began in the Cambrian period at the dawn of the Paleozoic era (540 million years ago) when living creatures developed the ability to build a solid skeleton based on calcium carbonate or silica. Contemporary techniques allow us to uncover fossils of much simpler organisms that do not possess a skeleton. Furthermore, these findings can nowadays be dated precisely and they push back the date of the emergence of life on Earth over 3.5 billion years ago (Schopf, 1999).

Charles Darwin in 1859 proposed a method that had great potential to reconstruct the history of life based on differences in selected features of living animal species and not extinct ones. In modern applications of this methodology, the most fundamental characteristics are the nucleotide sequences in the genomes of selected individual organisms (Dawkins, 1989). Based on the mathematically well defined distance between genomes, one tries to reconstruct, e.g., the history of the species *Homo sapiens* (Cann et al., 1987) or HIV 1 (Nieselt-Struve, 1997). Biochemistry and molecular biology (Darnell et al., 1999; Stryer et al., 2002), whose dynamic development has flourished since the mid-20th century, provide us with numerous examples of 'living fossils'. These are archaic metabolic pathways and more or less conserved domains in enzymes. The contemporary organization of animate matter reflects the history of its evolution and, conversely, the living structures that we encounter on the Earth today are products of the evolution of life.

Biophysics, when it attempts to describe the phenomenon of life using the conceptual framework of physics, only partially explains the structures of the elements of living systems, treating certain other components as given. Describing in this book the emergence of these 'given' components as a historical process, we will strive to provide the most precise answer possible to Erwin Schrödinger's famous question (Schrödinger, 1967): What is life?

4.2 Initiation

Planet Earth was formed about 4.6 billion years ago as a result of accretions (inelastic collisions and agglomerations) of larger and larger rocky fragments formed gradually from the dust component of the gaseous dusty cloud that was the original matter of the Solar System. The 'Great Bombardment' ended only 3.9 billion years ago when the stream of meteorites falling onto the surface of the newly formed planet reached a more or less constant intensity. The first well preserved petrified microstamps of relatively highly organized living organisms similar to today's cyanobacteria emerged about 3.5 billion years ago (Schopf, 1999), so life on Earth must have developed within a relatively short time of a few hundred million years.

Rejecting the hypothesis of an extraterrestrial origin of life [which may not appear so irrational, see arguments by Crick (1981)], we have to answer the question of the origin of the simplest elements of living organisms: amino acids, monosaccharides and nitrogenous bases. Three equally probable hypotheses have been put forward to explain their appearance (Cairns-Smith, 1990; Orgel, 1998). According to the first and oldest hypothesis, these compounds appeared as a result of electric discharges and ultraviolet irradiation of the primary Earth atmosphere containing CO_2 (the contemporary atmospheres of Mars and Venus are 98% composed of this gas) and H_2O, but also a certain amount of strongly reducing gases CH_4, NH_3 and H_2S. According to the second hypothesis, the basic elements of living organisms were formed in space outside the orbits of large planets and transferred to the Earth's surface via collisions with comets and, indirectly, carbon chondrites. The third and most recent hypothesis says that these elements appeared at the oceanic rifts where the new Earth's crust was formed and where water overheated to $400°C$ and containing strongly reducing FeS, H_2 and H_2S met cool water containing CO_2.

In fact the problem remains open and all three hypotheses have been seriously criticized. First, the primary Earth atmosphere might not have been sufficiently reducing. Second, organic compounds from outer space could have deteriorated while passing through the Earth's atmosphere. Third, the reduction of CO_2 in oceanic rifts requires non-trivial catalysts.

The three most important characteristics of life that distinguish it from other natural phenomena were aptly expressed by Charles Darwin, whose theory of evolution has appeared so crucial for biology. Taking into account the achievements of post-Darwinian genetics and biochemistry (Mayr and Provine, 1980) we can come up with the following definition:

> *Life* is a process characterized by continuous (1) *reproduction*, (2) *variability* and (3) *selection* (survival of the fittest). Using present-day language, to stay alive individuals must have a replicable and modifiable *program*, a proper *metabolism* (a mechanism of matter and energy conversion) and a capability of *self-organization*.

The emergence of molecular biology in the 1950s answered many questions about the structures and functioning of the three most important classes of biological macromolecules: DNA (deoxyribonucleic acid), RNA (ribonucleic acid) and proteins. However, in the attempts to develop a possible scenario of evolution from small organic elements to large biomolecules, a classical chicken-and-egg problem was encountered. What appeared first? The DNA that carried the coded information on enzymatic proteins controlling the physiological processes that determined the fitness of an individual, or the proteins that enabled the *replication* of DNA, its *transcription* into RNA and the *translation* of certain sequences of amino acids into new proteins (Fig. 4.1a)?

The question was resolved in the 1970s as a result of the evolutionary experiments carried out in Manfred Eigen's laboratory (Biebricher

(a)

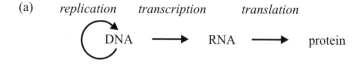

(b)

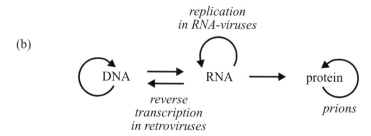

Fig. 4.1. Processing genetic information. (a) The classical dogma: the information is carried by DNA that undergoes *replication* during the process of biological reproduction and *transcription* into RNA when it is to be expressed; gene expression consists of *translation* of the information written in RNA onto a particular protein primary structure. (b) Modern corrections to the classical dogma. RNA can be replicated and transcribed in the opposite direction into DNA. Furthermore, the proteins themselves can also carry information, as is assumed to occur in prion diseases

and Gardiner, 1997). The primary macromolecular system undergoing Darwinian evolution may have been RNA. Single-stranded RNA is not only the information carrier, a program or *genotype*, but thanks to a specific spatial structure, it is also an object of selection or a *phenotype*. Originally, Manfred Eigen and coworkers used viral RNA replicase (Fig. 4.1b), a protein, to produce new generations of RNA in vitro. It soon appeared that the complementary RNA could polymerize spontaneously, without any replicase, on the matrix of the already existing RNA as a template. Consequently, we can imagine a very early 'RNA world' (Gesteland et al., 1999) composed only of ribonucleosides, their phosphates and their polymers – subject to Darwinian evolution, and thus alive according to the definition adopted.

A number of facts support the RNA world concept. Ribonucleoside triphosphates are a highly effective source of free energy. They fulfill this function as a relict in most chemical reactions of contemporary metabolism (Stryer et al., 2002). Diribonucleotides play the role of co-factors in many protein enzymes. In fact, RNA molecules themselves can serve as catalysts (Cech, 1986; Lilley, 2003) and it is becoming more common to talk about *ribozymes*. Contemporary ribosomes translating information from RNA onto a protein structure (see Fig. 4.1a) fulfill their catalytic function due to their ribosomal RNA content rather than their protein components (Ramakrishnan and White, 1998; Steitz and Moore, 2003). Finally, for a number of years now, we have known the *reverse transcriptase* that transcribes information from RNA onto DNA (Fig. 4.1b). It also appears that RNA may be a primary structure and DNA a secondary one, since modern organisms synthesize deoxyribonucleotides from ribonucleotides.

4.3 Origins of the Prokaryotic Cell Machinery

The RNA world could not exist any longer in conditions that occur on the contemporary Earth. The smallest present-day system that has the key function of a living object, namely reproduction, is a *cell*. There is a sharp distinction between simple prokaryotic cells (that do not have a cell nucleus) and much more complex eukaryotic cells (with a well-defined nucleus). There is a body of evidence pointing to an earlier evolutionary emergence of prokaryotic cells. Eukaryotic cells are believed to have resulted from mergers of two or more specialized prokaryotic cells. Unfortunately, little is known about the origins of prokaryotic cells themselves. The scenario that we present below is only an attempt to describe some key functional elements of the apparatus possessed by all prokaryotic cells (Harris, 1995) and does not represent a serious effort to reconstruct the history of life on Earth.

The world of competing RNA molecules must have eventually reached a point where a dearth of the only building material, nucleoside triphosphates, occurred. Molecules that were able to provide themselves with adequate supplies of the building materials gained an evolutionary advantage but they needed a bag, a container that could protect them and their supplies from the surroundings. In the liquid phase, such containers were formed spontaneously from phospholipids, amphiphilic molecules with one part attracted and another repelled by water. As a result of the movement of the hydrophobic part away from water and the movement of the hydrophilic part toward water, a lipid bilayer without a boundary, i.e., a three-dimensional vesicle, is formed (Fig. 4.2, Appendix C.4).

Since phospholipid vesicles can join to make bigger structures from several small ones, they are important for the RNA molecules that compete for food. Merging into bigger vesicles can be advantageous in foraging for food, while division into small vesicles can be seen as

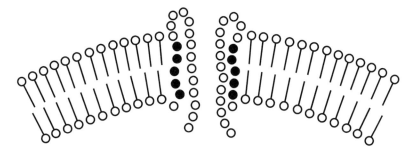

Fig. 4.2. In a water environment, amphiphilic molecules composed of a hydrophilic part (*white circle*) and a hydrophobic part (*black line segment*) organize themselves spontaneously into bilayers closed into three-dimensional vesicles. A protein, a linear polymer of appropriately ordered hydrophilic (*white circles*) and hydrophobic (*black circles*) amino acids, forms a structure that spontaneously builds into the bilayer and allows selectively chosen molecules, e.g., nucleoside triphosphates, to pass inside the vesicle

a type of reproduction. The phospholipid vesicle itself does not solve the problem, since there must also be a way of selective infusion of nucleotides into its interior. Employing a new type of biomolecule, amino acids, some of which are hydrophilic and some hydrophobic, solved this latter problem. Their linear polymers are called peptides and long peptides give rise to proteins. Proteins possess three-dimensional structures whose hydrophobicity depends on the order in which amino acid segments appear in a linear sequence (see Appendix C.5). Such proteins may spontaneously embed themselves in a lipid bilayer and play the role of selective molecular channels (see Fig. 4.2).

It appears, therefore, that the first stage in the development of a prokaryotic cell was probably the enclosure of some RNA molecules into phospholipid vesicles equipped with protein channels that enabled a selective transfer of nucleoside triphosphates into the interior (Fig. 4.3a). The second stage must have been the perfection of these channels and a link between their structure and the information contained in the RNA molecules. The latter property would have gained a selective significance. Selective successes could be scored by RNA molecules that could translate some of the information contained in the RNA base sequence into an amino acid sequence of a channel protein in order to synthesize it. This was the way to distinguish the so-called mRNA (*messenger* RNA) from tRNA (*transfer* RNA) and rRNA (*ribosomal* RNA). While mRNA carries information about the amino acid sequence in a protein, tRNA connects particular amino acids with the corresponding triples of bases. rRNA is a prototype of a ribosome, a catalytic RNA molecule that can synthesize amino acids, transported to it by molecules of tRNA, into proteins (Fig. 4.3b).

The analysis of the nucleotide sequences in tRNA and rRNA of various origins indicates that they are very similar and thus archaic. The genetic code based on sequences of triples is equally universal and archaic. Contemporary investigation of both prokaryotic and eukaryotic ribosomes has provided solid evidence that the main catalytic role is played by rRNA and not the proteins contained within them (Ramakrishnan and White, 1998; Steitz and Moore, 2003).

However, proteins have much better catalytic properties than RNA. A key property is their high specificity with regard to the substrate. In the form of polymerases, they soon replaced RNA in the process of self-replication. It was already possible on the RNA template to replicate sister RNA as well as DNA (deoxyribonucleic acid). DNA spontaneously forms a structure composed of two complementary strands (a double helix, see Appendix C.6) and is a much more stable carrier of information than RNA. This principle led to the current method of transferring genetic information (see Fig. 4.3c) which goes as follows. Genetic information is stored in double-stranded DNA. Protein *replicases* duplicate this information in the process of cell division. If necessary, protein *transcriptases* transcribe this information onto mRNA, which is used during the process of translation on *ribosomes* as a template to produce proteins. The transfer of information in the reverse direction from RNA to DNA via *reverse transcriptases* is a fossil remnant that has been preserved in modern retroviruses.

Protein enzymes can perform useful tasks. They can produce much-needed nucleoside triphosphates by recycling them from nucleoside diphosphates and inorganic orthophosphate with the use of organic compounds of a fourth type – saccharides as a source of free energy. In the now universal process of *glycolysis*, the oxidation of the most common monosaccharide, glucose, to pyruvate, two ATP molecules are reconstructed (see Sect. 4.6). The oxidizer (hydrogen acceptor) is NAD^+ – nicotinamide adenine dinucleotide. This is also a relict from the RNA world. The oxidizer NAD^+ is recovered in the process of fermentation of a pyruvate into a lactate (Fig. 4.3d).

The lactic fermentation process that accompanies phosphorylation of ADP to ATP with the use of sugar as a substrate has several drawbacks. In addition to its low efficiency (unused lactate), it leads to increased acidity of the cell. While sugars are neutral (pH near 7), lactate is a product of dissociation of lactic acid and, in the process of

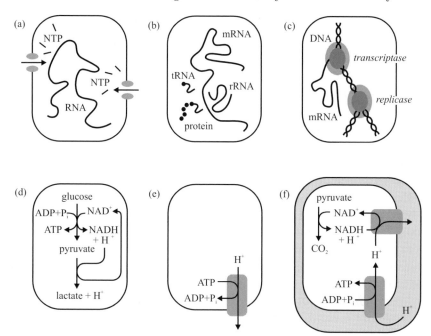

Fig. 4.3. Development of the prokaryotic cell machinery. (**a**) The self-replicating RNA molecule with a supply of nucleotide triphosphates (NTP) is enclosed in a vesicle bounded by a lipid bilayer with built-in protein channels that allow a selective passage of nucleoside triphosphates. (**b**) In an RNA chain, a distinction is made between mRNA and various types of tRNA and rRNA, the latter being a prototype of a ribosome that can synthesize proteins according to the information encoded in mRNA. Proteins produced in this way are more selective membrane channels and effective enzymes that can catalyze many useful biochemical processes. (**c**) Double-stranded DNA replaces RNA as a carrier of information. Protein replicases double this information during cell division and protein transcriptases transfer this information onto mRNA. (**d**) Protein enzymes appear to be able to catalyze the process of lactose fermentation of sugars as a result of which the pool of high energy nucleoside triphosphates (mainly ATP) can be replenished using low energy diphosphates (mainly ADP). The amount of oxidizer (hydrogen acceptor) NAD^+ remains constant, but the cell interior becomes acidic. (**e**) Proton pumps are created which are able to pump H^+ ions into the cell exterior via ATP hydrolysis. (**f**) Other proton pumps use hydrogen obtained from the decomposition of sugars through pyruvate to CO_2 as fuel. Due to the presence of a wall or a second cell membrane, pumped-out protons may return to the cell interior through the pumps of the first type that act in reverse to reconstruct ATP from ADP. Membrane phosphorylation becomes the basic mechanism of bioenergetics in all modern living organisms

sugar decomposition, hydrated protons H^+ (in the form of hydronium ions H_3O^+) are released. This is a well-known experience during hard physical work and exercise when blood is unable to deliver a sufficient amount of oxygen to the muscles. The lowering of the pH results in a significant slowdown or even a complete stoppage of the glycolysis reaction.

For the decomposition of sugars to be effectively used in the production of ATP, a cell must find a different mechanism of fermentation whose product has a pH close to 7, or a way for the protons H^+ to be expelled outside the cell. The new type of fermentation was discovered only by yeast, where it consists of the reduction of pyruvate to ethanol with a release of carbon dioxide. Before it took place, a *proton pump* had been found utilizing the hydrolysis of ATP as a source of energy (Fig. 4.3e). During the production of one molecule of ATP, one hydrated proton H^+ is released inside the cell, while the hydrolysis of one molecule of ATP results in the pumping outside the cell membrane of three hydrated protons H^+ (see Fig. 4.4a). Hence the process is still energetically favorable.

However, from the viewpoint of ATP production, a more efficient process is further oxidation of pyruvate to final products: carbon dioxide and water. This process takes place in the citric acid cycle of Krebs (see Sect. 4.6). Discussing the economy of the Krebs cycle makes sense only when a cell is able to utilize fuel in the form of hydrogen bound to the NAD^+ (and FAD – flavin adenine dinucleotide) for further phosphorylation of ADP to ATP. This became possible when a new generation of proton pumps was found. These pumps worked as a result of the decomposition of hydrogen into a proton and an electron instead of ATP hydrolysis. These particles were further transported along different pathways to the final hydrogen acceptor which, in the early stages of biogenesis, may have been an anion of an inorganic acid.

Primitive bacterial cells were endowed with cell membranes composed of peptidoglycan, a complex protein–saccharide structure, and later developed additional external cell membranes. This facilitated accumulation of protons in the space outside the original cell membrane, from which they could return to the cell interior using the proton pump of the first type (Fig. 4.3f). This pump, working in reverse, synthesizes ATP from ADP and an orthophosphate. This very efficient mechanism of membrane phosphorylation is universally utilized by all present-day living organisms.

A more detailed explanation of the proton pump that utilizes the oxidation of hydrogen is shown in Fig. 4.4b. In the original version,

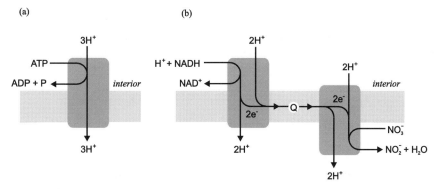

Fig. 4.4. Proton pumps transport free protons H^+ across the membrane from the cell interior to its exterior at the expense of the following chemical reactions: hydrolysis of ATP into ADP and an inorganic orthophosphate (**a**) or oxidation of hydrogen released in the decomposition of glucose to CO_2 and transported by NAD^+ (**b**). A derivative of quinone Q is an intermediary in hydrogen transport. This molecule is soluble inside the membrane and oxidation is finally accomplished, for example, through nitrate NO_3^- that is reduced to nitrite NO_2^-. If pumps of both types are located in the same membrane, the first protons passing in the reverse direction can phosphorylate ADP to ATP

the pump is composed of two protein trans-membrane complexes: a dehydrogenase of NADH and a reductase, for example, the one reducing nitrate NO_3^- to nitrite NO_2^-. In the first complex, two hydrogen atoms present in the NADH–hydronium ion pair are transferred to FMN (flavin mononucleotide). Later, after two electrons are detached, the hydrogens (as protons) are transferred to the other side of the membrane. The two electrons are accepted in turn by one and then the other iron–sulfur center. (Iron is reduced from the state Fe^{3+} to Fe^{2+}.) Subsequently, at a molecule of a quinone derivative Q, they are bound to another pair of protons that reached the same site from the interior of the cell. An appropriate derivative of quinone Q is well soluble inside the membrane and serves as an intermediary that ferries two hydrogen atoms between the two complexes. At the other complex, two hydrogen atoms are again split up into protons and electrons. The released protons are transferred to the exterior of the membrane and the electrons together with the protons from the interior of the membrane are relocated to a final acceptor site that can be a nitrate anion. The nitrite thereby created can be the oxidizer in another reaction used by another reductase:

$$NO_2^- \longrightarrow N_2 .$$

Alternatively, the nitrite can be involved with other inorganic anions such as an acid carbonate or sulfate in reactions leading to the formation of compounds with hydrogen: ammonia, methane, or sulfurated hydrogen:

$$NO_2^- \longrightarrow NH_3 , \quad HCO_2^- \longrightarrow CH_4 , \quad SO_4^{2-} \longrightarrow H_2S .$$

4.4 The Photosynthetic Revolution

The Earth is energetically an open system and a substantial flux of solar radiation has reached it since the moment of its creation. Coupled with the rotational motion of the planet, this flux has powered the machinery of oceanic and atmospheric motions. The primary energy sources for the newly emerged life on Earth were nucleoside triphosphates and small organic molecules such as monosaccharides that soon turned out to be exhaustible. Life became energetically independent only when it learned how to harness the practically inexhaustible solar energy or, more precisely, the fraction of it that reaches the surfaces of the oceans.

The possibility of utilizing solar energy by living cells is linked to the use of *chlorophyll* as a photoreceptor (Nitschke and Rutherford, 1991). The chlorophyll molecule contains an unsaturated carbon–nitrogen porphyrin ring (see Fig. C.4) with a built-in magnesium ion Mg^{2+} and phytol, a long saturated hydrophobic carbohydrate chain. The molecules of chlorophyll are easily excited in the optical range and equally easily transfer this excitation among each other, creating a light-harvesting system in an appropriate protein matrix. The last chlorophyll molecule in such a chain can become an electron donor and replace the $NADH + H^+$ fuel in a proton pump (see Fig. 4.4b).

The first organisms that found this possibility were most likely purple bacteria. Their proton pumps are also composed of two protein complexes built into the cell membrane (Fig. 4.5). In the protein complex called the type-II *reaction center* (RC), two electrons from the excited chlorophyll are transferred with two protons from the cell interior to a quinone derivative Q with a long carbohydrate tail. Q is soluble in the membrane. When reduced to quinol QH_2, it carries the two hydrogen atoms inside the membrane to the next complex that contains a protein macromolecule called cytochrome bc1. The macromolecule catalyzes the electron transfer from each hydrogen atom onto another macromolecule called cytochrome c, while the remaining proton moves to the extracellular medium.

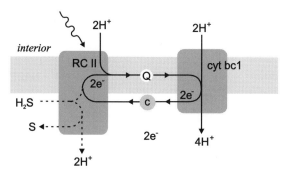

Fig. 4.5. A proton pump in purple bacteria utilizing solar radiation energy. In the first protein complex called a type-II reaction center (RC II), an electron from an excited chlorophyll molecule (the 'primary donor') is transferred to a molecule of a quinone derivative (Q) along with a proton taken out of the cell interior. The quinone derivative molecule carries two hydrogen atoms formed this way to another protein complex containing cytochrome bc1. In the latter complex hydrogen atoms are again separated. A proton is moved outside the cell, while an electron reduces a molecule of the water-soluble cytochrome c, which carries it back to the primary donor. An alternative source of electrons (*broken line*) for purple sulfur bacteria can be the molecule of sulfurated hydrogen H_2S

Cytochromes are proteins that contain a *heme*, a porphyrin ring with a built-in iron ion Fe^{2+}, that may also exist in a form oxidized to Fe^{3+}. Cytochrome c is a water soluble protein that drives electrons outside the cell membrane back to the reaction center. This completes the cyclical process during which two protons are carried from inside the cell to the outside. An alternative source of electrons needed to restore the initial state of the reaction center used, for example, in purple sulfur bacteria, may be molecules of sulfurated hydrogen H_2S. In contrast to the oxidation of NADH, oxidation of H_2S to pure sulfur is an endoergic reaction (consuming and not providing free energy) and it cannot be used directly for proton pumping.

The proton concentration difference on each side of the cell membrane is further used by purple bacteria to produce ATP in the same way that it is produced by non-photosynthetic bacteria. An alternative way of using solar energy was found by green bacteria (Fig. 4.6). In the protein complex called the type-I reaction center, an electron from photoexcited chlorophyll is transferred to a water-soluble protein of *ferredoxin*. The lack of electrons in the chlorophyll molecules is compensated uncyclically from sulfurated hydrogen decomposition. The electron carrier in ferredoxin is the iron–sulfur center composed

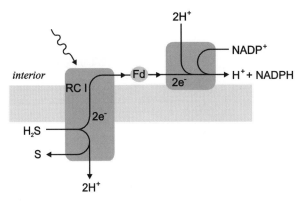

Fig. 4.6. The utilization of solar energy by green sulfur bacteria. In the first protein complex called the type-I reaction center (RC I), an electron from an excited chlorophyll molecule is transferred to the water-soluble protein molecule of ferredoxin (Fd) which carries it to the complex of $NADP^+$ (nicotinamide adenine dinucleotide phosphate) reductase. The deficit electron in the initial chlorophyll is compensated in the process of oxidation of sulfurated hydrogen H_2S. The reduced hydrogen carrier $NADPH + H^+$ serves as a fuel in the Calvin cycle, synthesizing sugar from water and carbon dioxide

of four Fe atoms directly bound to four S atoms. After the reduction of iron, ferredoxin carries electrons to the next protein complex where they bind to protons moving from the cell interior and reducing the molecules of $NADP^+$ (nicotinamide adenine dinucleotide phosphate) to $NADPH + H^+$. The entire system is not really a proton pump since there is no net proton transport across the cell membrane. The system transforms light energy into fuel energy in the molecules of NADPH, together with hydrated protons H^+ that carry the original charge of $NADH^+$. This fuel is used in the synthesis of glucose from CO_2 and H_2O in the Calvin cycle. This cycle is in a sense reverse to the Krebs cycle (compare Sect. 4.6). In the Calvin cycle, ATP is still being used. In the final balance, after the oxidation of glucose in the same way as for non-photosynthetic bacteria, an excess of ATP is produced.

A combination of the two methods of using solar energy offers the optimal solution. It was found by cyanobacteria. In cyanobacteria, cytochrome c1 was replaced by the slightly different cytochrome f, whereas cytochrome c was replaced by *plastocyanin* (PC) and used as an electron carrier between type-II and type-I reaction centers. The centers are now referred to as *photosystem* II (PS II) and *photosystem* I (PS I), respectively (Fig. 4.7). The electron carrier in plastocyanin

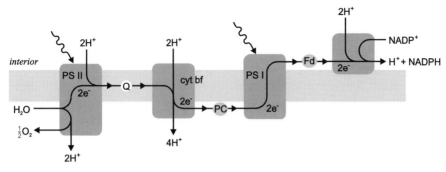

Fig. 4.7. A proton pump using solar energy in cyanobacteria can be thought of as a combination of the proton pump in purple bacteria (type-II reaction center, now called photosystem II or PS II) and the photosynthetic system of green bacteria (type-I reaction center, now called photosystem I or PS I). The coupling of the two systems is realized by a water-soluble molecule of plastocyanin (PC) with a copper ion serving as an electron carrier. The final electron donor is water H_2O which, after donating electrons and protons, becomes molecular oxygen O_2. The proton concentration difference between the two sides of the cell membrane is used to produce ATP via H^+ATPase (see Fig. 4.4a) working in the reverse direction. In principle, the photosynthetic system in the thylacoid membranes of chloroplasts that are organelles of eukaryotic plant cells are identical structures

is the Cu^{2+} copper ion reducible to Cu^+ and directly bound via four covalent bonds to surrounding amino acids.

However, the greatest breakthrough resulted not from the combination of the two photosystems, but from the utilization of water as the final electron donor (and a proton donor, hence a hydrogen donor). The dissociation of hydrogen atoms from a water molecule H_2O turned it into a highly reactive molecular oxygen O_2 gas that was toxic to the early biological environment. Initially, it oxidized only iron Fe^{2+} ions that were soluble in great amounts in ocean water at that time. As a result of this oxidation, poorly soluble Fe^{3+} was formed. This sedimented, giving rise to modern iron ore deposits. Simultaneously, the increased production of sugars from CO_2 and H_2O reduced ocean acidity and caused a transformation of acidic anions of HCO_3^- into neutral CO_3^{2-} ions. The latter reacted with the Ca^{2+} ions initially present in high concentrations, leading to sedimentation of insoluble calcium carbonate $CaCO_3$. The membranes of cyanobacteria captured the calcium carbonate and produced a paleobiological record of these processes in the form of fossils called stromatolites (Schopf, 1999).

The formation of calcified stromatolites depleted CO_2 from the atmosphere. When a deficit of compounds capable of further oxidation occurred, molecular oxygen O_2 started to be released into the atmosphere. Along with molecular nitrogen N_2 formed by the reduction of nitrates, they brought about the contemporary oxygen–nitrogen-based atmosphere containing only trace quantities of carbon dioxide. Life had to develop in a toxic oxygen environment from that point onward. The problem was solved by the mechanism of oxidative phosphorylation used by modern aerobic bacteria and all higher organisms. A proton pump that used inorganic anions as final electron acceptors (see Fig. 4.4b) was replaced by a pump in which the final electron acceptor is molecular oxygen (Fig. 4.8). Use has been made of the protein complex with cytochrome bc1 transferring electrons from quinone Q to the water-soluble cytochrome c (see Fig. 4.7), a mechanism found earlier by purple bacteria.

The source of electrons transferred to the quinone can be the NADH + H^+ fuel generated by glycolysis and in the Krebs cycle or, directly, $FADH_2$ (reduced flavin adenine dinucleotide) produced in one stage of the Krebs cycle. Electrons can also come from an inorganic source (chemotrophy). For example, nitrifying bacteria can oxidize ammonia to nitrite and further to nitrate:

$$NH_3 \longrightarrow NO_2^- \longrightarrow NO_3^- \ ,$$

using molecular oxygen.

Nature demonstrates here, as it has many times before and since, its ability to use environmental pollution to its advantage. It will be interesting to see, for example, what use it finds for the countless tons of plastic bottles deposited in modern garbage dumps.

4.5 Origins and Structure of the Eukaryotic Cell. Further Stages of Evolution

In its 19th century interpretation, Darwin's theory of natural selection favoring the survival of the fittest could be readily associated with the struggle for survival, tooth and nail, and all that was called the 'law of the jungle'. This, in turn, became the foundation of the ideological doctrine of many totalitarian regimes on the 20th century political landscape ('the struggle of classes' and 'the struggle of races'). To the credit of Lynn Margulis (1981, 1998), the third scientist we mention by name in this chapter, some emphasis has been given to the fact

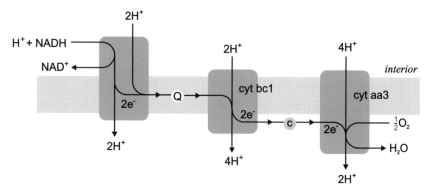

Fig. 4.8. Protein pump of heterotrophic aerobic bacteria. Electrons from the fuel in the form $NADH + H^+$ (produced in glycolysis and in the Krebs cycle) are transferred via a quinone derivative (Q) to the protein complex with cytochrome bc1 and then via cytochrome c to the complex with cytochrome aa3. The final electron acceptor is molecular oxygen O_2. During the transfer of two electrons along the membrane, eight protons are pumped across it. The proton concentration difference between the two sides of the membrane is used to produce ATP by H^+ATPase (see Fig. 4.4a) working in reverse. This is in principle identical to the mechanism of oxidative phosphorylation in the mitochondrial membrane, which is an organelle present in all eukaryotic cells

that survival can be accomplished not only through struggle but also through peaceful coexistence, called *symbiosis* in biology. Many clues support the significance of symbiosis in the formation of modern eukaryotic cells.

Figure 4.9 shows a simplified scheme of the phylogenetic tree of living organisms. There is a clear division between archaic bacteria (*Archaebacteria*) and true bacteria (*Eubacteria*) that may already have emerged in the earliest periods of life on Earth. The history of subsequent differentiation of prokaryotic organisms within these two groups, however, is not so clear. The modern phylogenetic tree is based on differences in DNA sequences coding the same functional enzymes or ribozymes. The more differences are found in the DNA sequences, the earlier the two branches of the compared species must have divided. However, the results obtained by comparing, for example, ribosomal RNA with the genes of the proteins in the photosynthetic chain show that they differ greatly from each other and hence lead to very dissimilar reconstructions of the history of the evolution of photosynthesis (Doolittle, 1999; Xiong et al., 2000). The reason for this ambiguity is the lateral gene transfer processes by which genes are borrowed by one organism from another. Branches can split away and merge over time.

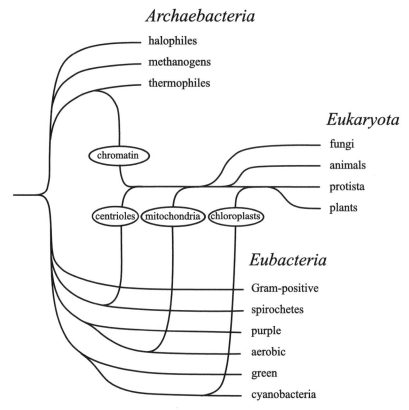

Fig. 4.9. Simplified and somewhat hypothetical phylogenetic tree of living organisms. The earliest are the two groups of prokaryotic organisms, *Archaebacteria* and *Eubacteria*. As a result of the merger of prokaryotic cells with different properties, eukaryotic cells (*Eukaryota*) were formed. The latter further evolved either in undifferentiated form as single-cell organisms (the kingdom of *protista*) or differentiated into multicellular organisms (the kingdoms of heterotrophic *fungi* with multinuclear cells, heterotrophic *animals* and phototrophic *plants*)

Therefore, the phylogenetic tree of *Eubacteria* shown in Fig. 4.9 must be viewed with caution, especially since only the taxons essential to our discussion are depicted.

Gram-positive bacteria (the name originates in the staining process proposed by Gram) have only a single external membrane and hence are more sensitive to antibiotics. Fortunately, this group includes most of the pathogenic bacteria. *Spirochetes* have developed mechanisms of internal motion for the entire cell. Photosynthetic *purple bacteria* with the type-II reaction center can be sulfuric or non-sulfuric. They must be evolutionarily close to *aerobic bacteria* because they utilize the same mechanism of reduction of cytochrome c through the protein complex with cytochrome bc1 (see Figs. 4.5 and 4.8). To be more precise, biologists do not distinguish a taxon of bacteria that have such a name. Most aerobic bacteria, including the common *Escherichia coli*, can survive in oxygen-deprived conditions. *Green bacteria* and *cyanobacteria* have in common the mechanism of sugar photosynthesis with the use of type-I reaction centers.

Lateral gene transfer can be fully accomplished when several simple prokaryotic cells merge into one supercell. According to Margulis, this is how eukaryotic cells first formed. Most probably, a thermophilic bacterium with a stable genomic organization whose DNA was protected by proteinaceous histones that combined to form a prototype of *chromatin* entered into a symbiotic arrangement with a spirochete containing a motile apparatus formed from microtubules (see Fig. 4.9). This combination gave rise to a *mitotic* mechanism of cell division. Chromatin with a doubled amount of genetic material organizes itself after replication into *chromosomes* pulled in opposite directions by a karyokinetic spindle formed from *centrioles* by self-assembling microtubules that consume GTP as fuel (see Sect. 4.4).

In the next stage, cells with *nuclei* that contained chromatin assimilated several aerobic bacteria (Fig. 4.9). The latter were transformed into *mitochondria*, the power plants of cells that synthesized ATP via the oxidative phosphorylation mechanism shown in Fig. 4.8. The *Eukaryota* cells thereby formed continued to evolve (Fig. 4.9) in undifferentiated forms as single-cell organisms (the *protista* kingdom) or in differentiated forms as multicellular organisms (*fungus* and *animal* kingdoms). All the above organisms were heterotrophs. The assimilation of prokaryotic cells of cyanobacteria as *chloroplasts* led to the formation of phototrophic single-cell organisms and multicellular *plants*.

So far, we have only discussed the symbiosis of *different* prokaryotic cells. An encounter of two organisms belonging to *the same* species will lead either to cannibalism or to symbiosis. According to Margulis, symbiotic encounters led to the emergence of *sex*. A symbiotic cell becomes diploidal, i.e., contains two slightly different copies of the same genome. Obviously, reproductive cells nurtured by a parent organism before entering into new symbiotic arrangements are *haploidal* and contain only one copy of the genetic material. A reduction of the genetic information took place in the process of generating reproductive cells when *meiotic* division replaced mitotic division. The evolutionary advantage of sexual reproduction is due to the *recombination* (*crossing over*) of maternal and paternal genes in meiotic division. As a result, the genetic material undergoes a much faster variability compared to random point *mutations* and such changes are seldom lethal.

Let us summarize the most important stages in the evolution of life on Earth. We know some of the elements of the puzzle quite well since they left visible traces that can be precisely dated. Some other elements are somewhat hypothetical and have not yet been backed by reliable precisely dated discoveries. They are indicated by question marks. We finish with a survey of events over the last billion years, a period in which life on Earth reached a supra-cellular level of organization (Cowen, 1990):

- 4.6 billion years ago – creation of Earth.
- 3.9 billion years ago – end of Big Bombardment. The surface temperature of Earth is lowered so that the early gaseous envelope of volcanic or cometic and meteoric origin differentiates into atmosphere (mainly CO_2) and ocean (mainly H_2O with the addition of simple organic compounds).
- ? – emergence of nucleosides and RNA allows the storage of information and self-replication subjected to Darwinian selection based on the survival of the fittest. Nucleoside triphosphates become the key source of free energy for the reactions of polymerization.
- ? – emergence of key elements of the prokaryotic cell machinery: membranes, membrane channels, ribosomes, DNA and RNA polymerases, and proton pumps. Phospholipid bilayers with selective protein channels isolate various types of RNA from their surroundings and protect a supply of the required nucleotides. Primitive ribosomes express the information contained in RNA in terms of the proteins produced. Proteins turn out to be more efficient as catalysts than RNA. The double-stranded DNA is found to be a more stable information carrier than RNA. Protein enzymes enable the use of sugars as a source of free energy in recycling nucleoside diphospates into nucleoside triphospates. This is first carried out through the process of fermentation and then in a more efficient process of membrane phosphorylation with inorganic anions as oxidizers.
- 3.5 billion years ago – discovery of photophosphorylation is linked to dissociation of H_2O. The resulting molecular O_2 oxidizes Fe^{2+} to the poorly soluble Fe^{3+}, while sugar synthesis from CO_2 and

H_2O lowers the acidity of the ocean, leading to the elimination of insoluble calcium carbonate.

- 2 billion years ago – molecular oxygen starts to accumulate in large quantities in the atmosphere. The chemical machinery of oxidative phosphorylation is discovered.
- ? – thermophilic bacteria possessing DNA protected by histone proteins enter into a symbiotic relationship with motile bacteria possessing microtubular cytoskeletons. The cell nucleus is formed and the mechanism of mitotic cell division is carried out.
- ? – sex is discovered and with it the amount of genetic information is doubled. A diploidal cell is formed and its meiotic division, reducing the amount of information, offers the possibility of genetic recombination.
- 1.4 billion years ago – symbiotic coexistence of cells containing nuclei with oxygen bacteria and possibly cyanobacteria is established. Oxygen bacteria play the role of mitochondria and cyanobacteria that of chloroplasts. The modern eukaryotic cell is born.
- 1.0 billion years ago – some cells that arose from cell division stop dividing, which leads to the emergence of an embryo that differentiates into a multicellular animal or plant organism.
- 540 million years ago – animals start developing skeletons from calcium carbonate or silicate, thus enabling the formation of fossils that give us a lasting chronological record. Beginning of Paleolithic era (the Cambrian explosion).
- 440 million years ago – symbiosis of plants and fungi allows their emergence on land.
- 100 million years ago – perfection of the most effective ways of protecting the embryo (angiospermous plants and mammals with placenta).
- 6 million years ago – first hominid forms are recorded.
- 150 thousand years ago – first *Homo sapiens* appears on Earth.

4.6 The Main Metabolic Pathways. Enzymes

Figure 4.10 illustrates schematically the main metabolic pathways of energy and matter processing, common to contemporary bacteria (prokaryotic cells) as well as animals, fungi and plants (*Eukaryota*). Only the most important biochemical reactions have been shown; today we know close to a hundred times as many reactions composing the metabolism (Stryer, 2002). For simplicity, substrates are represented by black dots and reversible or practically irreversible reactions by bi- or unidirectional arrows.

It is easy to see the vertical pathway of *glycolysis*, the oxidation of the most common monosaccharide, glucose, to pyruvate. It is equally easy to see the closed *citric acid cycle* of Krebs. The archaic origin of the main metabolic pathways is evident not only from their universality (from bacteria to man) but also from the presence in many of these reactions of nucleoside triphosphates, mainly ATP (adenosine triphosphate). If a given reaction is connected with the hydrolysis of ATP to ADP (adenosine diphosphate), it is marked in Fig. 4.10 by a letter P at the start of the reaction. If, on the other hand, a given reaction is linked to a synthesis of ADP and an orthophosphate group into ATP (a process called phosphorylation), it is marked by a letter P at the end of a reaction.

From the chemical point of view, the process of going from glucose, $C_6H_{12}O_6$, to pyruvate, $CH_3-CO-COO^-$, is an oxidation reaction and it consists in taking hydrogen atoms from glucose. NAD^+ (nicotinamide adenine dinucleotide) is a universal oxidant (an acceptor of hydrogen, i.e., simultaneously an electron and a proton). This is also a relict of the RNA world. The process wherein NAD^+ (or FAD in the case of one reaction of the Krebs cycle) accepts two hydrogen atoms is marked in Fig. 4.10 by a letter H at the end of a given reaction.

An overall balance of the glycolysis reaction, i.e., an oxidation of glucose to a pyruvate, takes the form

$$C_6H_{12}O_6 + 2\,NAD^+ + 2\,ADP + 2\,P_i \qquad\qquad (4.1)$$
$$\longrightarrow 2\,CH_3-CO-COO^- + 2\,NADH + 2\,H^+ + 2\,ATP + 2\,H_2O \ .$$

During this process two molecules of NAD^+ are reduced by four atoms of hydrogen:

$$C_6H_{12}O_6 + 2NAD^+ \longrightarrow 2CH_3-CO-COO^- + 2H^+ + 2NADH + 2H^+$$
$$(4.2)$$

(two protons are obtained from the dissociation of pyruvic acid into a pyruvate anion, whereas another two protons transfer the original positive charge of NAD^+) and the phosphorylation of two molecules of ADP to ATP takes place according to the equation

$$ADP + P_i + H^+ \longrightarrow ATP + H_2O \ . \qquad\qquad (4.3)$$

In a neutral water environment, ATP is present as an ion with four negative charges, ADP with three negative charges and an orthophosphate P_i with two.

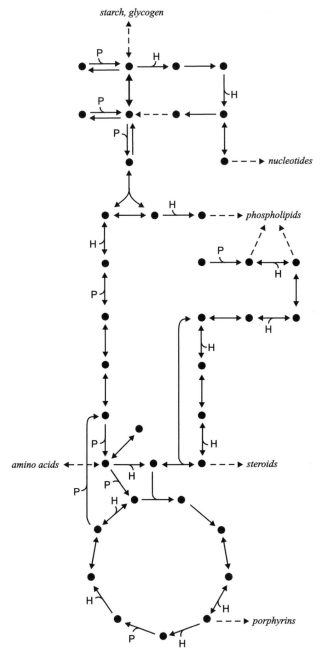

Fig. 4.10. Outline of the main metabolic pathways. Substrates of successive reactions are represented by *black dots*. Reversible or practically irreversible reactions catalyzed by specific enzymes are represented, respectively, by *bi-* or *unidirectional arrows*

The entry of pyruvate into the Krebs cycle requires its further oxidation to acetate and carbon dioxide:

$$CH_3-CO-COO^- + H_2O + NAD^+ \qquad\qquad (4.4)$$
$$\longrightarrow CH_3{-}COO^- + CO_2 + NADH + H^+ \ .$$

As a result, a reduction of one molecule of NAD^+ is made by two atoms of hydrogen. The net balance in the Krebs cycle is

$$[CH_3{-}COO^- + H^+ + 2H_2O] + [3NAD^+ + FAD] + [GDP + P_i + H^+]$$
$$\longrightarrow 2\,CO_2 + [\,3\,NADH + 3\,H^+ + FADH_2\,] + [\,GDP + H_2O\,] \ . \quad (4.5)$$

Acetate enters into it, bound to a so-called co-enzyme A (CoA) as acetyl CoA. During one turn of the Krebs cycle a further reduction takes place, of three molecules of NAD^+ and one molecule of FAD (flavin adenine dinucleotide) involving eight atoms of hydrogen and phosphorylation of a molecule of GDP (guanosine diphosphate) to GTP (guanosine triphosphate). For greater clarity of the overall reaction, we have used square brackets to indicate the component subprocesses.

Photosynthesizing organisms use hydrogen in the form of NADPH $+ H^+$ (see Fig. 4.7) in the synthesis of glucose from CO_2 and H_2O in the Calvin cycle, whose overall balance equation takes the form

$$[\,6\,CO_2 + 12\,NADPH + 12\,H^+\,] + [\,18\,ATP + 18\,H_2O\,] \qquad (4.6)$$
$$\longrightarrow [C_6H_{12}O_6 + 6H_2O + 12NADP^+] + [18ADP + 18P_i + 18H^+] \ .$$

This cycle is in a sense reverse to the Krebs cycle. As for the Krebs cycle, we have used square brackets to denote summary component reactions, in order to show the net reaction more clearly. In the Calvin cycle, ATP is still being used. In the final balance, after the oxidation of glucose in the same way as for non-photosynthetic bacteria, an excess of ATP is produced.

Each metabolic reaction is catalyzed by a specific protein *enzyme*. Besides *accelerating* reactions many millions of times, enzymes have two other important functions: *regulatory* – a given reaction is needed for the cell only at a certain space and time; and *coupling* – processes of biological free energy or signal transduction occur only when several reactions are catalyzed by the same multienzymatic complex.

Enzymes have been given a special nomenclature (*Enzyme Nomenclature*, 1973). The name of each enzyme ends with an -*ase* suffix. Here we list the six main classes of enzymes and give several typical examples:

1. *Oxidoreductases* catalyze electron transfer. If there is a simultaneous transfer of a proton (i.e., the whole hydrogen atom), then we call them *dehydrogenases* or *transhydrogenases*. If the hydrogen acceptor is molecular oxygen, we use the name *oxidase*, and if the hydrogen acceptor is hydrogen superoxide we use the name *peroxidase*.

2. *Transferases* catalyze the transfer of a radical or a molecular group from one compound to another. Here, an important subclass contains *phosphotransferases* or *kinaseses*, which phosphorylate various substrates.

3. *Hydrolases* catalyze the hydrolysis of bonds of ester type, in particular proper ester bonds (*esterases* or *lipases*), peptide bonds (*peptidases* or *proteases*) and phosphodiester bonds (*phosphodiesterases* or *deoxyribonucleases*). An important subclass here are the already mentioned *ATPases*, in which the hydrolysis of ATP to ADP and P_i is coupled with various processes that require a supply of free energy, e.g., active transport across ion channels in biological membranes or movement along cytoskeletal structures.

4. *Lyases* catalyze the breaking of various bonds in a non-hydrolytic way. Particular examples are *decarboxylases*, which free carbon dioxide CO_2 from substrates.

5. *Isomerases* catalyze unimolecular reactions of intra-molecular isomerization. When this accompanies a relocation of groups inside molecules, we use the name *mutases* (e.g., acylmutase, phosphomutase).

6. *Ligases* (*synthetases*)indexsynthetase catalyze the synthesis of new bonds in conjunction with the breaking (not hydrolysis) of the pyrophosphate bond in ATP to give AMP and PP_i. Important examples are DNA and RNA *polymerases, transcriptase* and *reverse transcriptase*.

5 Molecular Biology of the Eukaryotic Cell

5.1 The Eukaryotic Cell as a System of Compartments

Under an optical microscope with small magnification (distance scale ≥ 5 m), a eukaryotic cell is seen as a drop of *cytoplasm* with a *nucleus* inside, surrounded by a *cytoplasmic membrane* (Fig. 5.1a). The largest possible magnification of the optical microscope (distance scale ≥ 0.5 m, a wavelength of visible light) can differentiate the cytoplasm into various *organelles* suspended in semi-liquid *cytosol* (Fig. 5.1b).

On the magnification scale of an electron microscope (distance scale ≥ 0.5 nm, Fig. 5.2) a system of endoplasmic membranes can be seen that spatially border particular organelles (Solomon et al., 2004): the *nucleus, mitochondria, rough* and *smooth cytoplasmic reticulum, Golgi apparatus* and *lysosomes. Centromeres* consisting of two *centrioles* that organize structures responsible for cell division are the only organelles without membranes. Figure 5.2 shows the main compartments distinguished by endoplasmic membranes in an animal cell. Plant cells (Fig. 5.3) contain in addition a cellulose *wall* that can be considered as

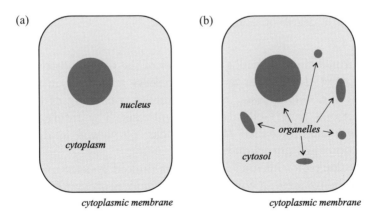

Fig. 5.1. A eukaryotic cell as seen under an optical microscope at small (**a**) and large (**b**) magnifications

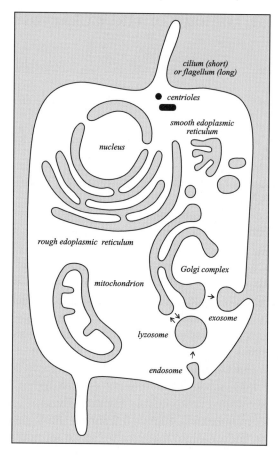

Fig. 5.2. Compartments of a eukaryotic cell seen under an electron microscope. *Solid lines* represent bilayer membranes and neighboring compartments differ in degree of shading

a transformed thick peptidoglycan layer of the Gram-positive bacteria, *vacuoles* that store water as well as reserve substance in the form of *granules*, and *chloroplasts*, large organelles consisting of three layers of membranes that facilitate photosynthesis. The internal flattened chloroplast bubbles called *thylacoids* can be viewed as disconnected *crista* (combs) of mitochondria.

The eukaryotic cell nucleus contains diploid genetic material organized into *chromatin*, a complex nucleic acid–protein structure. Three types of organelle have their own genetic material, different from the nuclear genetic material. As mentioned earlier, they probably originate from prokaryotic cells assimilated in the past. These are mitochondria

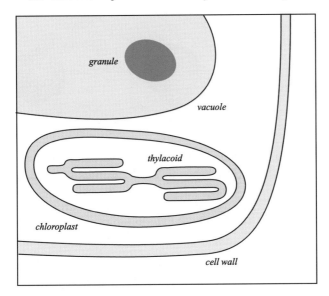

Fig. 5.3. Additional compartments of a plant cell. *Solid lines* represent bilayer membranes and neighboring compartments differ in degree of shading

originating from the former aerobic bacteria, chloroplasts originating from cyanobacteria, and centrioles originating from spirochetes, although the last statement still remains questionable.

Various supramolecular structures are of intermediate size between organelles and single protein or RNA macromolecules soluted in intracellular buffer. These are portions of a protein *cytoskeleton* and *multienzyme* or *enzyme–ribozyme complexes*. Figure 5.4 shows three examples of such supramolecular structures: a model DNA *polymerase* III complex (Stryer et al., 2002, Chap. 27), a *ribosome* (Stryer et al., 2002, Chap. 24), and a hypothetical and probably not very stable *metabolon* (Welch, 1985; Srere, 1987; Lyubarev and Kurganov, 1989).

Specific metabolic processes occur in their respective cell compartments or on membranes that enclose them. In particular:

- in the cell nucleus, *nucleic acid synthesis* takes place,
- in lysosomes, various *biopolymer hydrolysis* reactions occur,
- in the mitochondrial matrix, *Krebs cycle* and *fatty acid degradation* reactions occur,
- on the internal membranes of mitochondria, *oxidative phosphorylation* takes place,
- on the thylacoid membranes of chloroplasts, *photophosphorylation* takes place,

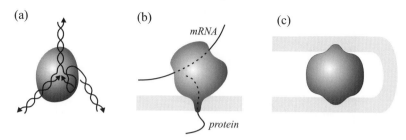

Fig. 5.4. Three examples of supramolecular structures with sizes exceeding 20 nm, surrounded by a homogeneous liquid medium and usually connected to a lipid membrane. (**a**) DNA polymerase III. (**b**) Ribosome. (**c**) Metabolon of tricarboxylic acid cycle. *Shaded regions* represent cross-sections of lipid membranes. Note that the membrane thickness is ~ 7 nm

- in the chloroplast stroma, *Calvin cycle* reactions take place,
- on membranes of the rough endoplasmic reticulum, *protein synthesis* takes place,
- on membranes of the smooth endoplasmic reticulum, *lipid synthesis* proceeds,
- in the interior of the Golgi apparatus, *polysaccharide synthesis* occurs.

After merging with membrane proteins, polysaccharides synthesized in the Golgi apparatus can be transported outside cells in a process called *exocytosis* (Fig. 5.2). The external cytoplasmic membrane armed with such *glycoproteins* recognizes various signals and selectively transports substances from the external environment to the cell interior in a process called *endocytosis* (Fig. 5.2). Transformations of monosaccharides, amino acids, and nucleotides take place in-between organelles in the cellular interior. This region is filled with supramolecular structures of the protein cytoskeleton that provide the cell with motile machinery.

5.2 Membrane Channels and Pumps

Transport across membranes that divide particular compartments of the eukaryotic cell is essential for its functioning. However, the purely lipid bilayer membranes are not permeable either for water molecules with their high electric dipolar moment or for ions endowed with an electric charge. For this purpose, various protein *channels* are necessary.

Although osmotic water transport across membranes has been the subject of intensive study for over a century, only recently have *aqua-*

(a)   (b)

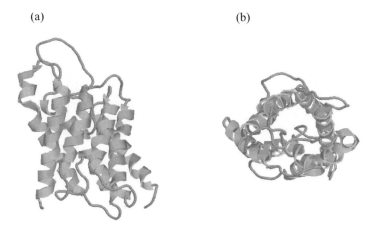

Fig. 5.5. Structure of aquaporin-1 water channel. View in directions parallel (**a**) and perpendicular (**b**) to the cytoplasmic membrane surface, showing the component secondary structure elements. In the present case, only α-helices occur. The drawing was made using the program Rasmol and Protein Data Bank (pdb) entry 1IH5 (Ren et al., 2001)

porin channels facilitating this transport been discovered (Agre et al., 2002; Murata et al., 2000). Figure 5.5 shows the spatial structure of the human aquaporin-1 water channel determined with the help of X-ray crystallography. The time of individual water molecule transitions through the channel is very short, of the order of 300 ps. The permeation mechanism, proposed on the basis of molecular dynamics simulations (de Groot and Grubmüller, 2001; 2005; Fujiyoshi et al., 2002), elucidates both the high water-permeation rate and the filtering properties with respect to protons. The former results from the highly collective behavior of the hydrogen bond network during individual water molecule transitions across the channel. The latter results from the form of a local electric field generated by some amino acid side chains in the middle of the channel, which forces the dipole moment of the passing molecule to rotate by 180°. This reorientation prevents the formation of a continuous network of hydrogen-bonded water molecules enabling proton transport via the Grotthus mechanism (see Fig. 5.17 below).

The concentration of three ions: K^+, Na^+ and Ca^{2+} is essential for many intra- and intercellular process. Formally, protein ion channels can be considered as typical enzymes, an ion on one or the other side of a membrane corresponding to two chemical states. Like enzymes, the ion channels have to be characterized by the three basic properties:

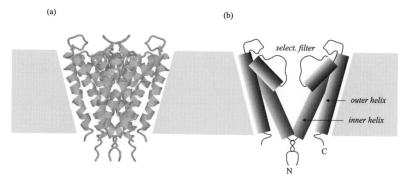

Fig. 5.6. Structure of the KcsA potassium channel. (**a**) Drawing made using the program Rasmol on the basis of pdb entry 1BL8 (Doyle at al., 1998), showing the component α-helices. (**b**) The particular functional elements are identified

high conduction rate, high selectivity, and possibility of control. The control of ion channels consists in 'gating' them, with a transition from closed to open state, and conversely. This can be realized either by applying an external *voltage* or by binding various *ligands* (Fuller and Shields, 1998, Chap. 9).

Most voltage-gated channels have a similar structure. They are tetramers consisting of four identical molecules (K^+ channels) or pseudotetramers consisting of four similar, almost identical domains (Na^+ and Ca^{2+} channels). Figure 5.6 shows the currently best known spatial structure of the K^+ channel of a particular type, KcsA, determined with the help of X-ray crystallography (Doyle et al., 1998). Each of the four component molecules (consisting of about 120 amino acids) contains three α-helices. One helix from each monomer enters a *selectivity filter* (Fig. 5.6) in the entrance pore which determines the maximum radius of an ion that can pass through the channel. The minimum radius is related to the hydration free energy (Morals-Cebrat, Zhou and MacKinnon, 2001). Gating by an applied external voltage results either from a counterclockwise rotation of two transmembrane helices in each monomer, with a nonzero dipolar moment (Doyle et al., 1998; Fig. 5.7) or from a movement of a charged NH_3^+ group at the end of a long flexible chain of an additional protein that behaves like a ball on a chain and blocks the channel (Zhou at al., 2001).

The spatial structure of the voltage-gated sodium channel is known to far less accuracy from cryo-electron microscopy studies (Sato et al., 2001). Its main α-subunit (mass $\sim$ 260 kda, which corresponds to about 2 300 amino acids) is composed of four homologous domains,

(a)   (b)

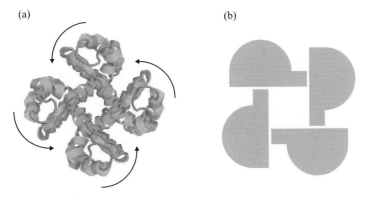

Fig. 5.7. KcsA potassium channel (**a**) closed and (**b**) open. The drawing was made using the program Rasmol with pdb entry 1BL8 (Doyle at al., 1998)

each of which contains six transmembrane α-helices and two longer loops that form extramembrane parts of the protein. It is interesting that the central pore is densely stained with ions from the bathing solutions and does not take part in the transport of ions from the extracellular to intracellular solution. Four actual transmembrane pores are located peripherally, one in each domain (Fig. 5.8). The voltage gating is related to a movement of the transmembrane helices.

The voltage-gated potassium and sodium channels play an essential role in the generation of an action potential in neuronal axons (see Sect. 8.6). However, for excitation of a nerve impulse in the axon,

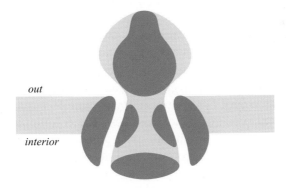

out

interior

Fig. 5.8. Schematic cross-section of the voltage-gated sodium channel as seen by cryo-electron microscopy (after Sato at al., 1999). The central pore (*lighter grey*) is densely stained with ions from the bathing solution and not involved in the transport of ions. The actual transmembrane pores are located peripherally in component domains

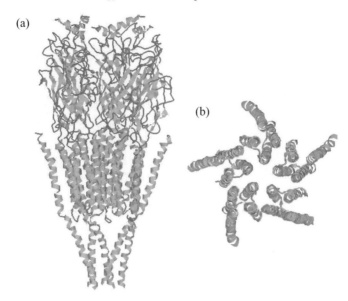

Fig. 5.9. Acetylcholine-gated sodium channel. The strands of β-structure are represented by *broad arrows* pointing to the C-terminal, and regions of the α-helix are shown as helical ribbons. (**a**) Drawing made using the program Rasmol on the basis of pdb entry 1QED (Miyazawa et al., 2003). (**b**) Rasmol drawing from pdb entry 2BG9 (Unwin, 2005)

ligand-gated sodium channels are necessary, controlled by neurotransmitters (Solomon, 2004). The best understood ligand-gated sodium channel is the nicotinic acetylcholine-gated channel, also called the *nicotinic acetylcholine receptor* (Brejc et al., 2001; Miyazawa et al., 2003). It is a glycoprotein with mass ~ 290 kda, composed of five similar subunits (Fig. 5.9a). Two of them are identical (the α-subunits) and bind acetylcholine, a small non-peptide hormone. All five subunits form a ring that spans the membrane and encircles a pore inside the membrane (Fig. 5.9b). Binding of acetylcholine to two α-subunits initiates their rotational movements. These are communicated to inner α-helices of other subunits shaping the pore and this opens the gate, making the whole pore hydrophilic and accessible to Na^+ ions.

All channels considered up to now conduct ions *passively* in the direction of lower concentration, i.e., the only driving force was the difference in chemical potential on either side of the membrane. However, there is a large group of channels that conduct the ions *actively*, against the concentration gradient. Such *active transport* in the direction contradicting the second law of thermodynamics is possible only

(a)   (b)

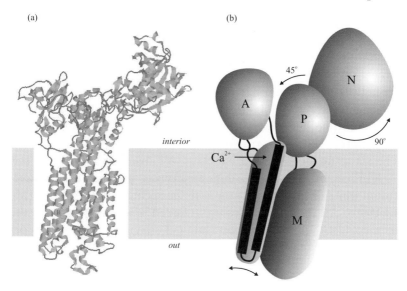

Fig. 5.10. Structure of the ATP-driven calcium pump, Ca^{2+}-ATPase. (**a**) Drawing made using the program Rasmol on the basis of pdb entry 1SU4 (Toyoshima et al., 2000), showing the particular secondary structure elements, α-helices and β-strands. (**b**) Schematic representation of component subunits. See text for details

at the expense of coupling to another chemical reaction, usually ATP hydrolysis, which serves as a free energy donor. Free energy transduction processes are one of the most characteristic and important thermodynamic processes in biological systems and will be considered in detail in Chaps. 8 and 9. Ion channels performing active transport are called *ion pumps*.

H^+, Na^+/K^+ and Ca^{2+} pumps (H^+-, Na^+/K^+- and Ca^{2+}-ATPases) are essential for biological functions, but the best known structure and mechanism of action is associated with the latter (Lee, 2002; Lykke-Moller et al., 2004). It is a protein monomer of mass ~ 110 kda (about 1000 amino acids) consisting of three cytoplasmic and two transmembrane domains (Fig. 5.10).

The *anchor* cytoplasmic domain A (about 50 amino acids) is connected rather rigidly to the transmembrane region. In contrast, the largest of the three cytoplasmic domains, the *nucleotide-binding* domain N (about 240 amino acids), has a considerable flexibility and can rotate up to 90° and penetrate the central, *phosphorylation* domain P (about 60 amino acids). Only after this penetration can the ATP molecule bound at the surface of the N domain phosphorylate one of the aspartate residues in the interior of the P domain. The two Ca^{2+} ions are bound cooperatively between transmembrane domains. The conformational changes that accompany the phosphorylation reaction move one of the transmembrane domains composed of two α-helices and close a cytosolic entrance for calcium ions, thereby preventing backflow before these ions are released on the other side of the membrane. Steroid inhibitors bound to the protein on this side can completely block the transport of ions.

The structures of H^+ and Na^+/K^+-ATPases are similar to that of Ca^{2+}-ATPase, but the first occurs as a dimer with two additional glycoprotein units of mass ~ 35 kda and the second as a hexamer (Moller et al., 1996). All these ATPases are referred to as P-type ATPases, because a covalent high-energy phosphoryl-enzyme intermediate occurs in their chemical reaction pathways.

5.3 Substrate, Oxidative, and Photo Phosphorylation

The ion pumps are one of many classes of biological molecular machines that consume free energy supplied by ATP. Other classes will be described before long. In Chap. 4, we indicated the three sources of ATP in the eukaryotic cells: substrate, oxidative, and photo phosphorylation. Here we present more details of each on the molecular level.

Substrate phosphorylation takes place in the next reaction after splitting the hexose molecule into two trioses in the glycolysis pathway (Fig. 4.10). It is a glyceraldehyde-3-phosphate dehydrogenation reaction. As a matter of fact, this reaction consists simultaneously of two processes: substrate dehydrogenation and substrate phosphorylation (Blumenfeld, 1974, Chap. 8; Darnell at al., 1999, Chap. 15). Both the processes take place on the same enzyme, *glyceraldehyde-3-phosphate dehydrogenase*, whose detailed structure is not yet known. If we write down the substrate formula in the form Ac–H, where the acyl group $Ac- \equiv PO_3^{2-}-O-CH_2-CHOH-CO-$, dehydrogenation can be considered to proceed through an enzyme acylation stage (Fig. 5.11a). With high probability, the acyl group is phosphorylated before an intermediate Ac–OH delivery. In this way, glyceraldehyde-3-phosphate dehydrogenase appears to be a chemochemical molecular machine that couples the dehydrogenation reaction donating the free energy with the phosphorylation reaction accepting the free energy (see Sect. 8.2). The delivery of the intermediate partly decouples the process of free energy transduction. The phosphate group with the high free energy bond is

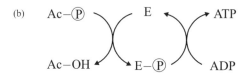

(a)   Ac−H + NAD⁺ 　　　　 E−SH 　　　　 Ac−Ⓟ

　　　　　　　　　　　　　　　Ac−OH

　　　NADH + H⁺ 　　　 E−S−Ac 　　　　 Pᵢ

(b)   Ac−Ⓟ 　　　　 E 　　　　 ATP

　　　Ac−OH 　　 E−Ⓟ 　　　 ADP

Fig. 5.11. (a) Since it proceeds on the same enzyme, the process of sub-strate phosphorylation is coupled to the process of substrate dehydrogena-tion. (b) In a subsequent reaction the phosphate group is transferred from the substrate to ADP which results in the creation of ATP. E denotes the free enzyme and Ac denotes the acyl group of the substrate (see text for more detailed explanation)

transferred from the substrate to ADP in a subsequent reaction, whose product is ATP (Fig. 5.11b).

Substrate phosphorylation is only a minor source of ATP. The main source is membrane phosphorylation. It took almost a decade until bio-chemists accepted the chemiosmotic concept of Peter Mitchell, stating that the free energy donor in membrane phosphorylation is a trans-membrane proton gradient (Darnell at al., 1999, Chap. 15; Stryer et al., 2002, Chap. 19). A molecular machine that transduces free energy in this case is ATP *synthase* (Stock et al., 2000; Capaldi and Aggeler, 2002).

In fact, ATP synthase consists of two machines: a water-soluble ro-tary motor called the F_1 *portion*, driven by ATP hydrolysis to ADP and P_i, and an intramembrane rotary motor called the F_o *portion*, driven by the transmembrane flow of protons. Both component ma-chines can operate reversibly in both directions and in ATP synthase are linked by a peripheral stalk that permits transmission of the two rotational motions (Fig. 5.12). In effect, proton flow drives the ATP synthesis.

The F_1 portion consists of five subunit types, labeled from α to ϵ, and has a molecular mass of about 380 kda. Three β subunits, each of mass 52 kda, i.e., containing about 450 amino acids, bind a nucleotide in a cleft between two domains that can move with respect to one another around a hinge. Three similar α subunits of mass 56 kda

(a)   (b)

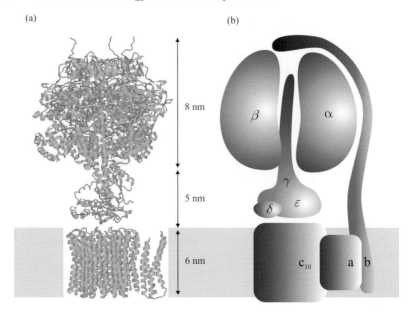

Fig. 5.12. Structure of ATP synthase. (**a**) Drawing made using the program Rasmol with pdb entry 1E79 (Gibbons et al., 2000) for the matrix portion F_1 end 1C17 (NMR data, Rastogi and Girvin, 1999) and for the intramembrane portion F_o, showing the particular secondary structure elements, α-helices and β-strands. (**b**) Schematic representation of component subunits. See text for details

(about 500 amino acids) are more rigid. They do not bind nucleotides but, together with the β subunits, form a muff within which a single γ subunit (34 kda, about 300 amino acids) linked to δ and ϵ subunits (14 and 6 kda, respectively) can rotate.

The F_o portion consists of a single a subunit, two b subunits, and 10 to 14 c subunits. The a subunit is a 'stator' and the c subunits, organized in a ring, form a 'rotor'. The b subunits form the periphelial stalk linking the F_o stator to the F_1 $3\alpha/3\beta$ muff. On the interface between the a subunit and the c subunit ring, there are two half-channels able to translocate protons. A proton transition from one half-channel linked to the outer membrane area to a second half-channel linked to the intra-membrane area takes place through the c subunits and requires a full rotation of the c ring. The F_o rotor and the F_1 rotor are connected in a noncovalent way and hence a torque resulting from the proton flow is transmitted to F_1 via the ϵ subunit and the γ shaft. The latter, pushing cyclically mobile domains of the three β subunits, facilitates nucleotide binding and rebinding, thus driving the

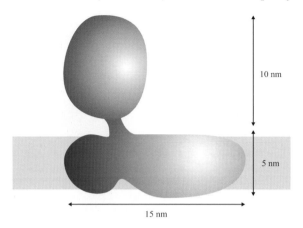

Fig. 5.13. Schematic structure of NADH:Q oxidoreductase as seen by cryo-electron microscopy (after Grigorieff, 1999)

ADP phosphorylation to ATP (Stock et al., 2000; Capaldi and Aggeler, 2002).

ATP synthase can operate in the reverse direction and it is then called H^+-ATPase. This is an example of an F-type ATPase, as the substrates and the products of the catalyzed reaction bind noncovalently to one of the F subunits of the enzyme.

When the enzyme operates in the membrane phosphorylation direction, the transmembrane proton gradient originates either from hydrogen oxidation in mitochondria or from use of light energy in chloroplasts.

The chain of charge transfer protein complexes in the mitochondrial inner membrane is shown in Fig. 4.8. The mitochondrial respiratory chain begins with NADF:Q oxidoreductase (Grigorieff, 1999). It utilizes $NADH + H^+$ formed during glycolysis and the citric acid cycle and transfers two electrons to quinone. This transfer is coupled to the translocation of two protons from the matrix to the intramembrane area. The X-ray structure is not yet known. Low-resolution electron microscopy reveals a characteristic L shape (Schultz and Chan, 2001). Two portions can be distinguished: a *membrane arm* of mass 370 kda, containing an Fe_2S_2 cluster, and a *matrix arm* of mass 520 kda, containing an Fe_4S_4 cluster (Fig. 5.13).

The second link of the electron transport chain in mitochondria is quinol:cytochrome c-oxidoreductase, the cytochrome bc_1 complex (Fig. 5.14). It is a protein dimer, each monomer consisting of some 2 150 amino acids (Darrouzet et al., 2001; Crofts and Berry, 1998;

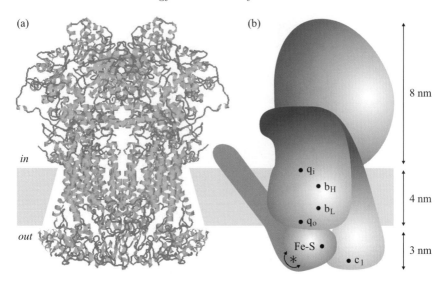

Fig. 5.14. Structure of quinol:cytochrome c-oxidoreductase, the cytochrome bc_1 complex. (**a**) Drawing made using the program Rasmol on the basis of pdb entry 2BCC (Zhang et al., 1998), showing the particular secondary structure elements, α-helices and β-strands, composing the supramolecular dimer. (**b**) Schematic representation of the component subunits of the monomer. See text for details. The *asterisk* indicates the position of a rotation axis of the moving domain of the iron–sulfur complex

Zhang et al., 1998). Functionally, four subunits can be distinguished in the monomer (Fig. 5.14b). The largest, intramatrix subunit (900 amino acids) is not involved in redox processes. The cytochrome b complex (570 amino acids, i.e., some 8.5 thousand atoms) transfers one electron from an outer quinol binding site q_o through two b-type heme groups b_L ('lower') and b_H ('higher') to an inner semiquinone binding site q_i. The iron–sulfur protein complex (270 amino acids, some 4.0 thousand atoms) transfers the second electron from the quinol binding site q_o to a c_1-type heme. The latter is a component of the cytochrome c_1 complex (440 amino acids, some 6.5 thousand atoms) that transfers this electron further to a cytochrome c molecule, soluble in the intramembrane mitochondrial medium.

A cycle of quinol to quinone reduction (see Sect. 8.4 and in particular Fig. 8.11) appears to be composed of two subcycles: that of quinol to semiquinone reduction proceeding on one monomer, and that of semiquinone to quinone reduction proceeding on the other (Fig. 5.15). There is a cave in-between the two monomers that facilitates diffusion

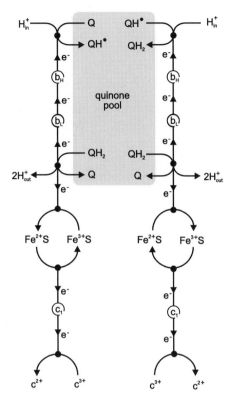

Fig. 5.15. Electron transfer processes taking place in two component monomers of quinol:cytochrome c-oxidoreductase. In each monomer, one electron is transferred from the outer quinol binding site through the lower and higher cytochromes b, b_L and b_H, respectively, to the inner binding site. The latter binds either the quinone Q releasing the semiquinone $QH^\bullet$, or the semiquinone $QH^\bullet$ releasing the quinol QH_2. The second electron in each monomer is transferred through the mobile Fe–S center to the cytochrome c_1, from which it is taken by the water soluble cytochrome c. Diffusion of various forms of quinone molecules takes place in a cave in-between the two monomers, linked to the whole membrane quinone pool. As a result, one quinol molecule with two hydrogen atoms is transferred within the cave from the inner to the outer side of the membrane in which the considered complex is built

of all forms of the quinone molecules between the appropriate binding sites (Zhang et al., 1998). The transfer of two electrons perpendicularly to the membrane, combined with the transfer of two hydrogens on a quinol molecule diffusing in the cave in the opposite direction, is equivalent to the transfer of two protons. The large-scale domain

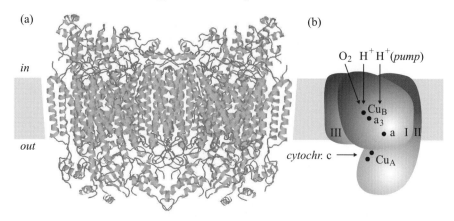

Fig. 5.16. Structure of the cytochrome c oxidase. (**a**) Drawing made using the program Rasmol on the basis of pdb entry 1OCC (Tsukihara et al., 1996), showing the particular secondary structure elements, α-helices and β-strands, of a dimer. (**b**) Schematic representation of three component subunits of the monomer. See text for details

movement in the iron–sulfur complex (Fig. 5.14b) makes it possible to transfer a single electron per cycle. This is the main mechanism minimizing energy losses (Darrouzet et al., 2001).

The third and terminal link of the electron transport chain in mitochondria is cytochrome c oxidase (Michel, 1998; Rottenburg, 1998; Bränden et al., 2001). It is a dimer, each monomer being a protein complex of molecular weight up to 200 kda, consisting of up to 13 subunits (Fig. 5.16a). Three of them, subunit I (60 kda, about 550 amino acids), subunit II (26 kda, about 230 amino acids) and subunit III (30 kda, about 270 amino acids), are conserved across all species, from bacteria to mammalians (Fig. 5.16b). The enzymatic complex catalyzes oxidation of four cytochrome c^{2+} molecules and four intra-matrix protons by a single O_2 molecule and couples this reaction to the transport of the next four protons from the outer to the inner side of the intramitochondrial membrane.

Subunit II contains the binuclear Cu_A center, which receives the electrons from cytochrome c. These are then transferred to heme a and the binuclear heme a_3-Cu_B center in subunit I, where the hydroxyl ions are first formed out of water and molecular oxygen. Two different pathways of proton transfer are possible within the protein body (Michel, 1998; Hofacker and Schulten, 1998; Rottenburg, 1998; Bränden et al., 2001). As in liquid water, the proton transfer in the protein matrix proceeds according to the Grotthus mechanism (Fig. 5.17).

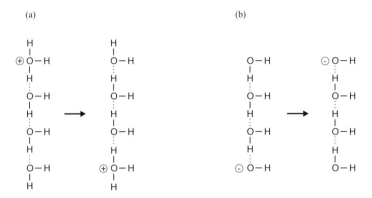

Fig. 5.17. The Grotthuss mechanism for proton conductivity in an acid (**a**) and a base (**b**)

The essence of this mechanism is that no long-distance motion of the hydrated proton H_3O^+ or the hydrated lack of proton OH^- needs to occur, but only a series of local proton jumps in the successive symmetrical hydrogen bonds between neutral water molecules and the charged hydronium or hydroxyl ions (see Appendix C.4).

The chain of photosynthetic charge transfer protein complexes in the thylakoid membrane of the chloroplast, identical to that in cyanobacteria, is shown in Fig. 4.7. It consists of two photosystems and two oxidoreductase complexes. The electron carrier diffusing within the membrane is quinone and those diffusing in the water phase outside the membrane are the small proteins, plastocyanin and ferredoxin.

The chain begins with photosystem II (PS II) in which two electrons are transferred from water to a single quinone molecule at the expense of the light absorption. PS II is a protein dimer (Fig. 5.18a), each monomer consisting of 19 protein or polypeptide molecules with total mass 210 kda, which corresponds to about 2 000 amino acids (Barber, 2002; Ferreira et al., 2004). Functionally, four subunits can be distinguished in each monomer (Fig. 5.18b). The core *reaction center* (RC) composed of D1, D2 and connecting proteins of masses 31, 36 and 28 kda, respectively, is structurally very similar to the RC of the photosynthesizing purple bacteria with component L, M and H proteins (Fig. 5.19a). The *oxygen evolving center* (OEC) is linked directly to the RC at the lumenal site. It consists of three proteins of masses 33, 23 and 17 kda, surrounding the Mn_4 cluster. Flanking the opposite sites of the RC are the CP43 and CP47 proteins of masses 43 and 47 kda, respectively, serving as *internal light-harvesting proteins*

(a)

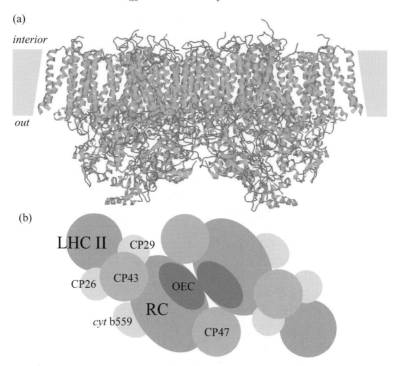

Fig. 5.18. Structure of cyanobacterial photosystem II. (a) Drawing made using the program Rasmol on the basis of pdb entry 1S5L (Ferreira et al., 2004), showing the particular secondary structure elements, α-helices and β-strands, of a dimer. (b) Sketch of the spatial organization of functional subunits composing the complex of the photosystem II dimer with external light-harvesting centers in higher plants (after Barber and Kühlbrandt, 1999). View along the membrane normal from the outer (lumenal) side. See text for details

(antennae). The *cytochrome* b559 *subunit*, consisting of two proteins α and β, probably protects the RC against photoinduced damage. Nine smaller polypeptides fulfill subsidiary functions.

PS II of the eukaryotic higher plants is surrounded by a system of *external light-harvesting centers* LHC-II that diffuse in the thylacoidal membrane (Kühlbrandt, 1994) and transiently enter into multisub-unit complexes with PS II (Barber and Kühlbrandt, 1999). In these complexes, two groups of CP29 and CP26 light-harvesting proteins transfer energy excitations between the LHC-IIs and the internal light-harvesting proteins flanking two central reaction centers of the PS II dimer (Fig. 5.18b). LHC-II is a trimer, each monomer being a protein molecule with mass 25 kda (232 amino acids) and containing 12 chlorophylls and 2 carotenoids (Kühlbrandt, 1994). Altogether, there are over 300 pigment molecules per PS II. The counterparts of the LHC-II complexes in higher plants are *phycobilisomes* (PBS) in cyanobacteria, LH-II *complexes* in purple bacteria, and *Fenna–Mathew–Olson* (FMO) *complexes* coupled to *chlorosomes* in green bacteria (Hu et al., 1998).

The light-harvesting antenna system essentially increases the efficiency of the process of light-to-chemical free energy transduction by photosynthesizing organisms. Under optimum conditions, over 90% of the absorbed light energy is transferred within a few hundred picoseconds from the antenna system to the reaction center. As said above, the RC of PS II is almost identical to the RC of the photosynthesizing purple bacteria (Fig. 5.19a), intensively studied in the past two decades (Hoff and Deisenhofer, 1997). In the RC, the excitation energy absorbed by a chlorophyll dimer results in a charge separation (Fig. 5.19b). The chlorophyll dimer (the special pair) is a primary electron donor from which the electron is transferred very quickly, within a few picoseconds, through a monomeric chlorophyll to pheophytin and then, within two hundred picoseconds, to a bound quinone molecule Q_A (Kriegl and Nienhaus, 2004). Owing to the strong irreversibility, the quantum yield of the process is greater than 99.5%. Within ten microseconds, Q_A^- reduces a second quinone Q_B which, after accepting a second electron and two protons from the interior (stromal) area, is liberated as a quinol molecule into the membrane interior, increasing the *quinone pool*. The electrons lacking on the primary donor are supplied by the oxygen evolving center where water is split into molecular oxygen, protons and electrons.

The structure of quinol:plastocyanin oxidoreductase (the cytochrome b_6f complex), the second link in the photosynthetic chain, is very similar to that of quinol:cytochrome-c oxidoreductase, i.e., the cytochrome bc_1 complex (Kurisu et al., 2003). Plastocyanin is a small water-soluble protein with mass 10 kda (92 amino acids) that operates in the lumen of the thylacoid. It contains a copper ion able to transfer a single electron from the cytochrome b_6f complex to PS I by diffusion. Under steady-state conditions, this process can repeat more than 1000 times per second (Gross, 1996).

From plastocyanin on the lumenal side of the thylacoid, electrons are further transferred to ferredoxin on the stromal side. This is an endoergic reaction and proceeds at the expense of the light energy on the photosystem I (PS I) complex. PS I is a protein trimer, each monomer consisting of 12 protein or polypeptide molecules with total mass 356 kda which corresponds to about 3 400 amino acids (Jordan et

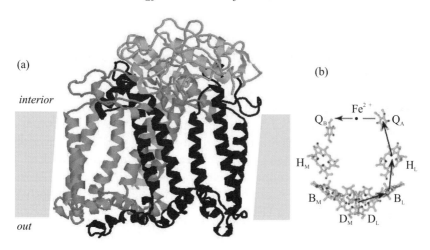

Fig. 5.19. (**a**) Structure of the photosynthetic reaction center of *Rhodobacter sphaeroides*. The drawing, made using the program Rasmol with pdb entry 1PCR (Ermler et al., 1994), shows the particular secondary structure elements, α-helices and β-strands. The three component domains L, M, and H are distinguished by different degrees of shading. (**b**) Pathway of electron transfer. M and L branches are shown. D (primary donor) is the special pair of chlorophylls, B is monomeric chlorophyll, H (secondary acceptor) is pheophytin, and Q_A and Q_B are strongly and weakly bound quinone, respectively

al., 2001). Six core subunits labeled A to F (Fig. 5.20) are involved in the electron transport. The three subunits A, B and C form the proper reaction center. Their organization shows striking similarities with the organization of D1, D2 and the connecting proteins of PS II, as well as L, M and H proteins of the purple bacteria RC (Heathcote, 2002). The electron transfer branches from the A1–B1 chlorophyll pair to Q_A and Q_B quinones in PS I correspond to the electron transfer chains from the special pair to Q_A and Q_B quinones in PS II and purple bacteria (see Fig. 5.19b). The only difference is that the AB entity in PS I is a real homodimer and both branches are active. The electrons from Q_A and Q_B are transferred through three Fe_4S_4 clusters F_X, F_A and F_B to the D subunit which binds the oxidized ferredoxin molecule. The lack of electrons on the primary A1–B1 donor is completed from the F subunit which binds the reduced plastocyanin molecule.

Subunits A and B of PS I (83 kda each) are larger then subunits D1 and D2 of PS II (31 and 36 kda, respectively), as they also collect light energy. Together with the smaller subunits I, J, K, L, M and X,

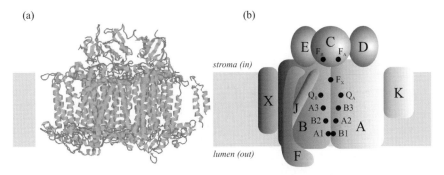

Fig. 5.20. Structure of cyanobacterial photosystem I. (**a**) Drawing made using the program Rasmol on the basis of pdb entry 1JBO (Nield et al., 2003), showing the particular secondary structure elements, α-helices and β-strands, of a monomer unit. (**b**) Schematic representation of the component subunits of the monomer. See text for details

the core light-harvesting system of PS I contains 90 chlorophylls and 22 carotenoids.

Ferredoxin is a small water-soluble protein with mass 11 kda (about 100 amino acids) that operates in the stroma of the chloroplast (Knaff, 1996). It contains an F_2S_2 cluster, ligated by four cysteine residues. Ferredoxin transfers two electrons from PS I to the terminal link of the photosynthetic chain, ferredoxin:NADP$^+$ oxidoreductase (Brunes and Karplus, 1995). It is a single peripheral protein molecule with mass 35 kda, bound to the stromal side of the thylacoid membrane. The electron carrier of that oxidoreductase is constantly bound flavin adenine dinucleotide (FAD).

5.4 Cytoskeleton and Cell Motility: Microfilaments

The protein cytoskeleton (Fuller and Shields, 1998, Chap. 7; Darnell et al., 1999, Chaps. 17 and 18, Stryer et al., 2002, Chap. 34) is composed of *microfilaments*, *microtubules*, and *intermediate filaments* connected by a three-dimensional *microtrabecular lattice*. The intermediate filaments and the microtrabecular lattice fulfill only constructive functions. They make the cytoplasmic membrane and intercellular junctions relatively rigid. In contrast, the microfilaments and microtubules play the main part in cell motility, besides fulfilling constructive functions. The present and the following sections will be devoted to them.

The microfilaments are made up of *actin* molecules. In monomeric form, actin is a globular protein composed of 375 amino acids (see

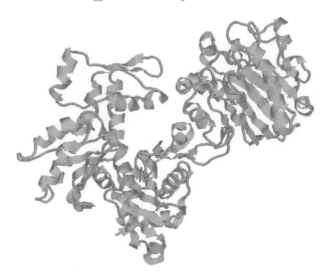

Fig. 5.21. Structure of the actin molecule. The drawing was made using the program Rasmol on the basis of pdb entry 1ATN (Kabsch et al., 1990)

Fig. 5.21). Its structure is strongly conservative: the actin of transversally striated skeletal muscles differs from that of smooth muscles by a few amino acids, and differences between various phylogenetically widely separated species do not exceed a dozen amino acids. The actin molecule binds ATP as well as Mg^{2+} and K^+ ions, and is an ATPase from the enzymatic point of view. The nucleotide binding site is a cleft between two similar domains. ATP hydrolysis results in a large conformational change in the actin molecule.

Under the conditions of a sufficiently high concentration of Mg^{2+} and K^+ ions, the ATP-bound actin molecules spontaneously polymerize to form microfilaments composed of two helically wound chains of total diameter ~ 9 nm and repetition period ~ 36 nm, which corresponds to 13 monomers placed every 5.5 nm along each component chain (see Fig. 5.22) (Squire, 1997). Just after completion of the polymerization process, ATP is hydrolyzed to ADP and P_i so that the products of a possible microfilament depolymerization are always actin molecules bound with ADP and not with ATP. The process of ADP to ATP exchange, necessary for repeated polymerization, is slow, which secures a large reserve of free actin molecules in the cell.

The actin microfilaments play a twofold role. Firstly, they can bind various proteins that link them either into bunches, stabilizing cellular appendices, or gel structures varying in time due to the polymeriza-

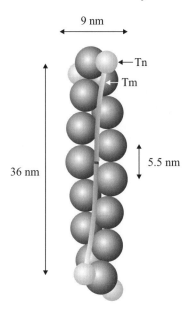

9 nm

← Tn

Tm

36 nm

5.5 nm

Fig. 5.22. Schematic structure of the actin microfilament. When it makes a track for the myosin molecule movement, access is forbidden to the actin molecules by two-strand superhelices of *tropomyosin* (Tm) in the conformation controlled by a complex of *troponins* (Tn). Binding of one of them, *troponin C*, to Ca^{2+} ions makes the actin molecules accessible for the myosin heads

tion at one end and the depolymerization at the other, and responsible for an amoeba-like movement of some cells (Marx, 2003). And secondly, they make tracks for the movement of special molecular motors, the myosin molecules. In the latter case the actin filaments are equipped with additional protein molecules that fulfill regulatory functions. Among these, *troponin C* plays the key role. It is a two-domain, Ca^{2+}-binding protein with structure similar to that of calmodulin (see Fig. 5.35). Moreover, its mechanism of action is similar to that of calmodulin: after binding two additional calcium ions, the C-end domain conformation changes essentially, and this influences the conformation of controlled neighboring proteins. The final effect is an unblocking of the myosin-to-actin binding site (Fig. 5.22).

The term *myosin* is ascribed to a wide class of proteins that perform various functions: motion of membranes, transport of macromolecules or whole liposomes and, in the most evolved form, a muscle contraction (Mermall et al., 1998). All myosins are molecular motors that move in a directed way along the actin microfilaments at the expense of ATP hydrolysis. The muscular cell myosin, called *myosin* II, is the most abundant animal protein (50% of all proteins in skeletal muscles). A single myosin II molecule (Fig. 5.23a) is composed of six protein chains with total mass 520 kda: two identical *heavy chains* (HC), each with mass 220 kda, two identical *essential light chains* (ELC), and two identical *regulatory light chains* (RLC), each with mass 20 kda. The light

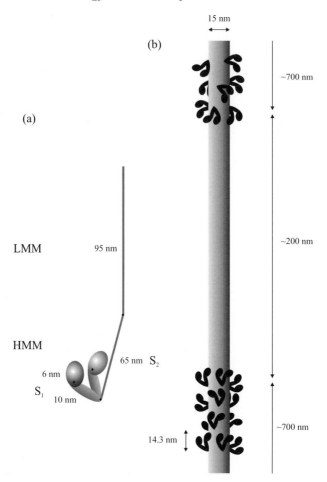

Fig. 5.23. (a) Structure of the myosin II molecule. The non-globular α-helices of two heavy chains are stranded in the long coiled-coil rod. Five mobile swivels are indicated. (b) Schematic structure of the thick filament composed of a number of myosin II molecules

chains have similar structure to troponin C or calmodulin and fulfill a similar function (a change of conformation after Ca^{2+}-ion binding). With the help of proteolytic enzymes, the myosin II molecule can be split into so-called *light meromyosin* (LMM) and *heavy meromyosin* (HMM), the latter then being split into two subfragments S1 (the myosin 'heads') and one subfragment S2 (the myosin 'tail').

Under the conditions of a sufficiently high concentration of appropriate ions, the myosin II molecules spontaneously polymerize to form structures called *thick filaments* (see Fig. 5.23b) (Squire, 1997).

The thick myosin filament has the form of a triple helix with repetition period ~ 43 nm. The slip of each component helix equals $3 \times 43 = 132$ nm and includes 9 myosin molecules. Together, every $132 : 9 = 43 : 3 \approx 14.3$ nm, there is a 'crown' consisting of 3 double myosin heads. An important role is played by five *swivels* of the myosin II molecule, one within the tail, two between the tail and two heads, and two within the heads (Fig. 5.23a). They admit of an independent and almost free motion of the myosin heads in the thick filament when looking for appropriate binding sites on the thin filament. The thick filament has diameter ~ 15 nm and consists of two oppositely oriented halves of length ~ 700 nm each with the protruding myosin heads, joined by a bare zone of length ~ 200 nm (Squire, 1997). Half of the thick filament contains $(700 : 14.3) \times 6 \approx 300$ myosin heads.

The proper molecular motor is subfragment S1, the *myosin head* (Fig. 5.24). It is a protein with mass 95 kda (a split fragment of the heavy chain) plus 2×20 kda (the two light chains), totalling 135 kda, i.e., about 1 200 amino acids or approximately 18 thousand atoms (Geeves and Holmes, 1999; Howard, 2001). The myosin head is an ATPase. Functionally, one can distinguish within it a catalytic subunit (630 amino acids) joined by a swivel with a regulatory subunit – an 8 nm long 'lever arm' (570 amino acids). Details deduced from X-ray crystallography studies are shown in Fig. 5.24.

Like all ATPases, the myosin head does not actually perform its enzymatic function until the conditions arise for the process to be biologically useful. The enzyme activator is the actin filament (hence the name 'actin'). Only after strong attachment to the actin filament at two sites is the myosin head able to bind and rebind substrates and products of the catalyzed reaction. The binding site is a pocket between the upper (U) and the lower (L) domains of the catalytic subunit (Fig. 5.24). This pocket can be in an *open* state (Rayment et al., 1993), affording a possibility for exchanging ADP with ATP through an outlet directed opposite to the filament, or in a *closed* state (Dominigues et al., 1998), affording a possibility for exchanging only P_i through an outlet directed toward the filament. The state of the pocket is transmitted onto the orientation of the lever-arm subunit through a long α-helix, referred to as the *relay* (Sablin and Fletterick, 2001), unless another α-helix, called the SH1–SH2 helix, is melted (Houdusse and Sweeney, 2001). Two orientations of the lever-arm subunit are seen, not only in X-ray crystallography, but also in cryo-electron micrography (Whittaker et al., 1995; Holmes et al., 2003) and in observations of the hyperfine structure of electron paramagnetic resonance (Baker et

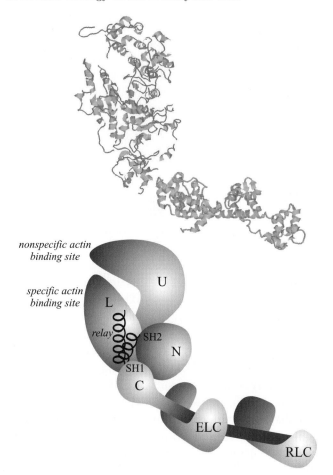

Fig. 5.24. Structure of the myosin head in the absence of a nucleotide. The *upper drawing* was made using the program Rasmol on the basis of pdb entry 2MYS (Rayment et al., 1993) for the catalytic unit and 1SCM (Xie et al., 1994) for the regulatory subunit. The *lower drawing* presents schematically the component domains of the catalytic subunit, i.e., the upper (U), lower (L) and amino end (N) domains, as well as those of the regulatory subunit, i.e., the globular converter (C) becoming a single α-helix stabilized by the essential light chain (ELC) and the regulatory light chain (RLC). The relay and the SH1–SH2 helices are shown. The swivel is close to the hydrosulfide group SH2

al., 1998; 1999; Baumann et al., 2004) or fluorescence energy transfer (Suzuki et al., 1998; Xiao et al., 1998; Xu and Root, 2000).

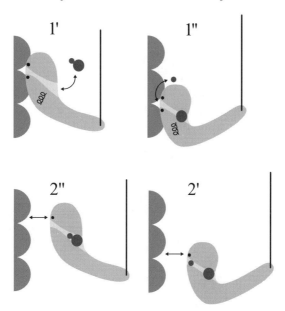

Fig. 5.25. Four distinguished conformational states of the myosin head attached to ($1''$ and $1'$) and detached from ($2''$ and $2'$) the actin filament in the ATPase cycle of the actomyosin motor. The cycle proceeds in the counterclockwise direction. A strong attachment of the myosin head to the actin at two sites (*black dots*) makes the nucleotide pocket (*weakly shaded*) relatively rigid. The SH1–SH2 helix transmits the motion of the lever-arm domain onto the pocket shape, which in the state $1'$ enables binding–rebinding of the ATP molecule (*two joined black disks*) and in the substate $1''$, the P_i molecule (*smaller disk*). In the detached state the SH1–SH2 helix is melted. For the myosin head to attach weakly (at one site) to the actin, an appropriate surface loop (*single black dot*) has to assume a suitable shape, different when the nucleotide is non-hydrolyzed (substate $2''$) or hydrolyzed (substate $2'$). This can, but need not, be related to the position of the lever-arm domain. Here we assumed that these positions coincide with the corresponding positions in the attached state

Both states of the myosin head, with the open and the closed pocket, are shown schematically in Fig. 5.25 as states $1'$ and $1''$, respectively. In the open state $1'$, the lever arm is oriented as in Fig. 5.24, making an angle of about $45°$ with respect to the actin filament axis. In the closed state $1''$, the lever arm is raised to make an angle of about $90°$. The transition from state $1''$ to $1'$, with the myosin head strongly attached to the actin filament and the myosin tail remaining strained, is related to the generation of a force that the myosin head exerts on the filament. This *swinging lever-arm* picture (Rayment et al., 1993; Spudich, 1994; Geeves and Holmes, 1999; Houdusse and Sweeney, 2001; Howard, 2001) refines H.E. Huxley's (1969) classical swinging cross-bridge hypothesis.

States of the myosin head with the third orthophosphate P_i included, bound to or hydrolyzed from the nucleotide, become more stable if the double attachment of the catalytic subunit to the actin filament is reduced to an unspecific single attachment of only the U domain (see Fig. 5.24), or if it is completely broken. Moreover, this *detached* state of the myosin head has been observed crystallographically (Houdussee at al., 2000). However, because of the melted SH1–SH2 helix (Volkmann and Hanein, 2000; Nitao and Reiser, 2000; Houdusse and Sweeney, 2001), the orientation of the lever-arm subunit is not well determined in true in vivo conditions (see the comment in the caption to Fig. 5.25).

In skeletal and heart muscle cells, the thick myosin filaments and the thin actin filaments are organized in structures called *sarcomeres* (Fig. 5.26) (Squire, 1997). Here, the two types of filament form two interpenetrating lattices with hexagonal symmetry. In a zone where the two lattices overlap, each myosin filament is surrounded by six actin filaments, recalling that the crown of myosin heads is formed out of six independent heads (see Fig. 5.23b), and each actin filament is surrounded by three myosin filaments. The distance between the myosin filament (diameter ~ 15 nm) and the actin filament (diameter ~ 9 nm) is about 13 nm. The actin filaments, with length ~ 1000 nm, are attached on one side to a membrane disk and the position of the myosin filaments with respect to the actin filaments is stabilized by *titin* molecules (Squire, 1997). The latter are huge proteins with mass 3.0 Mda which corresponds to about 27 000 amino acids. They contain some 300 linearly ordered domains of the immunoglobin and fibronectin III type, each of about 90 amino acids (see Fig. 5.40 below), and a single large PEVK domain (about 2 200 amino acids), rich in proline (P), glutamate (G), valine (V) and lysine (K), and playing the role of an elastic element (Fig. 5.26).

According to the *sliding-filament* model (A.F. Huxley, 1957; H.E. Huxley, 1969; A.F. Huxley and Simmons, 1971), the thick and thin filaments slide past each other during muscular contraction. Contrary to a widespread view, the myosin heads in sarcomere do not interact with the actin thin filaments in a cooperative way, but "like many out-of-time rowers in a boat, they attach asynchronously to actin, tilt their angles like oars stroking through the water, and then detach and

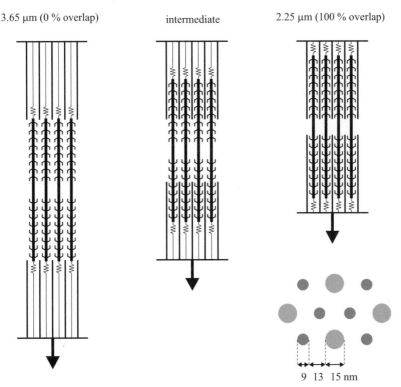

Fig. 5.26. Sarcomere in the loose (0% overlap), intermediate, and contracted (100% overlap) state. *Thick* and *medium thick lines* represent the myosin and actin filaments, respectively. *Thin lines* represent the titin molecules ('springs' correspond to the PEVK domains). *Bottom right*: Hexagonal symmetry of the sarcomere cross-section

recock, similar to pulling the oar out of the water at the end of the stroke" (Vale and Milligan, 2000). Each myosin head undergoes alternating *working* and *recovery strokes* in the attached and the detached state, respectively. Only 1–2% of myosin heads are doing their 'duty' at any given moment. The remaining behave like passengers (Howard, 2001).

Arranged one after the other, the sarcomeres form a *myofibril* with length 50 m or more. The myofibrils, surrounded by the *sarcoplasmic reticulum* which stores and supplies the controlling Ca^{2+} ions, are organelles of *muscle fibers*, the component cells of skeletal and heart *striated muscles*. *Smooth muscles* of internal organs other than the heart also make use of the sliding mechanism of the thick versus thin filaments, but show no sarcomere structure (Fuller and Shields, 1998, Chap. 7).

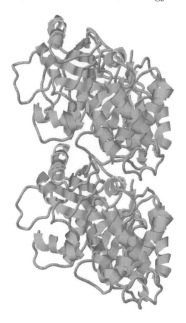

Fig. 5.27. Structure of the $\alpha\beta$ tubulin dimer. The drawing was made using the program Rasmol on the basis of pdb entry 1TUB (Nogales et al., 1998)

5.5 Cytoskeleton and Cell Motility: Microtubules

This detailed description of the actin microfilaments and myosin motors has been made because later on, in Sects. 9.4 and 9.5, we are going to consider a theory of their action. The other elements of the cytoskeleton, microtubules and the motors related to them will be described more superficially.

Microtubules are made up of dimers of α and β *tubulin* (Fig. 5.27), each with mass about 50 kda, comparable to the mass of actin (human α and β tubulin consist of 451 and 444 amino acids, respectively). In vitro and in vivo, as a result of polymerization, the tubulin dimers make up a hollow cylinder with outer diameter 25 nm and inner diameter 15 nm, consisting of 13 protofilaments (see Fig. 5.28) (Mandelkow and Mandelkow, 1994).

In contrast to actin, tubulin is not ATPase but GTPase. In physiological conditions, the GTP bound to α-tubulin is non-detachable whereas that bound to β-tubulin is hydrolyzed to GDP during microtubule assembly. As a consequence, the β-tubulin molecules in the interior of a microtubule contain GDP. The GTP hydrolysis provides the driving force for microtubule dynamics and the notorious variation in its length referred to as *dynamic instability* (Mandelkow and Mandelkow, 1994). The mechanism for switching between growth and

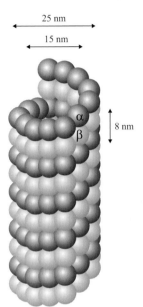

25 nm

15 nm

α
β

8 nm

Fig. 5.28. Microtubule structure (the so-called A-lattice)

shrinkage of the microtubules is still under discussion (Bolterauer et al., 1999).

The polymerization of microtubules starts at *centrioles* or *basal bodies* (Darnell et al., 1999, Chap. 18), each having a highly organized structure of 9-fold symmetry and consisting of several long-lived microtubular fragments[1] (Fig. 5.29). The microtubules are polar. The minus end is a germ at the organization center and it is the plus end that is subjected to growth or shrinkage. Two centrioles compose the *centrosome* (Fig. 5.2) which duplicates before mitosis. As mitosis proceeds from the two centrosomes on opposite sides of a dividing cell, the *mitotic spindle* grows up and joins with the chromosomes just formed from the nuclear chromatin. From the basal bodies, *cilia* and *flagella* grow up, responsible for cell locomotion. Both cilia and flagella have the same structure determined by the structure of the basal body (Fig. 5.29a). The difference is that cilia occur in large numbers and are short, whereas flagella are single and long.

Microtubules participate in an astounding variety of cell activities most of which are related to motion. In principle, one can distinguish three different mechanisms of microtubule-based motion, each making use of a specific kind of molecular motor. The first mechanism affords

---

[1]Centrioles and the basal bodies have declined in the case of higher plants. Here, the polymerization initiation mechanism remains unknown.

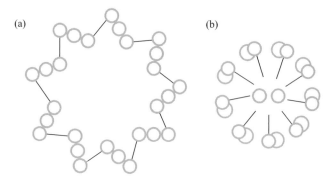

Fig. 5.29. Structure of microtubule organization centers. (**a**) Centrioles are built from a cylindrical array of 9 triples of microtubules connected by protein bridges. (**b**) Basal bodies are built from a cylindrical array of 9 double microtubules surrounding 2 single central microtubules. The circulating doublets are connected to the central microtubules by protein 'spokes'

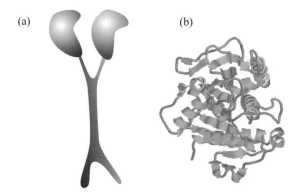

Fig. 5.30. (**a**) Schematic structure of the kinesin double-headed molecule. (**b**) More detailed structure of the kinesin head. The drawing was made using the program Rasmol on the basis of pdb entry 1BG2 (Kull et al., 1996)

possibilities for *transport* of various substances with the help of *kinesin* molecules. Kinesin is composed of two protein chains that form two heads (ATPases) and a tail (Fig. 5.30). It walks along a protofilament of the microtubule in a hand-over-hand manner with a step of 8 nm (Yildiz et al., 2004). A variety of cargo can be attached to the kinesin tail, including membranous organelles, mRNA, and signaling molecules (Goldstein and Philip, 1999). In this way microtubules can also take part in directed *signal transduction* processes.

The second mechanism affords possibilities for cell *locomotion*. It consists in a relative sliding of adjusted double microtubules in cilia or

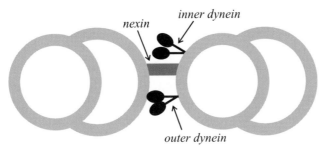

Fig. 5.31. Linking of two adjusted double microtubules in cilium or flagellum. Nexin as well as the inner and outer arms of dynein molecules are shown

flagella (see Fig. 5.29b). Here, two other kinds of molecular motor are used, *nexin* and *dyneins* (Darnell et al., 1999, Chap. 18). The former have extraordinary elastic properties, whereas the latter are very large complexes of mass 1 to 2 Mda, counting 1, 2 or 3 heads (Fig. 5.31). In contrast to sarcomere with actin filaments and myosins, sliding of dyneins fixed to one microtubule along the other does not result in contraction, but bends the whole structure.

The third mechanism of motion is used during cell *division* for chromosome segregation. Besides the already mentioned dynamic instability of the microtubules, this mechanism makes use of yet another kind of motor, i.e., *mitotic* motors (Heald and Walczak, 1999).

Each tubulin dimer included in a microtubule can switch between two or more conformational substates with different electrical properties. Interaction between individual dimers in the microtubular wall affords possibilities for information processing within axons of nerve cells, provided the level of noise is sufficiently low (Tuszyński et al., 1998). The problem is highly puzzling and various related questions have been considered on both the classical (Tuszyński et al., 1995) and the quantum level (Hagan et al., 2002).

5.6 Regulation of Enzyme Activity

Each biological molecular process is catalyzed by a specific enzyme. Since the cell requires these processes to happen only at a certain place and time, this must be controlled. Control of biological molecular processes occurs on two levels: expression of information recorded on genes (synthesis of appropriate enzymes) and regulation of the activity of appropriate enzymes. Most control of the first type consists in turning the gene transcription on or off, and will be considered in the next two sections. Here, we restrict our presentation to control of the second type.

Six mechanisms of enzyme activity regulation can be distinguished: (a) proteolytic precursor activation, (b) covalent precursor modification, (c) anchoring in membrane, (d) competitive inhibition, (e) feedback inhibition, and (f) allosteric control (Stryer, 2002, Chap. 10).

Just after synthesis on the ribosome, most proteins are only inactive precursors of enzymes. In order to convert to the active form, they must be subjected to additional chemical modifications. One possible modification involves cutting off some fragments of the main chain. A well-known textbook example is *chymotrypsin* (see Figs. C.28 and 7.4), an enzyme that hydrolyzes specific peptide bonds of proteins. It is synthesized within the chief cells in gastric glands as an inactive precursor, *chymotrypsinogen*, consisting of a single polypeptide chain. In this form, it does not attack the chief cell itself. Only in the medium of low pH outside the cell can a proteolytic reaction proceed in which accidentally present active chymotrypsin molecules hydrolyze specific peptide bonds in chymotrypsinogen molecules and transform them into active three-chain chymotrypsin molecules, thereby initiating the whole avalanche reaction (Stryer, 2002, Chap. 10).

Other possible modifications of inactive precursors involve either methylation of some charged side chains, which leads to charge neutralization, or phosphorylation of side chains ending with hydroxyl groups, which endows the originally neutral chain with a negative charge. The latter form of activity regulation is the most common for living organisms. There are three hydroxyl-containing amino acids: serine, threonine and tyrosine. The enzymes that catalyze the phosphate group transfer are called kinases. Here we present the structure of protein kinases A and C, the two important enzymes taking part in signal transduction cascades, discussed in the next section.

The *protein kinase* A (PKA) is composed of two regulatory subunits (about 450 amino acids each) and two catalytic subunits (about 350 amino acids each). It performs phosphorylation of specific serine and threonine groups in order to activate (a) enzymes taking part in glycogen to glucose transformation, (b) some ion channels, and (c) some transcription factors (see the next section). In an inactive form of PKA, the substrate binding site is occupied by a 'pseudosubstrate' which is part of a regulatory subunit and plays the role of a competitive inhibitor (Fig. 5.32). The binding of a messenger molecule to

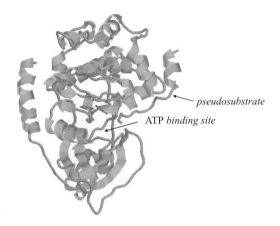

Fig. 5.32. Structure of the catalytic subunit of protein kinase A (PKA). The component secondary structure elements, α-helices and β-strands are shown. The drawing was made using the program Rasmol on the basis of pdb entry 1ATP (Zheng et al., 1993). A specially matched pseudosubstrate mimics either an actual substrate or a competitive inhibitor originating from the regulatory subunit

the regulatory subunits prevents the pseudosubstrate from associating with the catalytic subunit and activates the PKA itself.

The *protein kinase* C (PKC) contains one regulatory and one catalytic subunit but its activation mechanism is similar to that of PKA. PKC activates prostaglandins (lipid-soluble derivatives of arachidonic acid taking part in short-range signaling and stimulating inflammatory states) or some transcription factors. Besides the pseudosubstrate domain, the regulatory subunit of PKC contains a characteristic calcium-binding C2 domain, an eight-stranded antiparallel β-sandwich containing approximately 120 amino acids (Fig. 5.33a). Calcium binding induces a conformational change which enables domain anchoring in the eukaryotic cytoplasmic membrane. Similar structure characterizes bacterial toxins of phospholipase C activity (Fig. 5.33b), the C-terminal domain of which displays a strong structural and functional analogy to the C2 domain (Titball, 1993; Naylor et al., 1998). The membrane anchoring provides an additional form of enzyme activation mentioned at the beginning of this section.

We have already given an example of the *competitive inhibition* of enzyme activity due to replacement of an actual substrate by a pseudosubstrate. Creation of appropriate competitive inhibitors is a basic task in contemporary pharmacology. The kinetic mechanism of

(a)   (b)

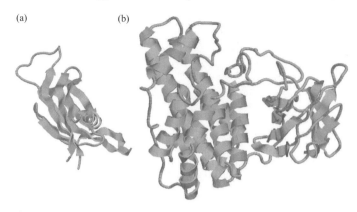

Fig. 5.33. (a) Structure of the C2 domain of the regulatory subunit of protein kinase C (PKC). (b) Structure of *Clostridium perfringens* α-toxin, the key determinant in gas gangrene. Drawings were made using the program Rasmol on the basis of pdb entry 1BDY (Pappa et al., 1998) and 1CA1 (Naylor et al., 1998) for (a) and (b), respectively

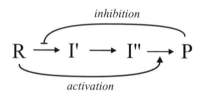

Fig. 5.34. Kinetic enzyme activity regulation: reagent R, if present in excess, activates the last enzyme in the metabolic pathway from R to P, whereas product P, if present in excess, inactivates the first enzyme in this pathway

competitive inhibition will be defined in detail in Sect. 7.3. In relation to kinetics, we would like to mention *feedback inhibition*. The main idea behind this form of enzyme activity regulation is illustrated in Fig. 5.34. It is worth mentioning that the opposite process of *feedback activation* can result in enzymatic oscillations (Sect. 7.6).

Physically, the most subtle mechanism of enzyme activity regulation is *allostery*, a change in the global conformation of the enzyme catalytic subunit caused by a weak local interaction. There are many examples of allosteric enzymes presented in textbooks (e.g., Stryer et al., 2002, Chap. 10). The kinetics of allosteric enzymes is considered in Sect. 7.5. Here, we shall only discuss *calmodulin*, a universal detector of calcium, already mentioned several times. This small (17 kda) protein consists of two similar globular lobes, each containing two Ca^{2+}-binding sites, joined by a long α-helix (Fig. 5.35). This α-helix winds

Fig. 5.35. Structure of calmodulin. The drawing was made using the program Rasmol on the basis of pdb entry 3CLN (Babu et al., 1988)

itself round another α-helix contained in a target protein in such a way that both lobes touch one another. The cooperative binding of more than three calcium ions results in serious conformational changes in both the calmodulin itself and the target protein.

5.7 Receptors

Signal transduction is no less important for the functioning of a living organism than matter transduction (metabolism) and energy transduction. However, the signal transduction system is much more complex than either of the other systems. Only intensive studies of the mechanisms of human transplant rejection and oncogenesis in the past two decades have given some insight into the way this system is organized on the intracellular level. The Human Genome Project has revealed that up to 20% of over 30 thousand human coding genes encode proteins involved in signal transduction (Blume-Jensen and Hunter, 2001).

Each cell of a living organism has to communicate with other cells and, indirectly, with the environment. There are three categories of molecules carrying extracellular signals (Darnell et al., 1999, Chap. 19). Category (a) includes *steroid hormones* and *nitric oxide* that freely diffuse across the cytoplasmic membrane and interact only with the receptors in the cytosol or nucleus membranes. Some amino acids or their derivatives known together as *neurotransmitters* form category (b). They open or close target channels allowing the appropriate ions to flow into or out of the cell interior. Category (c) consists of carriers that are ligands of *cell-surface receptors*. The carriers of category (a) will not be the subject of our considerations. Examples of ion channels gated by the carriers of category (b) have already been considered in Sect. 5.2. Here, we confine our attention only to the cascades of signal transduction impelled by the binding of carriers of the most extensive category (c) to their target receptors.

The cell-surface receptors are transmembrane proteins whose extra-cellular portion binds a signal carrier referred to as the receptor *ligand* and whose intracellular portion impels a cascade of reactions taking part in intramolecular signal transduction. Binding between a receptor and its ligand is like that between an enzyme and its substrate. It occurs by non-covalent forces and results in allosteric conformational change in the intracellular portion. This change triggers subsequent allosteric changes and/or effects chemical reactions taking place in molecules that bind to the intracellular portion of the receptor and are called generally adapter molecules or *adapters*. Either directly or through a *transducer*, they transmit the signal to an *effector* which, either directly or through a *second messenger*, transmits to the signal *target* (Fig. 5.36).

Four main classes of receptor can be distinguished (Fuller and Shields, 1998; Darnell et al., 1999, Chap. 19; Stryer et al., 2002, Chap. 15): (a) serpentine receptors, (b) receptor tyrosine kinases (RTKs), (c) tyrosine kinase-linking receptors (TKLRs) and (d) anti-gen receptors. The main links of the corresponding signal transduction pathways are listed in Table 5.1. RTKs are members of a more exten-sive class of receptors with intrinsic enzymatic activity, and TKLRs are members of a class of receptors that lack such activity but associate directly with cytosolic protein kinases. However, we shall not consider these more extensive classes of receptors and confine our attention to the mentioned subclasses.

The *serpentine receptors*, also called *seven-spanning receptors*, are the best and earliest studied. The names are related to the fact that the main protein chain of these receptors winds seven times back and forth through the cytoplasmic membrane. The ligands of the serpen-tine receptors are amino-acid or peptide hormones, e.g., *epinephrine* (*adrenaline*), *acetylcholine* or *glucagon*. The receptors can also be ex-cited by sensory stimuli such as light or odorants. Ligand binding or stimulation of the receptor activates a *G protein*. This is a trimeric protein whose serpentine subunits G_β and G_γ play the role of adapter proteins and whose subunit G_α is a GTPase that hydrolyzes GTP to GDP and the inorganic phosphate P_i (Fig. 5.37). The GTPase cycle of the G_α protein drives the whole signal transduction process (see Sect. 8.6).

The GTP-bound G_α protein dissociates from the receptor-bound G_β and G_γ proteins and activates an effector molecule. There are

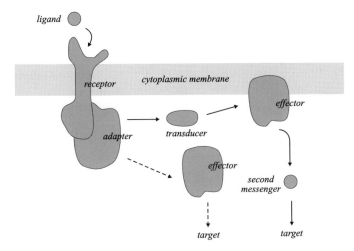

Fig. 5.36. General scheme of intramolecular signal transduction

Table 5.1. Main signal transduction pathways. Abbreviations: AC = adenylate cyclase, cAMP = cyclic AMP, DAG = diacylglycerol, IP$_3$ = inositol triphosphate, JAK = Janus kinase, LNK = either LAT (linker for activated T cells) or BLNK (B-cell linker protein), MAPK = mitogen-activated protein kinases (the entire pathway), PLC = phospholipase C, RTK = receptor tyrosine kinase, STAT = signal transducer and activator of transcription, TKLR = tyrosine kinase-linking receptor. For other abbreviations, see the comment in the text concerning proto-oncogenes

Ligand	Receptor	Adapter	Transducer	Effector	Second messenger
Hormone	Serpentine receptor	G_β–G_γ	G_α	AC	cAMP
		G_β–G_γ	G_α	PLC	IP$_3$ & DAG
Growth factor	RTK			PLCγ	IP$_3$ & DAG
		Grb–Sos	Ras	Raf-MAPK	
		JAK		STAT	
Cytokine	TKLR	JAK		STAT	
Antigen	Antigen receptor	LNK		PLCγ	IP$_3$ & DAG
		LNK–Grb–Sos	Ras	Raf-MAPK	

four different effectors depending on the kind of G_α protein involved (Neves et al., 2002). Here we mention only the adenylate cyclase (AC) and phospholipase C (PLC) pathways. AC catalyzes transformation of ATP into cyclic AMP (cAMP), which is a stable compound of high free

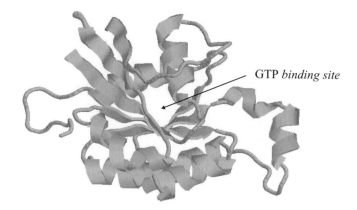

GTP *binding site*

Fig. 5.37. Structure of G_α protein. The drawing was made using the program Rasmol on the basis of pdb entry 1AN0 (Kongsaere et al., to be published)

energy that serves as a second messenger. Its target is protein kinase A (PKA), which catalyzes a number of metabolic reactions (see the last section). The other effector, PLC, cuts the phosphodiester bond in a phospholipid present in the cytoplasmic membrane, phosphatidyl inositol biphosphate, thereby generating two other second messengers: inositol triphosphate (IP_3) and diacylglycerol (DAG).[1] IP_3 opens calcium channels in the endoplasmic reticulum which liberates the Ca^{2+} ions stored there into the cytosol. A response can be, e.g., muscle contraction or glycogenolysis. DAG activates protein kinase C (PKC), which subsequently phosphorylates the serine or threonine residues of various target proteins (see the last section).

The ligands of *receptor tyrosine kinases* (RTK) are commonly called *growth factors*. They are peptides or small proteins, e.g., *insulin*, *epidermal growth factor* (EGF) or *platelet-derived growth factor* (PDGF). RTKs span the cytoplasmic membrane only once. Following ligand binding, the adjacent receptors dimerize and become active tyrosine kinases which autophosphorylate tyrosine residues on their own cytoplasmic portions. Various proteins can bind to the phosphorylated portions of RTKs provided that they have a characteristic SH2 domain (Src-homolog 2 domain, the name will be explained later on). One such protein is γ-type phospholipase C (PLCγ), which initiates a cAMP cascade identical to the one taking place in the the serpentine receptor case, but omitting the G protein intermediacy. An alternative

---

[1]Inositol is a six-carbon ring alcohol and diacylglycerol is glycerol doubly esterified by various fatty acids (see Appendix C.1). Diacylglycerols are precursors of *prostaglandins* that originate local inflammatory states.

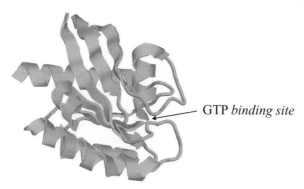

GTP *binding site*

Fig. 5.38. Structure of Ras protein. Note the similarity with the structure of G_α protein (Fig. 5.37). The drawing was made using the program Rasmol on the basis of pdb entry 1IOZ (Kigawa et al., 2001)

pathway is initiated by an adapter Grb–Sos complex that activates another GTPase, a Ras protein (Fig. 5.38). An effector of Ras protein is Raf protein, a kinase that initiates the entire pathway of *mitogen-activated protein kinases* (MAPKs). The final targets of MAPKs, just as for other second messengers like cAMP and DAG, are *transcription factors* (TF) that promote various gene expressions and contribute to cell proliferation or apoptosis (see the next section).

Names of many proteins taking part in the RTK and other signal transduction pathways have their origin in symptoms observed in the organism when the cellular gene coding a given protein is subject to mutation or a replacement by some retroviral counterpart (Fuller and Shields, 1998; Stryer et al., 2002, Chap. 15). Thus, the Src protein acquired its name from the src gene which, in the viral version, codes a protein that produces avian sarcoma. The name of the Ras protein originates in the gene which in its viral Ha-*ras* or Ki-*ras* versions produces Harvey or Kisten mouse sarcomas, respectively. The Ki-*ras* gene has a role in the origin of lung cancer. Cancer-producing genes are called *oncogenes* and their original, unaltered counterparts *proto-oncogenes*. The term 'Sos proteins' is an abbreviation for 'son-of-sevenless'. 'Sevenless' is the name of a *Drosophila* mutant with elementary eyes (omatidia), each containing six rather than seven receptor cells surrounding the eighth central cell. The origins of some other proteins are much more direct. The Grb protein simply grabs hold of the autophosphorylated RTK.

The *tyrosine kinase-linking receptors* (TKLR) lack intrinsic tyrosine kinase activity but can bind a special kind of cytosolic tyrosine

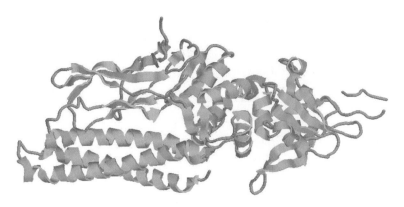

Fig. 5.39. Structure of signal transducer and activator of transcription (STAT) protein. The characteristic SH2 domain, containing some 100 amino acids, is to the right of the Ras molecule. The drawing was made using the program Rasmol on the basis of pdb entry 1UUR (Soler-Lopez et al., 2004)

kinases, called *Janus kinases* (JAK, the name comes from the Roman god with two faces). The TKLR ligands are called *cytokines*. Among them, we distinguish *interferons, interleukins, tumor necrosis factors* and *colony stimulating factors* of various kinds. TKLRs are dimers or trimers of identical or different cell-surface transmembrane proteins. Binding of the ligand stimulates the dimer or trimer formation which then activates JAK (Aaronson and Horvath, 2002). The activated JAKs phosphorylate *signal transducer and activator of transcription* (STAT) molecules. Moreover, they have a characteristic SH2 domain (Fig. 5.39). The phosphorylated STAT molecules form dimers that translocate to the nucleus and activate proteins that are involved in transcription of genes into mRNA. The cell response to a specific cytokine bound to a complementary specific TKLR are specific proteins produced using such transcribed genetic information.

The action of *antigen receptors* shows some features of the action of TKLRs and some of RTKs. For a better understanding, we present this final class of receptors together with a basic review of immunology on the subcellular level (Solomon et al., 2004; Stryer et al., 2002, Chap. 35; Perelson and Weisbuch, 1997).

The task of the immune system is to defend a multicellular organism against foreign pathogenic agents (pathogens) as well as its own cells when they have undergone virus infection or malignant transformation. The immune system of the vertebrates is a complex system of organs, cells and molecules diluted in intercellular fluids, but here we confine our attention to the organization of this system at the subcellular level.

Pathogens and transformed cells differ from one another by specific macromolecular components called *antigens*. Two kinds of cell present in blood are involved in the specific immune response to attack by pathogens: *macrophages*, developed from *monocytes*, and *lymphocytes*. Among the latter, we distinguish *B cells*, produced and maturing in the bone marrow, and *T cells*, produced in the bone marrow and maturing in the thymus. There are three types of T cell: *helper* T cells, *cytotoxic* T cells, and *regulatory* T cells. Each B cell and T cell is specific for a particular antigen, which means that they are provided with specific receptors exposed at the cell surface and made before the cell ever encounters an antigen.

The specificity of the immune response is related to special protein domains which, being the main component of immunoglobins (see below), are called *immunoglobin-like* (Ig-like) *domains*. They are composed of some 100 amino acids and contain two antiparallel β-pleated sheets forming a β-sandwich (Fig. 5.40). The Ig-like domains enter into the composition of many proteins, among which the already mentioned titin (Sect. 5.4). Besides *constant* versions of Ig-like domains, there are also somewhat larger *variable* versions with three hypervariable regions localized in three loops (Fig. 5.40). The genetic mechanism of the variable Ig-like domain diversity is truly fascinating (Stryer et al., 2002, Chap. 35). It is the variable Ig-like domains that determine the specificity of T-cell and B-cell receptors.

The Ig-like domains are also basic components of *major histocompatibility complex* (MHC) molecules that afford possibilities for the discrimination of 'self' from 'non-self'. There are two classes of MHS molecules. Class-I MHS molecules are present on the surface of almost all the cells of the organism. Class-II MHS proteins are found only on the surface of specialized antigen-presenting cells (macrophages after fagicitosis and B cells after endocitosis of antigens). All MHS molecules contain the two Ig-like domains and the two other domains that form together a binding site for peptides, characteristic of the content of a given cell and called *epitopes*. Protein molecules that help to operate the antigen receptors are called *cluster of differentiation* (CD) proteins. The class-I MHS molecules are recognized by CD8, whereas the class-II MHS molecules are recognized by CD4 molecules on the surface of T lymphocytes. They also involve Ig-like domains. Generally, CD8-containing T cells are cytotoxic (killer) cells whereas CD4-containing T cells are helper cells.

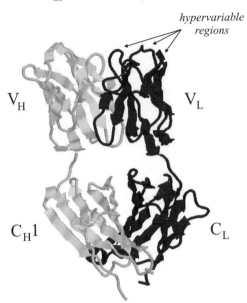

Fig. 5.40. Structure of Ig-like domains. In the presented fragment of immunoglobin G, originating from a B-cell receptor, four such domains are shown: constant and variable domain of the light chain (C_L and V_L, respectively), as well as the first constant and the variable domain of the heavy chain ($C_H 1$ and V_H, respectively). Three hypervariable regions are localized in the indicated loops. Concerning the component secondary structure elements, only β-strands occur in the present case. The drawing was made using the program Rasmol on the basis of pdb entry 1FDL (Fischmann et al., 1991)

The T-cell receptor (TCR) consists of two protein chains, each containing a variable and a constant Ig-like domain as well as an α-helix crossing the membrane, noncovalently linked to the cluster of differentiation 3 (CD3) complex (Fig. 5.41a). The recognition of a specific epitope presented by the class-II molecule is facilitated by the cluster of differentiation 4 (CD4). Alternatively, the recognition of a specific epitope presented by the class-I molecule is facilitated by the cluster of differentiation 8 (CD8). The consequence of epitope recognition and binding is activation of the Lck molecule, a protein tyrosine kinase that changes its place of binding from CD4 or CD8 to CD3. This creates the docking site for another protein tyrosine kinase, the Syk molecule which finally activates the *linker of activated T cell* (LAT). Due to the latter molecular complex, the TCR is effectively endowed with the tyrosine kinase function (Singer and Koretzky, 2002). Further signal transduction pathways are the same as for typical RTKs (Table 5.1).

(a)   (b)

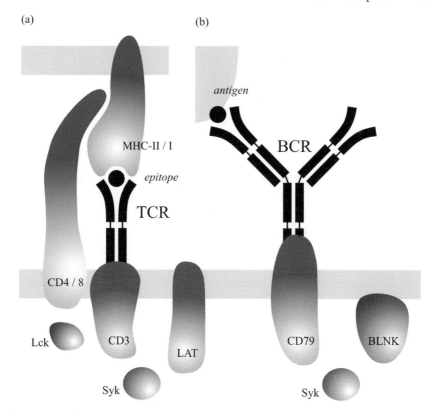

Fig. 5.41. (a) Schematic structure of the T cell receptor (TCR) and more important accompanied proteins. (b) Same for the B cell receptor (TBR). See text for details

The B-cell receptor consists of two to five identical subunits. Each subunit consists of two chains, light and heavy (see Fig. 5.40). The light chain contains one variable and one constant Ig-like domain, and the heavy chain contains one variable and three constant Ig-like domains as well as an α-helix crossing the membrane. The latter are linked to the *cluster of differentiation* 79 (CD79) complex (Fig. 5.41b). The consequence of antigen recognition and binding is activation of a Syk-like molecule, a protein tyrosine kinase that activates the *B-cell linker* (BLNK) molecular complex. Like TCR, the BCR is in this way effectively endowed with the tyrosine kinase function (Gauld at al., 2002).

General immune response mechanisms are quite well known today (Solomon et al., 2004; Stryer et al., 2002, Chap. 35). Small, soluble antigens (e.g., bacterial toxins) are engulfed directly into B lymphocytes via receptor-mediated endocytosis. Large antigens (e.g., viruses or the whole bacteria) are swallowed up by macrophages via fagocytosis. Both types of cell digest the antigen into fragments and display a characteristic portion of antigen called an *antigenic determinant* or *epitope* at the surface, nestling inside a class II histocompatibility molecule. Helper CD4-containing T cells specific for this structure then bind to these *antigen-presenting cells* and activate themselves.

There are two types of specific immunologic response: *humoral* and *cellular*. The humoral response occurs in the case when the antigen-presenting cell is a B cell. Then the activated helper T cell secretes some cytokines and stimulates the antigen-presenting B cell with the competent antigen receptor to enter the cell cycle. After multiple mitosis, a *clone* of the B cells with identical receptors is developed. These differentiate into *plasmatic* and *memory* cells. The former secrete the receptors as soluble *immunoglobins* (*antibodies*) into the plasma. Antibodies bind specifically to antigens and precipitate the pathogens. A small number of antibodies remain for a longer time and constitute the origin of an early immunologic response to repeated infection by the same pathogen. There are five classes of immunoglobins. They differ only in kind and number of the heavy chains. IG, IgD and IgE are dimers, IgA has a variable number of subunits, and IgM is a pentamer.

If the antigen-presenting cell is a macrophage, then the activated helper T cell secretes cytokines that stimulate both the macrophage and itself. The activated macrophage secretes other cytokines that attract other cells to the infection area and an inflammation state develops. Simultaneously, the stimulated helper T cells enter the cell cycle. After multiple mitosis, a *clone* of the T cells with identical receptors is developed. Those differentiate into *plasmatic* and *memory* cells. Most of the plasmatic cells are now the cytotoxic T cells, which secrete molecules that destroy the recognized pathogen.

Those of the organism's own cells that have undergone virus infection or malignant transformation present their epitopes nestling inside class I histocompatibility molecules, where they are bound by the specific *cytotoxic* CD8-containing T cells which secrete molecules that destroy the recognized cell. The human immunodeficiency virus (HIV) infects CD4-containing T cells. Therefore, the specific cytotoxic CD8-containing T cells, when destroying the latter cells, disorganize the whole immune system. To avoid autoagression against the organism's own regular cells, those T cells that react with the self-antigens are eliminated. However, this process is incomplete, and some *regulatory* T cells are helpful for deactivating remaining improper T cells.

5.8 The Cell Cycle

Besides its own small supply of genetic material in the form of mitochondria, chloroplasts and possibly centrosomes, the essential part of the genetic material of the eukaryotic cells is stored in the nucleus. There, it is organized in *chromosomes* (Darnell et al., 1999, Chap. 8; Stryer et al., 2002, Chap. 32) containing single molecules of DNA complexed with proteins, mainly basic histones that neutralize acidity and the negative charge of DNA. Chromosomes occur in pairs carrying two slightly different copies of particular sections of the genome. Only a part of the eukaryotic genome represents genes that encode proteins. The genes are split: the coding stretches of DNA, *exons*, are divided by the noncoding stretches, *introns*.

Gene *expression* proceeds in two steps: *transcription* from DNA to RNA and *translation* from RNA onto a given protein. The transcription starts by binding of a protein *transcription factor* to the *promoter* site of a gene, usually on the 5' end of the template. The signal transduction pathways that lead to the activation of appropriate transcription factors were discussed in the last section. In the present section, we consider signal transduction pathways controlling the second process involving DNA, i.e., genome *replication* followed by cell *reproduction* and *proliferation*.

The full *cell cycle* consists of two phases necessary for reproduction: that of *synthesis* (S) during which the genome is doubled, and that of *mitosis* (M) during which the genome is halved and the cell is divided. The phase between M and S is called G_1 (*gap* 1) and that between S and M, G_2 (*gap* 2). From phase G_1, the cell can enter a stable, *quiescent* phase G_0 during which it performs all its functions in the organism apart from reproduction (Fig. 5.42). The key point of the cycle is the *restrictive point* R of the late G_1 phase. Only after transition through this point can the cell enter the whole cycle.

The cell cycle must be controlled to ensure that cells divide only when necessary and the cycle proceeds correctly. The main control switches of the cycle preparing the cell to enter the succeeding phases are *cyclines*, whose concentration varies when the particular phases develop (Fig. 5.42). Cyclin A is present during the whole phase G_2, and cyclin D determines transition through the restrictive point R. Cyclin E initiates phase S, and cyclin B initiates phase M. Cyclins activate *cyclin-dependent kinases* (CDKs). Simultaneously, CDKs are inhibited by signal proteins in response to various factors. The whole system of cell cycle control forms a complex network with many positive and

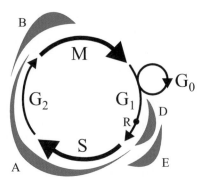

Fig. 5.42. The cell cycle. The time scale is not preserved. M lasts about 2 hours, S about 10 hours, and G_1 and G_2 about 4 hours each. The change in concentration of particular cyclins with the progress of succeeding phases is shown

negative feedbacks and is still far from being completely understood (Weinberg, 1996; Fuller and Shields, 1998, Chap. 6; Darnell et al., 1999, Chap. 12). A few better known CDKs are listed in Table 5.2, together with the cyclins and inhibitors controlling them.

The active CDKs phosphorylate one of the two main restrainers of the cycle, the *protein* pRB. (Its name comes from *retinablastoma*, an eye tumor that first proved to be related to damage of this protein.) pRB liberates an appropriate transcription factor enabling synthesis of proteins needed for the continuation of the cell cycle. The second restrainer of the cycle is the *protein* p53, a transcription factor that activates synthesis of a protein called p21 which blocks activity of all CDKs (see Table 5.2). If DNA damage is so extensive that it cannot be repaired, p53 also triggers synthesis of proteins leading to controlled cell death, or *apoptosis*.

Perturbations to the cell cycle manifest themselves as *cancer*, an unscheduled and uncontrolled cellular proliferation. Rapid progress over the last quarter of a century has revealed cancer to be a disease involving dynamic changes in the genome. Despite the apparently impenetrable thicket of complexity, there are serious premises indicating

Table 5.2. Main cyclin dependent kinases

Kinase	Cyclin	Inhibitor
CDK2	D, E, A, B	p21, p27
CDK4, CDK6	D	p21, p15, p16

that the emergence of all cancers from normal precursor tissues is governed by a few common mechanisms (Weinberg, 1996; Hanahan and Weinberg, 2000; Evan and Vousden, 2001; Hahn and Weinberg, 2002).

The first mechanism involves *self-sufficiency in growth signals*. No type of normal cell can divide in the absence of special external mitogenic growth factors. As mentioned in the last chapter, cancer cells can do it by transformation of their signaling molecules into a form that remains active in the signal transduction network without requiring initial receptor activation. Genes coding such transformed molecules, kinases as a rule, are called *oncogenes* (Blume-Jensen and Hunter, 2001). An alternative to oncogenes is production by cancer cells of their own growth factors, the *tumor growth factor*.

The second hallmark of cancer is *insensitivity to antigrowth signals*. Normal cells cease to divide as a result of their DNA damage. From the considerations of the present section, we know that the molecular mechanism blocking the cell cycle consists in inhibition of CDKs. All inhibitors of CDKs, including pRB and p53, are *tumor suppressors*. Both the key proteins pRB and p53 are activated by proteins originating through alternative splicings of the same locus (Quelle et al., 1995). An alternative to damage of the tumor-suppressor genes is production by the cancer cells of their own growth factors, the *transforming growth factor*.

The third hallmark of cancer is *evading apoptosis*. Normal cells die when subjected to genotoxic stress or a strong radiation that result in DNA damage. Controlled death, called *apoptosis*, follows the action of p53 and when it is inactive, the cells with damaged DNA can proliferate.

The fourth capability of cancer cells is their *limitless replicative potential*. Each time chromosomes of normal cells replicate, their DNA ends called *telomers* shorten. The length of telomeric DNA is a molecular marker of the number of divisions a given cell has passed since it originated in the embryo. After shortening below a critical length the telomers lose their ability to protect chromosomal ends and the cell is submitted to apoptosis. Cancer cells are immortal since, besides evading apoptosis, most of them activate *telomerase*, an enzyme of the reverse transcriptase type that extends telomeric DNA (Greider and Blackburn, 1996).

The remaining hallmarks of cancer are related to a supracellular organization and will not be discussed here.

6 Chemical Reactions

6.1 Single Unimolecular Reactions. The Chemical Equation of State

The compartmental model of the cell formulated in Chap. 5 makes it possible to treat biological processes at the subcellular level as a system of many coupled chemical reactions. Hence, we devote the present chapter to the thermodynamics of chemical reactions, not confining ourselves for a while to typical biochemical reactions. When speaking about thermodynamics, we have in mind in particular nonequilibrium thermodynamics, as in Chap. 3. In the case considered this means that our subject of interest will be essentially *chemical kinetics* (Kondepudi and Prigogine, 1998).

Let us start with the simplest system possible, an ideal gas or an ideal liquid solution comprising molecules each of which can occur in two states R and P, and let us assume that a transition takes place:

$$R \rightleftharpoons P . \tag{6.1}$$

The states R and P can be various chemical states (isomers) of the molecule or its various physical states, e.g., the occurrence on one or another side of a lipid membrane with a built-in channel that enables transport of the molecule across the membrane. In both cases the formal description of the process (6.1) is the same as the transition across a wall that divides the system into two subsystems (Fig. 6.1, see also Fig. 3.2). The molecule in state R will be called symbolically a *reagent*, and the molecule in state P a *product* of the reaction, although in general the reaction can proceed in both directions.

As a rule, chemical reactions take place not only under isothermal conditions, $T = \text{const.}$, but also under isobaric conditions, $P = \text{const.}$ Then the free energy (3.52), referred to more exactly as the Helmholtz free energy, should be replaced by the Gibbs free energy (*free enthalpy*) equal to the difference between the *enthalpy* H (energy increased by the product PV) and the bound energy TS:

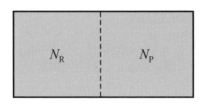

Fig. 6.1. Chemical reaction as a process of transition across a wall

$$G = G(T, P, N) = H - TS = E + PV - TS \ . \tag{6.2}$$

There is a formal derivation in Appendix A.2.

Let N_R denote the number of molecules in state R and N_P the number in state P. The total free energy G is the sum of the component free energies:

$$G = G_R(T, P, N_R) + G_P(T, P, N_P) \ . \tag{6.3}$$

As the total number of molecules does not change during reaction (6.1),

$$N_R + N_P = N = \text{const.} \ , \tag{6.4}$$

only one of N_R and N_P is an independent thermodynamic variable that characterizes a state of chemical nonequilibrium. Let us choose $N_P \equiv X$ as the independent thermodynamic variable.

The thermodynamic force conjugate to the variable X is called the *chemical affinity* and it is equal to the difference between the chemical potential of the reagent and the product in this reaction (Hill, 1989; Kondepudi and Prigogine, 1998):

$$A = -\left(\frac{\partial G}{\partial X}\right)_{T,P,N} = -\frac{\partial G_P}{\partial N_P} + \frac{\partial G_R}{\partial N_R} = -\mu_P + \mu_R \ . \tag{6.5}$$

For ideal gases or perfect solutions the energy per molecule does not depend on the kind of molecule [see (3.30)]. Hence, for a mixture of ideal gases or perfect solutions, the chemical potential of molecules in state R takes the form (Atkins, 1998, Chap. 7, see also Appendix A.1):

$$\mu_R = \mu_R^\circ + k_B T \ln \frac{N_R}{N_R^\circ} \ , \tag{6.6}$$

and similarly for molecules in state P. N_R° is the number of *all* molecules in the region occupied by molecules R, whence the fraction N_R/N_R° determines the probability that a given molecule is found in state R.

It appears that the second term in (6.6) has the meaning of minus the *entropy of mixing* multiplied by temperature [see (2.28)].

If the distribution of molecules is spatially homogeneous, then the numbers N_R and N_P can be expressed by *molar concentrations* denoted by the symbol of the molecular state in a square bracket:

$$[R] \equiv \frac{N_R}{N_A V_R}, \qquad [P] \equiv \frac{N_P}{N_A V_P}. \qquad (6.7)$$

N_A is equal to the *Avogadro number*, the number of molecules in one mole, $N_A = 6.0 \times 10^{23}$, whereas V_R and V_P are the volumes of the regions occupied by molecules R and P, respectively. Those regions may be identical but need not be (e.g., in the case of dissolution or transport across a membrane). The numbers N_R° and N_P° are usually chosen so that

$$\frac{N_R^\circ}{N_A V_R} = \frac{N_P^\circ}{N_A V_P} = 1 \, \text{mol/dm}^3 \equiv 1 \, \text{M}. \qquad (6.8)$$

Then (6.6) takes the form

$$\mu_R = \mu_R^\circ + k_B T \ln \frac{[R]}{M}. \qquad (6.9)$$

The unit of molar concentration $1 \, \text{M}$ has been adopted as the *standard* or *normal* concentration that determines the *standard chemical potential* μ_R° ($\mu_R = \mu_R^\circ$ in the case when $[R] = 1 \, \text{M}$). Similar relations are obtained for μ_P.

Equation (6.9) can be considered to be always satisfied independently of the assumption that the gas is ideal or the solution is perfect, i.e., regardless of the statistical independence of the molecules of the system. This relationship should then be treated as a definition of the quantity $[R]$, which is in general called the *activity*. Chemists have adopted a convention (Atkins, 1998, Chap. 7) that the activity of a substance in the pure solid phase is equal to unity (it has then to be $M = 1$ for $\mu^\circ = 0$), while the activity of a substance in the gas phase is expressed by pressure measured in bars ($M = 1 \, \text{bar} = 10^5 \, \text{Pa}$). The activity of substances in dilute liquid solutions is approximated by their molar concentrations.

Substituting (6.9) and a similar expression for the chemical potential μ_P into (6.5), we obtain the following expression for the chemical affinity:

$$A = -\frac{\Delta G^\circ}{N_A} + k_B T \ln \frac{[R]}{[P]}, \qquad (6.10)$$

where the quantity

$$\Delta G^\circ \equiv N_A(\mu_P^\circ - \mu_R^\circ) \tag{6.11}$$

is called the *free energy of reaction*. It has the physical meaning of the free energy change resulting from the transition of one mole of molecules from state R to state P.

Equation (6.10) can be rewritten as

$$\frac{[P]}{[R]} = K\,e^{-A/k_B T} , \tag{6.12}$$

where

$$K \equiv e^{-\Delta G^\circ / RT} . \tag{6.13}$$

The gas constant R is the product of Avogadro's number and the Boltzmann constant:

$$R = N_A k_B = 8.3 \times 10^{-3}\,\mathrm{kJ\,mol^{-1}deg^{-1}} . \tag{6.14}$$

In chemical equilibrium, the chemical affinity vanishes, $A = 0$, since the chemical potentials of the molecules in both states are equal. Hence the ratio of the equilibrium concentrations of the molecules in states P and R is given by

$$\frac{[P]^{eq}}{[R]^{eq}} = K . \tag{6.15}$$

The quantity K is called the *chemical equilibrium constant*. Equation (6.13) relates it to the free energy of the reaction ΔG°.

For a homogeneous mixture of molecules, when $V_R = V_P = V$, the conservation law (6.4) can be expressed via the concentrations:

$$[R] + [P] = [R]_0 = \mathrm{const.} . \tag{6.16}$$

The *mole fractions*,

$$P_R = \frac{[R]}{[R]_0} = \frac{X}{N} , \qquad P_P = \frac{[P]}{[R]_0} = \frac{N - X}{N} , \tag{6.17}$$

are interpreted as the probabilities that a given molecule is found in state R and P, respectively. From (6.12) and (6.16), we obtain a unique relationship between the thermodynamic variable $X = N_P$ and its conjugate force A:

$$X = \frac{N}{1 + K^{-1}e^{A/k_B T}} . \tag{6.18}$$

This is the *chemical equation of state* for the unimolecular reaction (Fig. 6.2).

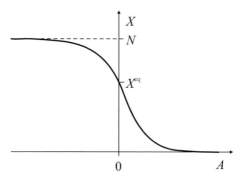

Fig. 6.2. Dependence of the variable $X = N_P$ on the conjugate force (chemical affinity) A for a unimolecular reaction

The free energy of reaction can be divided into enthalpy and entropy components according to [see (6.2)]

$$\Delta G^\circ = -RT \ln K = \Delta H^\circ - T\Delta S^\circ . \tag{6.19}$$

In a single chemical reaction that proceeds in a closed reactor under isothermal and isobaric conditions, no useful work is performed. As a consequence, the quantity ΔH° corresponds to the *heat of reaction* (see Fig. 3.8 in which the energy E is replaced by the enthalpy H and the Helmholtz free energy F by the Gibbs free energy G). By differentiating (6.19) with respect to temperature and using the thermodynamic relation (see, e.g., Appendix A.2)

$$\left(\frac{\partial H}{\partial T}\right)_P = T\left(\frac{\partial S}{\partial T}\right)_P , \tag{6.20}$$

we get the *van't Hoff equation*

$$\Delta H^\circ = RT^2 \frac{\partial \ln K}{\partial T} = -R\frac{\partial \ln K}{\partial (1/T)} . \tag{6.21}$$

This relates the reaction heat ΔH° to the chemical equilibrium constant K.

The heat generated during a reaction spontaneously proceeding from R to P may be given off ($\Delta H^\circ < 0$) in an *exothermic reaction*, or taken up from the environment ($\Delta H^\circ > 0$) in an *endothermic reaction*. Similarly, the free energy of a reaction can be both negative ($\Delta G^\circ < 0$, $K > 1$) in an *exoergic reaction*, or positive ($\Delta G^\circ > 0$, $K < 1$) in an *endoergic reaction*. If the free energy of reaction ΔG° does not depend on temperature, by differentiating (6.19) and using the relation (6.20), we conclude that $\Delta S^\circ = 0$, i.e., $\Delta G^\circ = \Delta H^\circ$. As a consequence, an exoergic reaction automatically becomes an exothermic reaction while an endoergic reaction becomes endothermic.

The direction of a reaction does not depend only on the sign of ΔG° but also on the initial value of the variable $X = N_P$ which, together with ΔG°, determines the sign of the chemical affinity A [see (6.10)]. Let us recall that spontaneous thermodynamic processes, without work being performed on or by the environment, proceed in such a way that the change in the *total* free energy of the system,

$$\Delta G = -A\Delta X , \tag{6.22}$$

is negative, i.e.,

$$A\Delta X \geq 0 , \tag{6.23}$$

whence ΔX has the same sign as the force A.

The time variation of the numbers of molecules $N_P = X$ and $N_R = N - X$ is determined by the *kinetic equation*

$$\dot{N}_P = k_+ N_R - k_- N_P = -\dot{N}_R . \tag{6.24}$$

This can be derived on the basis of statistical physics (see Sect. 3.7 and Appendix B.3). In spite of the fact that (6.24) is linear, it applies to situations arbitrarily far from the state of chemical equilibrium. Parameters k_+ and k_- are called *forward* and *reverse reaction rate constants*, respectively. They appear in the reaction equation

$$R \underset{k_-}{\overset{k_+}{\rightleftarrows}} P . \tag{6.25}$$

For the homogeneous mixture of molecules, (6.24) written in terms of the concentrations [R] and [P] takes the form

$$\frac{d}{dt}[P] = -k_-[P] + k_+[R] = -\frac{d}{dt}[R] . \tag{6.26}$$

The equilibrium solution to (6.24) or (6.26) satisfies the condition [see (6.15)]

$$\frac{N_P^{eq}}{N_R^{eq}} = \frac{[P]^{eq}}{[R]^{eq}} = \frac{k_+}{k_-} = K , \tag{6.27}$$

linking one of the rate constants with the other through the equilibrium constant K.

Taking into account the conservation law (6.4), we can rewrite the kinetic equation (6.24) in terms of the only independent variable X :

$$\dot{X} = -\tau^{-1}(X - X^{\text{eq}}) , \qquad (6.28)$$

where

$$\tau^{-1} \equiv k_+ + k_- \qquad (6.29)$$

and

$$X^{\text{eq}} \equiv \frac{k_+ N}{k_+ + k_-} . \qquad (6.30)$$

This is the simplest possible differential equation. It has a solution

$$X(t) - X^{\text{eq}} = \left[X(0) - X^{\text{eq}}\right] e^{-t/\tau} , \qquad (6.31)$$

which represents an exponential decay of the initial value $X(0)$ to the equilibrium one X^{eq}, with a *relaxation time* τ.

The kinetic equation (6.24) and the relation (6.12) written in terms of the variable X lead to the equation

$$\dot{X} = k_- \left(e^{A/k_{\text{B}}T} - 1\right) X . \qquad (6.32)$$

As $A(e^{A/k_{\text{B}}T} - 1) \geq 1$ always holds, (6.32) guarantees the fulfillment of the second law of thermodynamics [see (6.23)]:

$$A\dot{X} \geq 0 , \qquad (6.33)$$

independently of whether the system is closed or open. The chemical equation of state (6.18) is satisfied in both the closed and the open reactor, in steady state conditions. Expressing X in (6.32) in terms of the force A, we obtain a one-to-one relationship between the reaction flux per molecule and the thermodynamic force (Fig. 6.3):

$$J \equiv \frac{\dot{X}}{N} = \frac{1 - e^{-A/k_{\text{B}}T}}{k_+^{-1} + k_-^{-1}e^{-A/k_{\text{B}}T}} . \qquad (6.34)$$

Close to chemical equilibrium, if the chemical affinity A is lower than the thermal energy $k_{\text{B}}T$, (6.32) or (6.34) can be linearized:

$$\dot{X} \approx (k_- X^{\text{eq}}/k_{\text{B}}T)A \equiv LA . \qquad (6.35)$$

The proportionality coefficient L is interpreted as Onsager's kinetic coefficient (Sect. 3.7). The dependence of the flux on the force, as sketched in Fig. 6.3, shows directly the limits of applicability of linear nonequilibrium thermodynamics to chemical reactions.

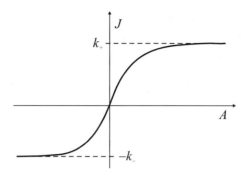

Fig. 6.3. Dependence of the reaction flux $J = \dot{X}/N$ on the chemical affinity (force) A for a unimolecular reaction

6.2 Transport Across Membranes

As already mentioned, the unimolecular reaction formalism applies equally to transport processes across biological membranes. These processes take place through the intermediary of protein channels and involve both neutral and charged solute molecules as well as solvent (water) molecules (see Sect. 5.2).

By assumption, if the membrane closes a distinguished region of a cell: cytosol in the case of the cytoplasmic membrane, matrix in the case of the inner mitochondrial membrane, or lumen in the case of the thylacoid membrane, the transported molecule *outside* this region is considered to be in state R, and the molecule *inside* this region to be in state P. In other words, the forward reaction corresponds to the transport process in towards the distinguished region.

From the chemical point of view, the molecules outside and inside the region are identical,

$$\mu_P^\circ = \mu_R^\circ \,, \qquad \Delta G^\circ = 0 \,, \tag{6.36}$$

whereupon the force that drives the transport process is determined directly by the ratio of concentrations

$$A = \mu_R - \mu_P = k_B T \ln \frac{[\text{R}]}{[\text{P}]} \,. \tag{6.37}$$

In thermodynamic equilibrium, $A = 0$, this ratio is equal to unity and the concentrations are equal:

$$[\text{R}]^{\text{eq}} = [\text{P}]^{\text{eq}} \,. \tag{6.38}$$

Because the molecules in states R and P do not occupy the same volume, the conservation law (6.16) should be replaced by the more general relation (6.4). As a consequence, the equation of state (6.18) and Fig. 6.2 remain valid provided that the equilibrium constant K is replaced by a volume ratio V_P/V_R. Similarly, the flux–force dependence (6.34) and Fig. 6.3 remain valid provided that the forward and reverse rate constants are related through the redefined constant K [see (6.27)]. If the outside region is open, we have $V_R \to \infty$, $K \to 0$, and in the corresponding formulas and figures, one must put $N \to \infty$ and $k_- \to \infty$, although it is meaningful only for low negative values of the force A.

In the case of transport processes involving charged molecules, i.e., ions (Hille, 2001), different values of the electric potential on either side of the membrane may have to be taken into account (see Fig. 6.4a) and the chemical potential (6.9) should be replaced by the *electrochemical potential*

$$\tilde{\mu}_R = \mu_R^\circ + ze\phi_R + k_B T \ln \frac{[R]}{M} \, . \tag{6.39}$$

Here ϕ_R is the electric potential of the membrane on the R side and ze is the electric charge of the molecule expressed as a multiple of the elementary charge $e = |e|$ taken with an appropriate sign. A similar expression can be written for the P side of the membrane. In this way the equation for the affinity (6.37) takes the form of the *Nernst equation*:

$$A = \tilde{\mu}_R - \tilde{\mu}_P = -ze\Delta\phi + k_B T \ln \frac{[R]}{[P]} \, , \tag{6.40}$$

where

$$\Delta\phi \equiv \phi_P - \phi_R = \phi_{in} - \phi_{out} \tag{6.41}$$

represents the *membrane potential*.

In electrochemical equilibrium, $A = 0$, there is no ion transport across the membrane and the membrane potential balances the concentration difference:

$$\Delta\phi^{eq} = \frac{k_B T}{ze} \ln \frac{[R]^{eq}}{[P]^{eq}} = \frac{RT}{zF} \ln \frac{[R]^{eq}}{[P]^{eq}} \, . \tag{6.42}$$

In the second equation we introduced the *Faraday constant* $F = N_A e = 0.96 \times 10^5$ C, which represents the charge of one mole of elementary positive charges. Table 6.1 lists the concentrations of the biologically most important ions outside and inside a typical cell, together with the corresponding equilibrium membrane potentials $\Delta\phi^{eq}$.

(a)   (b)

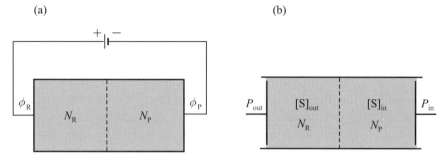

Fig. 6.4. (a) Transport of ions across a membrane is influenced by the difference of electric potentials ϕ_P and ϕ_R on either side of the membrane. (b) Transport of water (solvent) across a membrane is influenced by the osmotic pressure, which is the difference of pressures P_{in} and P_{out} on either side of the membrane. $[S]_{in}$ and $[S]_{out}$ denote the concentrations of a solute inside and outside the cell, respectively

It should be noted that the values of the concentrations quoted here are not in fact equilibrium values, but rather steady state values kept fixed by molecular pumps remaining in a slippage state at the expense of a constant ATP hydrolysis (see Sect. 8.4).

The equality of concentrations of electrically neutral molecules expressed by (6.38) or the constant concentrations of charged molecules given in the relation (6.42) are achieved either by solute transport across the membrane or as a result of solvent transport. In biological cells, the solvent is water. The phenomenon of water transport across membranes, driven by a difference in the solute concentration, is called *osmosis* (Kondepudi and Prigogine, 1999).

Changes in water content result in changes in pressure, which have to be taken into account when calculating the Gibbs free energy G. From (A.47) we find that, for a small change in the pressure value from P_0 to P, the change in the free energy is

Table 6.1. Concentration of key ions outside and inside a typical cell and the corresponding equilibrium membrane potential

	[R]/mM (out)	[P]/mM (in)	$\Delta\phi$/mV
K^+	5	140	−84
Na^+	145	5 to 15	+85 to +57
Cl^-	110	4 to 15	−83 to −50
Ca^{2+}	1.5	0.0001	+120

$$G(P) - G(P_0) = \left(\frac{\partial G}{\partial P}\right)_T (P - P_0) = V(P - P_0) \, , \qquad (6.43)$$

where V is the thermodynamic variable conjugate to pressure, i.e., the volume.

In order to describe the osmosis phenomenon, let us assume that $N_R = N_{out}$ and $N_P = N_{in}$ determine the numbers of water molecules outside and inside the cell, respectively, and that only one type of electrically neutral molecule S is dissolved in the water (Fig. 6.4b). For dilute solutions, the ideal gas approximation applies to both solute and solvent. As a consequence, the chemical potential of water outside the cell can be written as

$$\mu_{out} = \mu^* + k_B T \ln \frac{N_{out}}{N_{out} + N_S} + v(P_{out} - P_0) \, , \qquad (6.44)$$

and likewise for the chemical potential of water inside the cell. In (6.44), we used the relation (6.43) and took into account the fact that the pressure P_{out} can differ from the reference pressure P_0.

The quantity μ^* is the chemical potential of pure water ($N_S = 0$) for $P_{out} = P_0$, and v is the mean volume per single water molecule at P_0. With the help of the quantity v, the argument of the logarithm in (6.44) can be expressed in terms of the molar concentration $[S]_{out}$ of molecules S outside the distinguished region [see the definition (6.7)]:

$$\mu_{out} = \mu^* - k_B T \ln \left(1 + N_A v[S]_{out}\right) + v(P_{out} - P_0) \qquad (6.45)$$
$$\approx \mu^* - v[S]_{out} RT + v(P_{out} - P_0) \, ,$$

where N_A is the Avogadro number and $R = N_A k_B$ is the gas constant. The approximation assumed corresponds to a linear expansion of the logarithm in the neighborhood of unity and is justified by the small value of the ratio $N_S/N_{out} = N_A v[S]_{out}$.

The thermodynamic force that drives the water transport across the membrane is the chemical potential difference

$$A = \mu_{out} - \mu_{in} = -v\pi + v\left([S]_{in} - [S]_{out}\right)RT \, , \qquad (6.46)$$

where the pressure difference inside and outside the distinguished region, viz.,

$$\pi \equiv P_{in} - P_{out} \, , \qquad (6.47)$$

is called the *osmotic pressure*. In osmotic equilibrium, $A = 0$, there is no water transport across the membrane and the osmotic pressure balances the concentration difference:

$$\pi = ([S]_{in} - [S]_{out})RT \ . \tag{6.48}$$

This equation, similar to Clapeyron's ideal gas equation of state, is called the *van't Hoff equation*.

6.3 Bimolecular Reactions

The vast majority of chemical reactions proceed with the participation of two rather than one molecule. Two types of such reactions can be distinguished: association–dissociation and exchange reactions.

In *association–dissociation reactions* of the form

$$A + C \underset{k_-}{\overset{k_+}{\rightleftarrows}} AC \ , \tag{6.49}$$

three thermodynamic variables N_A, N_C and N_{AC} (the numbers of particular molecules) undergo changes. Only one variable is independent, a consequence of the two conservation laws:

$$N_A + N_{AC} = \text{const.} \ , \qquad N_C + N_{AC} = \text{const.} \ . \tag{6.50}$$

The latter can be rewritten in terms of concentrations:

$$[A] + [AC] = [A]_0 = \text{const.} \ , \qquad [C] + [AC] = [C]_0 = \text{const.} \ . \tag{6.51}$$

The chemical affinity corresponding, for example, to the variable $X = N_{AC}$ takes the form

$$A = -\mu_{AC} + \mu_A + \mu_C \tag{6.52}$$

$$= -\mu_{AC}^\circ + \mu_A^\circ + \mu_C^\circ - k_B T \ln \frac{[AC]M}{[A][C]} \ ,$$

and on introducing the free energy of the reaction,

$$\Delta G^\circ \equiv N_{Av} (\mu_{AC}^\circ - \mu_{Av}^\circ - \mu_C^\circ) \ , \tag{6.53}$$

exceptionally denoting the Avogadro number in the present section by N_{Av} instead of N_A, to avoid confusion with the number of A molecules, we obtain the relation

$$\frac{[AC]}{[A][C]} = M^{-1} e^{-\Delta G^\circ/RT} e^{-A/k_B T} \equiv K \, e^{-A/k_B T} \ . \tag{6.54}$$

The corresponding kinetic equation is of the form

$$\frac{d}{dt}[AC] = -\frac{d}{dt}[A] = -\frac{d}{dt}[C] = k_+[A][C] - k_-[AC] \,, \qquad (6.55)$$

and its equilibrium solution is given by

$$\frac{[AC]^{eq}}{[A]^{eq}[C]^{eq}} = \frac{k_+}{k_-} = K = M^{-1}e^{-\Delta G^\circ/RT} \,. \qquad (6.56)$$

The latter determines what is called the *law of mass action*. It states that the addition or removal of one substance from the closed reactor is followed by changes in the concentration of the other substances, in such a way that the value of the parameter K remains constant. The parameter K depends only on temperature and, through ΔG°, on pressure.

The equation of state and the flux–force relation for the association–dissociation reaction are rather complex, but if one of the reactants is in excess, e.g.,

$$[A]_0 \gg [C]_0 \,, \qquad (6.57)$$

then

$$[A] \approx [A]_0 \,, \qquad (6.58)$$

and the kinetic equation (6.55) becomes linear, so that the equation of state and the flux–force relation are identical to (6.18) and (6.34), respectively, with the forward unimolecular rate constant k_+ replaced by the *pseudo-unimolecular* rate constant $k_+[A]_0$.

In *exchange reactions* of the form

$$A + CB \; \underset{k_-}{\overset{k_+}{\rightleftarrows}} \; AC + B \,, \qquad (6.59)$$

the four thermodynamic variables N_A, N_B, N_{AC} and N_{CB} undergo changes, while only one variable is independent as a consequence of the three conservation laws:[1]

$$N_A + N_{AC} = \text{const.} \,, \qquad N_{CB} + N_B = \text{const.} \,,$$

$$N_{CB} + N_{AC} = \text{const.} \,. \qquad (6.60)$$

The conservation laws can be rewritten in terms of concentrations:

$$[A] + [AC] = [A]_0 = \text{const.} \,, \quad [CB] + [B] = [B]_0 = \text{const.} \,,$$

$$[CB] + [BC] = [C]_0 = \text{const.} \,. \qquad (6.61)$$

---

[1] The general rule is: one reaction – one independent thermodynamic variable, m reactions – m independent thermodynamic variables.

The chemical affinity corresponding to the variable $X = N_{AC}$ takes the form

$$A = -\mu_B - \mu_{AC} + \mu_A + \mu_{CB}$$

$$= -\mu_B^\circ - \mu_{AC}^\circ + \mu_A^\circ + \mu_{CB}^\circ - k_B T \ln \frac{[AC][B]}{[A][CB]} , \qquad (6.62)$$

and on introducing the free energy of the reaction

$$\Delta G^\circ \equiv N_{Av} \left(\mu_B^\circ + \mu_{AC}^\circ - \mu_A^\circ - \mu_{CB}^\circ \right) , \qquad (6.63)$$

we obtain the relation

$$\frac{[AC][B]}{[A][CB]} = e^{-\Delta G^\circ / RT} e^{-A/k_B T} \equiv K e^{-A/k_B T} . \qquad (6.64)$$

The kinetic equation for the exchange reaction is of the form

$$\frac{d}{dt}[B] = \frac{d}{dt}[AC] = -\frac{d}{dt}[A] = -\frac{d}{dt}[CB] = k_+[A][CB] - k_-[AC][B] , \qquad (6.65)$$

and its equilibrium solution (the law of mass action) is given by

$$\frac{[AC]^{eq}[B]^{eq}}{[A]^{eq}[CB]^{eq}} = \frac{k_+}{k_-} = K = e^{-\Delta G^\circ / RT} . \qquad (6.66)$$

The equations of state and the flux–force relation for the exchange reaction are even more complex than for the association–dissociation reaction, but if two of the reactants are in excess with respect to the third,

$$[A]_0 , \quad [B]_0 \gg [C]_0 , \qquad (6.67)$$

then

$$[A] \approx [A]_0 , \qquad [B] \approx [B]_0 , \qquad (6.68)$$

and the kinetic equation (6.65) becomes linear. In this case the equation of state and the flux–force relation are identical to (6.18) and (6.34), respectively, with the forward and reverse unimolecular rate constants k_+ and k_- replaced by the corresponding pseudo-unimolecular rate constants $k_+[A]_0$ and $k_-[B]_0$.

6.4 Protolysis Reactions

An example of the association–dissociation reaction taking place in an aqueous environment is the *electrolytic dissociation* of neutral molecules into ions (Atkins, 1998, Chap. 9):

$$AB \rightleftharpoons A^- + B^+ , \tag{6.69}$$

in particular, the electrolytic dissociation of water itself:

$$H_2O \rightleftharpoons H^+ + OH^- . \tag{6.70}$$

The equilibrium constant for the latter reaction is

$$K = \frac{[H^+]^{eq}[OH^-]^{eq}}{[H_2O]^{eq}} = 1.8 \times 10^{-16} \text{ M} . \tag{6.71}$$

One mole of H_2O has a mass of 18 g, and hence the molar concentration of pure water is

$$[H_2O] = 55 \text{ M} . \tag{6.72}$$

In dilute solutions, this value well approximates $[H_2O]^{eq}$ in (6.71). Using the value of the dissociation constant from (6.71), we obtain

$$[H^+]^{eq} = [OH^-]^{eq} = 10^{-7} \text{ M} . \tag{6.73}$$

It is commonly accepted to use a negative decimal logarithm of the molar concentration of hydrogen ions expressed in units of M, referred to as the pH:

$$\text{pH} \equiv -\log_{10} \frac{[H^+]^{eq}}{M} . \tag{6.74}$$

For pure water, we have

$$\text{pH} = 7 . \tag{6.75}$$

Actually, the electrolytic dissociation reaction of water does not proceed according to the scheme (6.70), as hydrogen ions cannot occur as protons H^+ of diameter five orders of magnitude smaller than the diameter of the remaining ions. Much more realistic is the scheme of an exchange reaction

$$H_2O + H_2O \rightleftharpoons H_3O^+ + OH^- , \tag{6.76}$$

in which hydrated protons (*hydronium ions*) H_3O^+ take part. Reaction (6.76) represents a particular case of a proton-transfer reaction in an aqueous environment, i.e., a *protolysis* reaction:

$$H_2O + HA \rightleftharpoons H_3O^+ + A^- , \qquad (6.77)$$

$$A^- + H_2O \rightleftharpoons HA + OH^- . \qquad (6.78)$$

The compound HA which is a proton donor is called an *acid*. The compound A^- which is a proton acceptor is called a *base*, coupled to HA. We have assumed here that its molecules have a negative charge but in general this does not have to be so. It can be assumed that $A^- = B$ and $HA = BH^+$.

The equilibrium constants for the two reactions define the ratios of the equilibrium concentrations:

$$K_a' = \frac{[H_3O^+]^{eq}[A^-]^{eq}}{[H_2O]^{eq}[HA]^{eq}} \equiv \frac{K_a}{[H_2O]} , \qquad (6.79)$$

$$K_b' = \frac{[HA]^{eq}[OH^-]^{eq}}{[A^-]^{eq}[H_2O]^{eq}} \equiv \frac{K_b}{[H_2O]} . \qquad (6.80)$$

Since the concentration of water $[H_2O]$ is practically unchanged during the reaction, it is convenient to use the constants K_a and K_b instead of K_a' and K_b'. These sets are related via

$$K_a K_b = [H_3O^+]^{eq}[OH^-]^{eq} = 10^{-14} \ M^2 . \qquad (6.81)$$

Defining the quantities

$$pK_a \equiv -\log_{10} \frac{K_a}{M} , \qquad pK_b \equiv -\log_{10} \frac{K_b}{M} , \qquad (6.82)$$

we obtain the relationship

$$pK_a + pK_a = 14 . \qquad (6.83)$$

The *degree of dissociation* of acid molecules HA is defined by the ratio

$$P \equiv \frac{[A^-]^{eq}}{[A^-]^{eq} + [HA]^{eq}} . \qquad (6.84)$$

Equation (6.79) leads to the relationship

$$\frac{P}{1-P} = \frac{K_a}{[H^+]^{eq}} , \qquad (6.85)$$

and after applying the logarithm function on both sides of this relationship, to the *Henderson–Hasselbalch equation*:

$$pH = pK_a + \log_{10} \frac{P}{1-P} . \qquad (6.86)$$

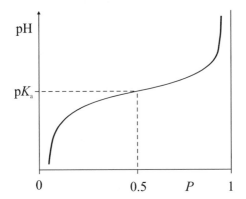

Fig. 6.5. Titration curve corresponding to the Henderson–Hasselbalch equation

The latter equation forms the quantitative basis for *titration*, i.e., an experimental method for determining the values of pK_a or pK_b from the value of pH at the inflection point (see Fig. 6.5). In the neighborhood of the inflection point, the pH value varies only slightly and the system behaves like a *buffer*.

An acid is considered to be strong if $pK_a \ll 7$ and a base is considered to be strong if $pK_b \gg 7$. From (6.83), it follows that a strong acid is coupled to a weak base and vice versa. In the case of dissociation of one acid, the inflection point of the titration curve corresponds to $P = 0.5$, which can be changed by mixing a strong acid with an uncoupled weak base or a strong base with an uncoupled weak acid. Such a technique offers the possibility of buffering practically any value of the pH.

A particular case of protolysis reactions are processes of proton transport through channels or pumps in biological membranes (see Sect. 5.3). Replacing natural logarithms by decimal logarithms (using $\ln 10 \approx 2.3$), we can rewrite the Nernst equation (6.40) for the proton channels or pumps ($z = +1$) to take the form

$$E_p = \Delta\phi - \frac{2.3\,RT}{F}\Delta pH\,, \qquad (6.87)$$

where R is the gas constant, F is the Faraday constant, and

$$\Delta pH = pH_P - pH_R\,. \qquad (6.88)$$

The above expression represents the difference in pH between the interior and exterior of the distinguished region. In the transition from

(6.40) to (6.87), the chemical affinity A has been replaced by the *proton-motive force*:

$$E_\mathrm{p} = -\frac{A}{e} \ . \tag{6.89}$$

(The two thermodynamic forces balance one another under dissipationless conditions).

Proton transfer reactions proceed very quickly. Their rate constants are comparable to the reciprocal Debye relaxation time of electric dipole moments in water. This time is defined by the rate of reorganization of the hydrogen bond system and is of the order of $1 \ \mathrm{ns}^{-1} = 10^9 \ \mathrm{s}^{-1}$ (Eigen, 1964).

6.5 Redox Reactions

The second important class of exchange reactions are *reduction–oxidation (redox) reactions* (Atkins, 1998, Chap. 10). These are chemical processes during which one or more electrons are transferred from molecule to molecule, for example:

$$\mathrm{Zn} + \mathrm{Cu}^{2+} \ \rightleftharpoons \ \mathrm{Zn}^{2+} + \mathrm{Cu} \tag{6.90}$$

or

$$\mathrm{Fe} + \frac{1}{2}\mathrm{O}_2 + (\mathrm{H}_2\mathrm{O}) \ \rightleftharpoons \ \mathrm{Fe}^{2+} + 2\mathrm{OH}^- \ , \tag{6.91}$$

where the second example exhibits an excess of water. The substance that accepts electrons is called an *oxidizer* and the one that donates electrons is called a *reducer*. A reducer undergoes oxidation while an oxidizer is reduced.

Reactions of reduction and oxidation are mutually coupled and, as a rule, occur in the same place. They can be spatially separated in an *electrochemical cell*. Figure 6.6 illustrates the classical Daniell cell in which the reaction (6.90) is separated into the following two reactions:

$$\begin{aligned} \mathrm{Zn} \ &\rightleftharpoons \ \mathrm{Zn}^{2+} + 2\mathrm{e}^- \ , \\ 2\mathrm{e}^- + \mathrm{Cu}^{2+} \ &\rightleftharpoons \ \mathrm{Cu} \ . \end{aligned} \tag{6.92}$$

Each cell is composed of two *half-cells* with *electrodes*, which are electronic conductors, submerged in *electrolytes* which are ionic conductors. Electrolytes must be connected through an electrolytic key that enables the flow of ions. Electronic currents flowing between the two electrodes can be used to perform useful work, whence the cell is a chemoelectrical machine (see Sect. 8.1). Oxidation takes place at

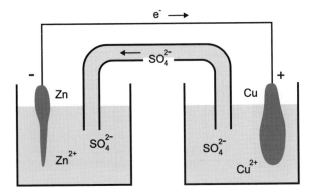

Fig. 6.6. The Daniell electrochemical cell

the *anode*, while reduction occurs at the *cathode*. In the case of the Daniell cell shown in Fig. 6.6, the anode is a zinc electrode. The atoms of Zn are oxidized to Zn^{2+} ions and move to the electrolyte, leaving electrons behind at the anode. The anode therefore reduces its mass and becomes negatively charged. The cathode is a copper electrode that accumulates Cu^{2+} ions after neutralization as a result of electron absorption. The cathode therefore increases its mass and becomes positively charged.

The chemical reactions taking place spontaneously at the two electrodes continue as long as ionic concentrations in the electrolyte do not reach their equilibrium values. Later on, the reactions can be induced by applying an external source of electric current. The cell then becomes an electrochemical machine called an *electrolyzer*, in which electrons flow in the same direction as in the cell and the processes at the two electrodes are the same as earlier. However, the cathode must now be negatively polarized in order to attract positive ions of Cu^{2+}, while the anode must be positively polarized so that it can accept electrons from the Zn atoms and turn them into positively charged ions of Zn^{2+}.

Redox processes can take place at the boundary between phases: liquid and solid or liquid and gas. They can also occur entirely within the liquid phase. We thus distinguish three types of half-cell:

- those with solid electrodes (Fig. 6.7a),
- those with gas electrodes (Fig. 6.7b),
- those with neutral electrodes, also called redox half-cells (Fig. 6.7c).

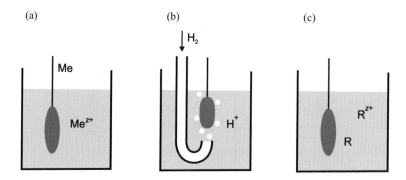

Fig. 6.7. Three types of half-cell. (**a**) Half-cell with a metallic electrode. (**b**) Half-cell with a gas electrode. Gas at pressure P is adsorbed at an electrode made from porous nickel or covered with platinum. These substances often act as catalysts. (**c**) Redox half-cell. The electrode is made from a chemically neutral substance, e.g., platinum. Both oxidized and reduced forms of molecules are found in the solution

In the latter case the active substances can be found in the solution in both the oxidized and the reduced form.

In general, the redox reaction with electron transfer can be written as

$$D_{red} + A^{z+}_{oxy} \rightleftharpoons D^{z+}_{oxy} + A_{red} . \qquad (6.93)$$

We need not assume that molecules D and A in the reduced state D_{red} and A_{red}, respectively, are electrically neutral. Each substance taking part in the reaction can be a positive or negative ion. In an electrochemical cell, the reaction (6.93) divides into two processes:

$$D_{red} \rightleftharpoons D^{z+}_{oxy} + ze^- \qquad (6.94)$$

and

$$A^{z+}_{oxy} + ze^- \rightleftharpoons A_{red} . \qquad (6.95)$$

The chemical equation of state for the reaction (6.93) links the activities of the substances taking part in it with the driving force:

$$A = -\frac{\Delta G^\circ}{N_A} + k_B T \ln \frac{[D_{red}][A^{z+}_{oxy}]}{[D^{z+}_{oxy}][A_{red}]} . \qquad (6.96)$$

Recall that the activity of a substance in a pure solid phase is equal to unity, the activity in a gaseous phase is determined by the pressure expressed in bars, and the activity in a dilute liquid solution can be approximated by the value of its molar concentration in M.

In a cell, which becomes a chemoelectrical machine after replacing a conductor linking the anode to the cathode with an electric energy receiver, the free chemical energy per molecule A is transformed into the free electrical energy of the charge transfer involving z electrons between electrodes at a potential difference (voltage)

$$E = E^\circ + \frac{RT}{zF} \ln \frac{[D_{red}][A^{z+}_{oxy}]}{[D^{z+}_{oxy}][A_{red}]} . \tag{6.97}$$

The voltage E, defined by the equation

$$E = \frac{A}{ze} , \tag{6.98}$$

is called the *electromotive force* of the cell [compare with (6.89) and ensuing comments, taking into account the fact that the charge of one proton equals $+e$ and the charge of z electrons equals $-ze$]. E° is the electromotive force of a cell working under *standard* conditions, where all the activities are equal to unity and the logarithm in (6.97) vanishes. Measurement of E° allows one to determine directly the free energy of the redox reaction, i.e., $\Delta G^\circ = -zFE^\circ$, where F is the Faraday constant. It should be emphasized that the electromotive forces E and E° are voltages between the electrodes of the unloaded cells with no additional voltage drops on an internal resistivity, i.e., with no free energy dissipation. The relationship (6.97), similar to the relationships (6.40) and (6.87), is also called the *Nernst equation*.

The electromotive forces of different half-cells can be compared with respect to the same reference half-cell. As a convention, the *hydrogen half-cell*, which is a gas cell (Fig. 6.7b), has been adopted as such a standard reference point. In this half-cell, a layer of oxidized platinum that covers the electrode is saturated with gaseous hydrogen under a pressure of 1 bar (10^5 Pa), while the electrolyte has a standard hydrogen ion activity of $[H^+] = 1$ M, i.e., it has a pH equal to 0. For the reaction (6.94) applied to a hydrogen half-cell,

$$\frac{1}{2}H_2 \rightleftarrows H^+ + e^- , \tag{6.99}$$

the reaction (6.93) adopts the more specific form

$$A^{z+}_{oxy} + \frac{z}{2}H_2 \rightleftarrows A_{red} + zH^+ . \tag{6.100}$$

E° for this reaction is called the *standard reduction potential*. Table 6.2 lists the values of E° for several selected reduction reactions (6.95)

Table 6.2. Standard reduction potential for chosen redox reactions at 25°C

$A_{oxy}^{z+} + ze^- \longrightarrow A_{red}$	E°/V
$Li^+ + e^- \longrightarrow Li$	-3.05
$K^+ + e^- \longrightarrow K$	-2.93
$Ca^{2+} + 2e^- \longrightarrow Ca$	-2.87
$Na^+ + e^- \longrightarrow Na$	-2.71
$Mg^{2+} + 2e^- \longrightarrow Mg$	-2.36
$H_2O + e^- \longrightarrow \frac{1}{2}H_2 + OH^-$	-0.83
$Zn^{2+} + 2e^- \longrightarrow Zn$	-0.76
$Fe^{2+} + 2e^- \longrightarrow Fe$	-0.44
$H^+ + e^- \longrightarrow \frac{1}{2}H_2$	0.00
$Cu^{2+} + 2e^- \longrightarrow Cu$	$+0.34$
$\frac{1}{2}O_2 + H_2O + 2e^- \longrightarrow 2OH^-$	$+0.40$
$Fe^{3+} + e^- \longrightarrow Fe^{2+}$	$+0.77$
$\frac{1}{2}O_2 + 2H^+ + 2e^- \longrightarrow H_2O$	$+1.23$
$\frac{1}{2}Cl_2 + e^- \longrightarrow Cl^-$	$+1.36$
$Au^{3+} + 3e^- \longrightarrow Au$	$+1.69$
$\frac{1}{2}F_2 + e^- \longrightarrow F^-$	$+2.87$

(Atkins, 1998, Chap. 10). They vary from below -3.0 V for the reduction of the lithium cation Li^+ to metallic Li, up to almost $+3.0$ V
for the reduction of molecular fluorine F_2 to the fluorine anion F^-.
The electromotive force of a cell working under standard conditions is
equal to the difference between standard reduction potentials for the
reactions taking place at the cathode and the anode, respectively. The
standard reduction potential for the reaction from Cu^{2+} to Cu equals
$+0.34$ V, and for the reaction from Zn^{2+} to Zn, it equals -0.76 V.
Therefore, the electromotive force of the Daniell cell equals $+0.34$ V
$-(-0.76$ V$) = 1.10$ V.

6.6 Fuel Cells and Photocells.
Biological Processes of Electron and Proton Transport

Any two half-cells can in principle be combined. This applies also to
half-cells with gas electrodes. In particular, combining the hydrogen
half-cell with the oxygen half-cell, we obtain a machine that directly
transforms hydrogen fuel energy into electrical energy, i.e., the *fuel cell*.
Figure 6.8 presents two versions of such a cell. In the cell with a base
electrolyte using the hydroxyl groups OH^- (Fig. 6.8a), the reaction

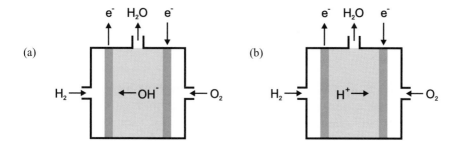

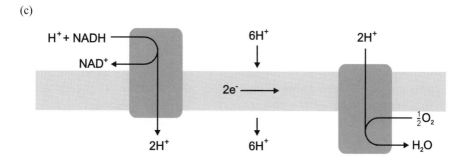

Fig. 6.8. Hydrogen–oxygen fuel cell with a base electrolyte (**a**) and an acid electrolyte (**b**), and a system of respiratory chain complexes in the internal membrane of mitochondrium (**c**)

$$H_2 + 2OH^- \longrightarrow 2H_2O + 2e^- \tag{6.101}$$

takes place on the anode and the reaction

$$\frac{1}{2}O_2 + H_2O + 2e^- \longrightarrow 2OH^- \tag{6.102}$$

on the cathode. On the other hand, in the cell with an electrolyte using hydrated hydrogen cations H^+ (Fig. 6.8b), the reaction

$$H_2 \longrightarrow 2H^+ + 2e^- \tag{6.103}$$

takes place on the anode and the reaction

$$\frac{1}{2}O_2 + 2H^+ + 2e^- \longrightarrow H_2O \tag{6.104}$$

on the cathode.

Is easy to check in Table 6.2 that the electromotive force in both cases equals +1.23 V. The total reaction,

$$H_2 + \frac{1}{2}O_2 \longrightarrow H_2O \ , \tag{6.105}$$

is the same as the reaction of hydrogen gas burning in an internal combustion engine, where chemical energy is first turned into thermal energy, and then transformed to mechanical energy of rotational motion. Just combining such an organized system with a generator allows one to obtain electrical energy. In comparison to the internal combustion engine, the fuel cell is silent and much more efficient. Important technological progress in the construction of catalysts able to decompose hydrogen gas into protons and electrons opens the way to considering a mass application of fuel cells for driving motor cars in the near future (Burns et al., 2002). Animate nature made a similar invention more than three billion years ago. Indeed, a system of enzymatic complexes of the respiratory chain in the internal membrane of mitochondria (see Fig. 4.8 and Sect. 5.3) is a highly efficient fuel cell (Fig. 6.8c). The only difference consists in the fact that hydrogen is supplied in a much safer way, bound to NAD^+, and instead of protons, electrons are transported along the membrane, resulting in the creation of proton-motive rather than electromotive transmembrane forces.

Similarly, a system of photosynthetic chain complexes in the thylacoid membrane (see Fig. 4.7 and Sect. 5.3) is a highly efficient photocell (Fig. 6.9a). Here animate nature solved the problem in a better way than man. Besides a transmembrane proton-motive force, the hydrogen is produced safely, bound to $NADP^+$. Later on, it is used in the Calvin cycle for synthesizing glucose (see Sect. 4.6). Present technology offers photocells that use semiconducting materials (Fig. 6.9b). The best-studied semiconductors in which photons easily create electron–hole pairs are cadmium selenite (CdSe), cadmium telluride (CdTe), gallium arsenide (GaAs), or microcrystalline silicon in the form of a thin film (Shah at al., 1999). The idea is to use the liberated electrons to produce molecular hydrogen in an acid electrolyte (Fig. 6.9c), but it appears that this is not an easy task.

From the point of view of biochemical reactions, the assumed definition of the standard reduction potential is not quite realistic, since the reference system is the electrolyte with pH = 0, i.e., a very strong, biologically destructive acid. The Nernst equation (6.97) also describes the case where the two half-cells are made from identical substances

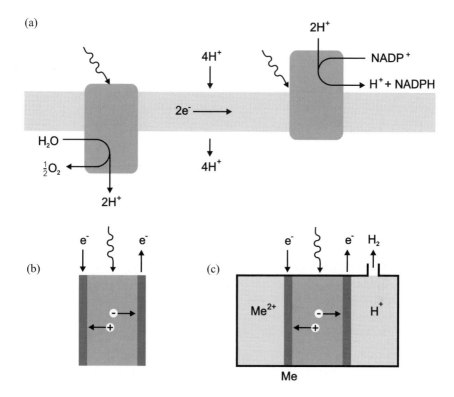

Fig. 6.9. (a) The system of photosynthetic chain complexes in the thylacoid membrane can be considered as a photocell. **(b)** Photocell with a semiconducting material. **(c)** Possible use of electrons created in the semiconductor photocell to produce hydrogen

(D = A, $E^\circ = 0$) and only differ in the concentrations of the electrolyte. Such a cell is called a *concentration cell*. In particular, for a cell composed of a hydrogen half-cell at an arbitrary pH and a standard hydrogen half-cell at pH = 0,

$$E' = -\frac{k_\mathrm{B}T}{e}(\ln 10)\,\mathrm{pH} = -0.059\ \mathrm{V\,pH}\ ,\qquad (6.106)$$

at 25°C.

For pH = 7, $E' = -0.42$ V. The value of this additional potential must be taken into account for biochemical redox reactions which take place in a buffer environment at a pH close to the neutral value. Electron transfer in such reactions is often linked to a proton transfer, and a whole hydrogen atom is then transferred. In measurements of standard reduction potentials for biochemical reactions in an environment with pH $= 7$, one does not determine the value of E°, but instead the value of

$$E^{\circ\prime} = E^\circ - 0.42 \text{ V} . \tag{6.107}$$

Table 6.3 lists the values of the potential $E^{\circ\prime}$ for the most important biochemical redox reactions (Stryer et al., 2002, Chap. 19). Note that the reference reaction potential now equals just -0.42 V.

Table 6.3. Standard reduction potential for key biochemical redox reactions at temperature 25°C and pH $= 7$

$A_{oxy}^{z+} + ze^- \longrightarrow A_{red}$	$E^{\circ\prime}/V$
ferredoxin$_{oxy}^+ + e^- \longrightarrow$ ferredoxin$_{red}$	-0.43
$H^+ + e^- \longrightarrow \frac{1}{2}H_2$	-0.42
$(NAD^+ + 2H^+) + 2e^- \longrightarrow (NADH + H^+)$	-0.32
$(NADP^+ + 2H^+) + 2e^- \longrightarrow (NADPH + H^+)$	-0.32
$(FAD + 2H^+) + 2e^- \longrightarrow FADH_2$	-0.18
$(\text{fumarate} + 2H^+) + 2e^- \longrightarrow \text{succinate}$	$+0.03$
$(\text{quinone} + 2H^+) + 2e^- \longrightarrow \text{quinone } H_2$	$+0.10$
cytochrome $c^{3+} + e^- \longrightarrow$ cytochrome c^{2+}	$+0.22$
$\frac{1}{2}O_2 + 2H^+ + 2e^- \longrightarrow H_2O$	$+0.82$

6.7 Two Successive Reactions.
The Steady State Approximation

Consider the following set of two consecutive unimolecular reactions:

$$R \; \underset{k_{-1}}{\overset{k_{+1}}{\rightleftharpoons}} \; I \; \underset{k_{-2}}{\overset{k_{+2}}{\rightleftharpoons}} \; P . \tag{6.108}$$

The corresponding kinetic equations take the form

$$\frac{d}{dt}[R] = -k_{+1}[R] + k_{-1}[I] ,$$

$$\frac{d}{dt}[I] = -(k_{-1} + k_{-2})[I] + k_{-1}[R] + k_{-2}[P] , \tag{6.109}$$

$$\frac{d}{dt}[P] = -k_{-2}[P] + k_{+2}[I] .$$

However, only two of them are independent in view of the conservation law

$$[R] + [I] + [P] = [R]_0 = \text{const.} . \tag{6.110}$$

Indeed, adding (6.109), we obtain

$$\frac{d}{dt}\Big([R] + [I] + [P]\Big) = 0 . \tag{6.111}$$

The reaction rate constants satisfy *detailed balance conditions* (in thermodynamic equilibrium the number of forward and reverse transitions is balanced for each reaction separately):

$$k_{+1}[R]^{eq} = k_{-1}[I]^{eq} , \qquad k_{+2}[I]^{eq} = k_{-2}[P]^{eq} , \tag{6.112}$$

from which it follows that

$$\frac{k_{+1}k_{+2}}{k_{-1}k_{-2}} = \frac{[P]^{eq}}{[R]^{eq}} . \tag{6.113}$$

Since only two of the three differential equations (6.109) are linearly independent, it is easy to find their analytical solutions, although they are not very transparent. In practice, one often uses the *steady state approximation*. If the equilibrium concentration of the intermediate I is sufficiently small, or if its free energy is relatively large, in view of the detailed balance conditions (6.112), we have

$$\frac{k_{+1}}{k_{-1}} = \frac{[I]^{eq}}{[R]^{eq}} = e^{-(G_I^\circ - G_R^\circ)/RT} \ll 1 \tag{6.114}$$

and

$$\frac{k_{+2}}{k_{-2}} = \frac{[P]^{eq}}{[I]^{eq}} = e^{-(G_P^\circ - G_I^\circ)/RT} \gg 1 . \tag{6.115}$$

It follows from the above inequalities that

$$k_{-1} + k_{+2} \gg k_{+1}, k_{-2} . \tag{6.116}$$

From the second equation of (6.109), we conclude that a weakly occupied state under equilibrium conditions is also a short-lived state. Its concentration [I] is a fast variable compared to [R] and [P], and after a short *transient* period

$$\tau_{tr} = (k_{-1} + k_{+2})^{-1} , \tag{6.117}$$

it reaches the *steady state*:

$$\frac{d}{dt}[I] = 0 . \tag{6.118}$$

A steady state is constant on a short time scale. On a long time scale, however, the value of the concentration [I] follows the slowly varying concentrations [R] and [P] (Haken, 1990). From (6.118), we can deduce a relationship between the stationary value of the *fast* variable [I] and the values of the *slow* variables [R] and [P]:

$$[I] = \frac{k_{+1}[R] + k_{-2}[P]}{k_{-1} + k_{+2}} . \tag{6.119}$$

Substituting it into the first or the third of (6.109), we find

$$\frac{d}{dt}[P] = -\frac{d}{dt}[R] = k_{+}[R] - k_{-}[P] , \tag{6.120}$$

where

$$k_{+} = \frac{k_{+2}k_{+1}}{k_{-1} + k_{+2}} , \qquad k_{-} = \frac{k_{-2}k_{-1}}{k_{-1} + k_{+2}} . \tag{6.121}$$

Equation (6.120) can be considered as a kinetic equation for the effective reaction

$$R \underset{k_{-}}{\overset{k_{+}}{\rightleftarrows}} P , \tag{6.122}$$

which, over a long time scale, well approximates the system (6.108). Applying the detailed balance conditions (6.112), we can rewrite the relationships (6.121) as

$$k_{+} = \frac{[I]^{\mathrm{eq}}}{[R]^{\mathrm{eq}}} \left[(k_{-1})^{-1} + (k_{+2})^{-1} \right]^{-1} \tag{6.123}$$

and

$$k_{-} = \frac{[I]^{\mathrm{eq}}}{[P]^{\mathrm{eq}}} \left[(k_{-1})^{-1} + (k_{+2})^{-1} \right]^{-1} . \tag{6.124}$$

The effective reaction rates are determined by the equilibrium occupation probability for the intermediate state relative to the initial and final states, respectively, multiplied by the inverse of the sum of the average transit times from the intermediate to the initial and final states.

6.8 Phenomenological Theory of Reaction Rates

The steady state approximation applies to an arbitrary number of mutually coupled chemical reactions provided the fast- and slowly-varying concentrations can be clearly distinguished. In particular, it can be applied to the following sequence of reactions:

$$
R \; \underset{k_{-1}}{\overset{k_{+1}}{\rightleftarrows}} \; R^{\ddagger} \; \underset{k_{-0}}{\overset{k_{+0}}{\rightleftarrows}} \; P^{\ddagger} \; \underset{k_{-2}}{\overset{k_{+2}}{\rightleftarrows}} \; P \;,
\tag{6.125}
$$

with the two short-lived intermediates $R^{\ddagger}$ and $P^{\ddagger}$. The effective single reaction (6.122) is described by the kinetic equation (6.120) with reaction rate constants

$$
k_{+} = \frac{k_{+1}k_{+0}k_{+2}}{k_{-1}k_{-0} + k_{-1}k_{+2} + k_{+0}k_{+2}}
\tag{6.126}
$$

and

$$
k_{-} = \frac{k_{-1}k_{-0}k_{-2}}{k_{-1}k_{-0} + k_{-1}k_{+2} + k_{+0}k_{+2}} \;.
\tag{6.127}
$$

Making use of the three detailed balance conditions for the three reactions (6.125), we rewrite (6.126) and (6.127) as

$$
k_{+} = \left[\left(\frac{[R^{\ddagger}]^{eq}}{[R]^{eq}} k_{+0} \right)^{-1} + (k_{+1})^{-1} + \frac{[R]^{eq}}{[P]^{eq}} (k_{-2})^{-1} \right]^{-1}
\tag{6.128}
$$

and

$$
k_{-} = \left[\left(\frac{[P^{\ddagger}]^{eq}}{[P]^{eq}} k_{-0} \right)^{-1} + (k_{-2})^{-1} + \frac{[P]^{eq}}{[R]^{eq}} (k_{+1})^{-1} \right]^{-1} \;.
\tag{6.129}
$$

The intermediates $R^{\ddagger}$ and $P^{\ddagger}$ can be interpreted as forming together a *transition state* of the reaction (6.122) (see Fig. 3.2). The reciprocal rate constants $(k_{+0})^{-1}$ and $(k_{-0})^{-1}$ can then be thought of as the mean passage times through this state in the forward and reverse directions, respectively. Without loss of generality, we can assume both times to be equal, whence

$$
k_{+0} = k_{-0} \equiv \nu \;, \qquad [R^{\ddagger}]^{eq} = [P^{\ddagger}]^{eq} \;.
\tag{6.130}
$$

In accordance with (6.13) and (6.15), we define the free energy differences $\Delta G_{R}^{\ddagger}$ and $\Delta G_{P}^{\ddagger}$ between the transition and the initial and final states, respectively:

$$\frac{[R^{\ddagger}]^{eq}}{[R]^{eq}} = e^{-\Delta G_R^{\ddagger}/RT} \ , \qquad \frac{[P^{\ddagger}]^{eq}}{[P]^{eq}} = e^{-\Delta G_P^{\ddagger}/RT} \ . \qquad (6.131)$$

The exponentials are treated as probabilities of transition state occupation under the assumption that it is in a partial thermodynamic equilibrium with the initial or final state, respectively.

In terms of the newly introduced quantities, the reciprocals of the reaction rate constants (6.128) and (6.129) can be rewritten in the form

$$(k_+)^{-1} = \left(\nu e^{-\Delta G_R^{\ddagger}/RT}\right)^{-1} + (k_{+1})^{-1} + K^{-1}(k_{-2})^{-1} \qquad (6.132)$$

and

$$(k_-)^{-1} = \left(\nu e^{-\Delta G_P^{\ddagger}/RT}\right)^{-1} + (k_{-2})^{-1} + K(k_{+1})^{-1} \ , \qquad (6.133)$$

where K is the equilibrium constant (6.15). Both (6.132) and (6.133) represent a sum of three characteristic times that determine the resultant reaction rate. The first component determines the time required to cross the transition state under the assumption that it is in a partial thermodynamic equilibrium with the initial (but not the final) or the final (but not the initial) state. As a result of the transition, this equilibrium is obviously disturbed. The second and third contributions in (6.132) and (6.133) describe processes of restoring the local equilibrium from the side of the initial and final states.

If the limiting factor of the reaction is crossing the transition state, i.e., the first time components in (6.132) and (6.133) are the longest, the remaining contributions can be ignored, yielding

$$k_+ = \nu e^{-\Delta G_R^{\ddagger}/RT} \ , \qquad k_- = \nu e^{-\Delta G_P^{\ddagger}/RT} \ . \qquad (6.134)$$

The assumption of an infinitely fast process of reaching partial equilibrium between the initial state R and $R^{\ddagger}$ or P and $P^{\ddagger}$ forms the basis of the *transition state theory* (Atkins, 1998, Chap. 27). Equating the mean passage rate across the transition state ν with the mean frequency of thermal vibrations and using the relation

$$h\nu = k_B T \ , \qquad (6.135)$$

where h is Planck's constant and k_B is Boltzmann's constant (which yields $\nu = 0.5 \times 10^{13} \ s^{-1}$ at temperature $T = 300$ K), this theory is limited only to calculations of the Gibbs free energy differences $\Delta G_R^{\ddagger}$ and $\Delta G_P^{\ddagger}$.

Conversely, if the processes of restoring partial equilibrium in the transition state are slower than the processes of transition state crossing, a detailed knowledge of microscopic dynamics is required to compute the requisite expressions (see Appendix B.3). This involves intramolecular dynamics for isomerization reactions and, additionally, intermolecular dynamics for exchange reactions. In such cases we say that a reaction is *controlled* by dynamics.

One can always express the complete reaction rates (6.128) and (6.129) in terms of an *Arrhenius formula* like (6.134):

$$k_+ = \nu_+ e^{-\Delta G_{\mathrm{R}}^{\ddagger}/RT} \;, \qquad k_- = \nu_- e^{-\Delta G_{\mathrm{P}}^{\ddagger}/RT} \;. \tag{6.136}$$

The quantities $\Delta G_{\mathrm{R}}^{\ddagger}$ and $\Delta G_{\mathrm{P}}^{\ddagger}$ are called *activation free energies*, while ν_+ and ν_- are referred to as *preexponential factors*. From the detailed balance conditions for the three reactions (6.125) and the relations (6.130), it follows that

$$\nu_+ = \nu_- = \left[\nu^{-1} + (k_{-1})^{-1} + (k_{+2})^{-1} \right]^{-1} \;. \tag{6.137}$$

The value of (6.137) can in general be much lower than 0.5×10^{13} s^{-1}.

As the rate constants k_{-1} and k_{+2} themselves display an Arrhenius temperature dependence, the reaction rates (6.128) and (6.129) can be rewritten in yet another form:

$$k_+ = \nu e^{-(\Delta G_{\mathrm{R}}^{\ddagger} + \Delta G_{\mathrm{eff}}^{\ddagger})/RT} \;, \qquad k_- = \nu e^{-(\Delta G_{\mathrm{P}}^{\ddagger} + \Delta G_{\mathrm{eff}}^{\ddagger})/RT} \;. \tag{6.138}$$

Here the preexponential factor ν is identical to the one occurring in the relation (6.135) used in the standard transition state theory, and $\Delta G_{\mathrm{eff}}^{\ddagger}$ is an effective correction to the activation free energy. Evaluation of this correction is the subject of the so-called *generalized transition state theory* (Garcia-Viloca et al., 2004).

7 Enzymatic Catalysis

7.1 Chemical Mechanisms of Enzymatic Catalysis

In accordance with the considerations of Sect. 6.8, the highest rate of a chemical reaction allowed by thermodynamics is determined by the transition state theory. The expression for the mean reaction time (the reciprocal rate constant) given by this theory is

$$k^{-1} = \nu^{-1} e^{\Delta G^{\ddagger}/RT} \ . \tag{7.1}$$

The exponent above represents the reciprocal equilibrium occupation probability of the transition state. $\Delta G^{\ddagger}$ denotes the free energy of activation, i.e., the difference between the free energy of the transition state and that of the initial state, and RT for the temperature $T = 300$ K corresponds to the energy 2.5 kJ/mol. The first factor, the reciprocal mean frequency (the mean period) of thermal vibrations, can be estimated using the relationship (6.135) to yield the value 2×10^{-13} s. For comparison, life has existed on Earth for over 3.5 billion years, i.e., almost 10^{17} s, which means a factor of 10^{30} longer than this characteristic time scale. Since $10^{30} \approx e^{70}$, the reaction time would reach this astronomical value for a free energy of activation that is only 70 times larger than the mean thermal energy, i.e., for 175 kJ/mol. For comparison, the bond energy in the carbon–carbon case is approximately 350 kJ/mol (see Table C.1), and hence is twice as large. Most reactions linked to the reorganization of covalent bonds do not go through complete bond breaking, but their characteristic free energy of activation is not much lower than 175 kJ/mol. Hence, at physiological temperatures, spontaneous occurrence of such reactions would require hours, years or even millennia.

These time scales are too long for living organisms. Living systems accelerate the rates of almost all relevant reactions using specific *catalysts*, i.e., compounds that take part in chemical reactions but are recycled after their completion. Biochemical reactions are catalyzed by protein *enzymes* or, currently very seldom, by the more archaic

RNA-based *ribozymes*. A typical time for an enzymatic reaction is 10^{-3} s, whence enzymes accelerate biochemical reactions by a factor of at least 10 million. To accelerate a reaction by a factor of ten, the activation barrier has to be lowered by 5.7 kJ/mol. As a consequence, enzymes reduce the activation barrier by at least 40 kJ/mol.

Besides *accelerating* chemical reactions, enzymes fulfill two other important functions. First, they *control* reactions, which means that a given reaction takes place in a cell only at an appropriate moment and at the desired location. Second, enzymes *couple* reactions. To make use of a chemical reaction in a process of biological free energy transduction, it must occur simultaneously with another reaction at the same multi-enzymatic complex. Therefore, enzymes must be characterized by high *specificity*. Each metabolic reaction is catalyzed by its own enzyme (Sect. 4.6).

Before we move to the main topic of this chapter, i.e., the kinetics of enzymatic reactions including their regulatory functions, let us first consider the purely chemical aspects of accelerating chemical reactions by enzymes. As we cannot cover the whole area of enzyme biochemistry here, we will limit our attention to just one well-known class of enzymes, namely *proteases* which catalyze the hydrolysis of peptide bonds. A general presentation of this topic can be found, e.g., in the book by Fersht (1999) and a review by Garcia-Viloca et al. (2004).

The reaction of hydrolysis of a peptide or an ester bond,

$$\text{R--CO--X} + \text{H}_2\text{O} \rightleftarrows \text{RCOOH} + \text{H--X} , \qquad (7.2)$$

where $-\text{X} = -\text{NH--R}'$ or $-\text{O--R}'$, while R$'$ and R are arbitrary molecular groups, occurs via a very short-lived tetrahedral intermediate state (see Fig. 7.1) which can be identified with the transition state. In this state, an electrostatic charge separation takes place over a relatively long distance, and this is the main reason for the significant energy increase of the system.

The transition state is stabilized (its energy is lowered) as a result of bringing an *electrophilic* molecule (a cation and/or a lone electron pair acceptor – see Appendixes C.3 and C.4) to its negative end and a *nucleophilic* molecule (an anion and/or a lone electron pair donor) to its positive end. An example of an electrophilic molecule is an acid molecule that detaches hydrogen in the form of a proton keeping a lone electron pair from the hitherto existing bond (*general acid catalysis*, see Fig. 7.2a). An example of a nucleophilic molecule is a base that attaches hydrogen in the form a proton, bringing its own electron pair into the bond thereby created (*general base catalysis*, see

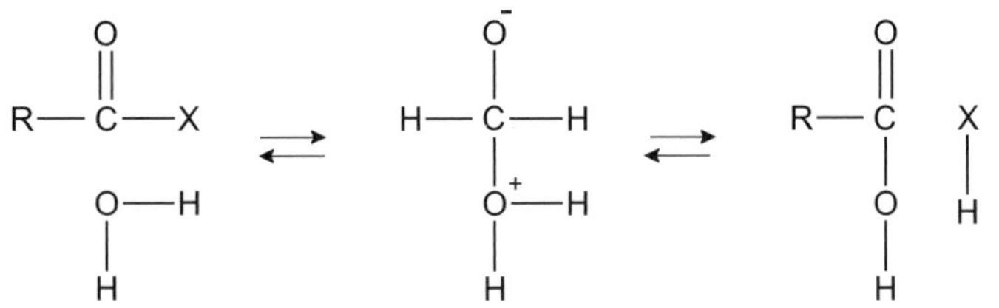

Fig. 7.1. Noncatalyzed reaction of hydrolysis of a peptide bond $(-X = -NH-R')$ or an ester bond $(-X = -O-R')$. R and R′ are arbitrary molecular groups. Electrostatic charge separation takes place in the transition state

Fig. 7.2b). Metal cations (Fig. 7.2c) play a purely electrostatic role as electrophiles, while nucleophiles often perform nucleophilic substitutions and make themselves tetrahedral intermediates. A newly created bond hydrolyzes faster than the original one, changing the reaction pathway (*covalent catalysis*, see Fig. 7.2d).

Adding an electrophile or a nucleophile to a solution containing a hydrolyzed peptide lowers the energy contribution to the free energy of activation and simultaneously increases its entropic contribution. Stabilization of a tetrahedral intermediate requires a collision of two molecules with an adequate mutual orientation which is very improbable and thus carries low entropy. The entropic contribution can be neglected when the two molecules are parts of a larger molecule within

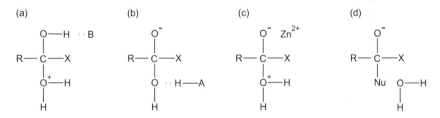

Fig. 7.2. Four main types of chemical catalysis stabilizing a tetrahedral intermediate. (**a**) General acid catalysis: an acid molecule HB^+ donates a proton to a negatively charged molecule of carboxyl oxide. The *dotted line* denotes a hydrogen bond. (**b**) General base catalysis: a base molecule A^- accepts a proton from a positively charged hydronium group. (**c**) Electrophilic catalysis taking place with the help of a metal cation, e.g., Zn^{2+}. (**d**) Nucleophilic catalysis: a nucleophile Nu^- forms a tetrahedral intermediate more stable than a molecule of water

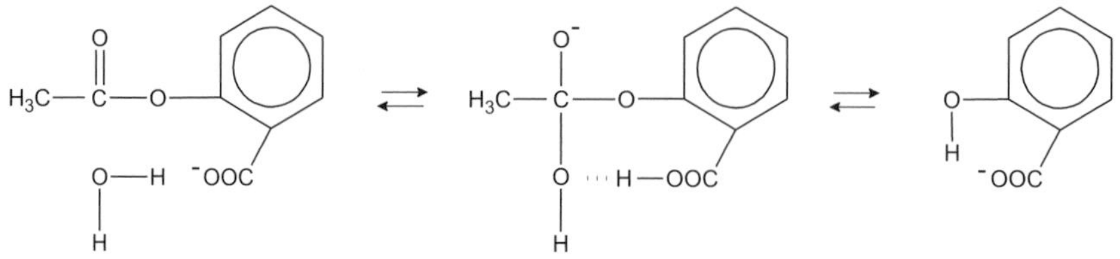

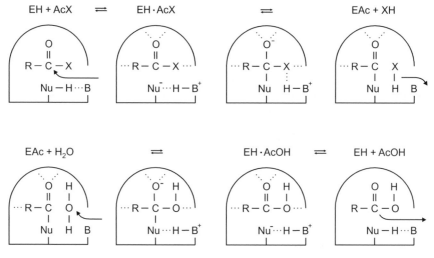

Fig. 7.3. Example of intramolecular catalysis. Hydrolysis of acetylsalicylic acid (aspirin). The basic catalytic group $-COO^-$ belongs to the same molecule as the acetyl group CH_3-CO- which, by contact with water, is transformed into a tetrahedral intermediate and then detached

which they are already properly oriented for bonding. A well-known example of such *intramolecular catalysis* is the hydrolysis of aspirin (acetylsalicylic acid) to salicylic acid (see Fig. 7.3).

Fig. 7.4. Hydrolysis of a peptide bond via serine and cysteine proteases proceeds in two stages. In the first stage, a covalent intermediate compound is formed, an acyl-enzyme EAc (the acyl group Ac referred to is the group R−CO−). In the second stage, a bond between the acyl group and the enzyme undergoes hydrolysis. The various functional groups in the active center are shown. Nu denotes a nucleophile, an oxygen of a side chain of serine or a sulfur of cysteine. The base B is a histidine. The binding of the radical R of the acyl group is specific while the binding of the amine group X in serine proteases is not. The latter can be readily replaced by an alcohol group. In fact, most investigations on the mechanism of functioning of serine proteases have been carried out using esters as substrates

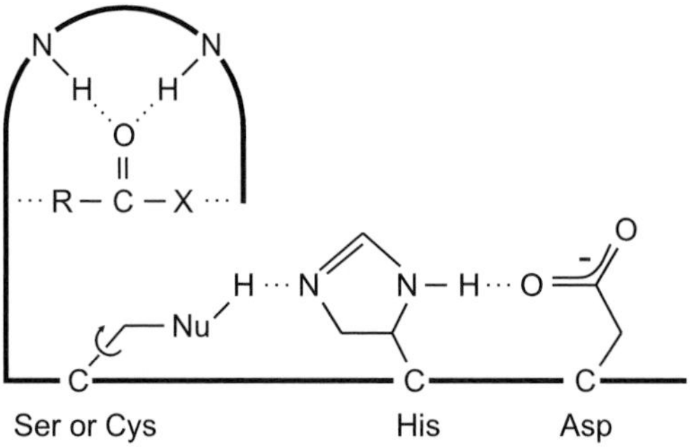

Fig. 7.5. Pathway for proton transfer from a nucleophile (a hydroxyl group of serine or a sulfhydryl group of cysteine) via a histidine to asparagine (the so-called catalytic triad). After freeing from the proton, the side chain of serine or cysteine rotates around the bond C^α–C and the nucleophile attacks the carboxylic carbon of the substrate

Usually in *enzymatic catalysis*, all elements of chemical catalysis are present. As an example, consider in more detail the mechanism underlying the action of *serine proteases* (e.g., pancreatic enzymes of mammals such as trypsin, chymotrypsin, see Fig. C.28 in Appendix C, elastase, or subtilisin in bacteria) or *cysteine proteases* (e.g., papain, a plant enzyme). In both classes of enzymes, we deal with covalent catalysis which changes the reaction pathway. First, a nucleophile (in this case a hydroxyl group of one of the serins or a sulfhydryl group of one of the cysteins) covalently bonds to the substrate forming an intermediate compound, acyl-enzyme, which then undergoes hydrolysis, as shown in Fig. 7.4 (Fersht, 1999). The nucleophile becomes active only after it donates a proton, and covalent catalysis must therefore be linked with a general base catalysis. In both classes of enzymes, a proton is transferred via a histidine onto an aspartate, as in Fig. 7.5 (Phillips and Fletterick, 1992; Dodson and Wlodawer, 1998). All the catalytically active molecular groups must have proper spatial orientation with respect to each other and with respect to a specifically bound substrate (Fig. 7.4). Therefore, enzymatic catalysis is to a large degree intramolecular catalysis.

7.2 Steady-State Kinetics of Enzymatic Reactions with One Intermediate

For a unimolecular biochemical reaction of the form

$$R \rightleftarrows P,$$ (7.3)

the simplest catalytic counterpart is

$$E + R \underset{k'_-}{\overset{k'_+}{\rightleftharpoons}} M \underset{k''_-}{\overset{k''_+}{\rightleftharpoons}} E + P . \tag{7.4}$$

This reaction contains two well-defined steps: the binding of a reagent with an enzyme and the release of the product. E denotes a free enzyme and M denotes the *enzyme–substrate complex*. During this reaction, the enzyme is not exhausted (see Fig. 7.6). The first approximate analysis of the kinetic equations for the scheme (7.4) was carried out by Michaelis and Menten in 1913. The full analysis was performed by Haldane in his monograph on enzymes published in 1930 (Cantor and Schimmel, 1980, Chap. 16).

Four concentrations occur in the kinetic equations describing the two steps in (7.4). Only two are independent due to the conservation laws

$$[R] + [M] + [P] = [R]_0 = \text{const.} , \tag{7.5}$$

and

$$[E] + [M] = [E]_0 = \text{const.} . \tag{7.6}$$

Figure 7.6 represents the reaction (7.4) from the viewpoint of the enzyme macromolecules. To represent it from the viewpoint of the substrate molecules, we rewrite it as

$$R \underset{k'_-}{\overset{k'_+[E]}{\rightleftharpoons}} M \underset{k''_-[E]}{\overset{k''_+}{\rightleftharpoons}} P . \tag{7.7}$$

In general, the molar concentration of the enzyme is much lower than that of the substrate:

$$[E]_0 \ll [R]_0 . \tag{7.8}$$

Hence, the concentration [M] is much lower than the concentrations [R] and [P]. After a short *prestationary* stage of the reaction (see Sect. 6.7),

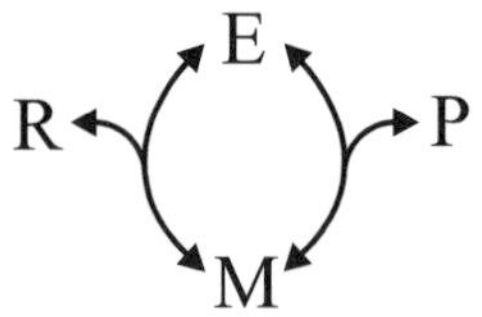

Fig. 7.6. Enzymatic cycle with one intermediate. In the first component reaction, the enzyme is used up, and in the second, it is recovered

we can apply the steady state approximation treating the enzyme concentration $[E]$ as a constant. Figure 2.3 in Sect. 2.3 illustrates the accuracy of this approximation, even for relatively large values of the fraction $[E]_0/[R]_0$. Utilizing (6.121) for the scheme (6.108) in which the constants k_{+1} and k_{-2} are replaced by the constants $k'_+[E]$ and $k''_-[E]$, respectively, we obtain

$$\frac{d}{dt}[P] = -\frac{d}{dt}[R] = \frac{k'_+ k''_+[R] - k'_- k''_-[P]}{k'_- + k'_+}[E] . \tag{7.9}$$

The value of the enzyme concentration $[E]$ in its free state is determined by the total enzyme concentration $[E]_0$. An appropriate relationship is found from (7.6) and the stationarity condition:

$$\frac{d}{dt}[M] = -(k'_- + k''_+)[M] + k'_+[E][R] + k''_-[E][P] = 0 , \tag{7.10}$$

which allows us to eliminate the fast-varying variable $[M]$. We thus obtain an equation that describes the steady state rate of product formation or reagent consumption:

$$\frac{d}{dt}[P] = -\frac{d}{dt}[R] = \frac{k_+ K_+^{-1}[R] - k_- K_-^{-1}[P]}{1 + K_+^{-1}[R] + K_-^{-1}[P]}[E]_0 , \tag{7.11}$$

where

$$k_+ = k''_+ , \qquad K_+ = (k'_- + k''_+)/k'_+ , \tag{7.12}$$

and

$$k_- = k'_- , \qquad K_- = (k'_- + k''_+)/k''_- . \tag{7.13}$$

It follows from (7.12), (7.13) and the detailed balance conditions for the two reactions (7.4) that the parameters k_+, K_+, k_-, and K_- are related by the *Haldane equation*:

$$\frac{k_+}{K_+}\frac{K_-}{k_-} = \frac{[P]^{eq}}{[R]^{eq}} \equiv K . \tag{7.14}$$

It is more convenient to define the enzymatic reaction rate, i.e., the *reaction flux*, with respect to a single molecule of enzyme [and not substrate as in (6.34)]:

$$J \equiv \frac{d}{dt}[P]/[E]_0 . \tag{7.15}$$

Using the Haldane equation (7.14), the approximate formula

$$[R] + [P] = [R]_0 , \tag{7.16}$$

and the definition of chemical affinity for the noncatalyzed reaction (7.3), viz.,

$$\frac{[P]}{[R]} = K e^{-\beta A} , \tag{7.17}$$

we can rewrite (7.11) in a form analogous to (6.34):

$$J = \frac{1 - e^{-\beta A}}{J_+^{-1} + J_-^{-1} e^{-\beta A}} , \tag{7.18}$$

where $\beta = (K_B T)^{-1}$ and

$$J_\pm = \frac{k_\pm [R]_0}{K_\pm + [R]_0} . \tag{7.19}$$

Under the condition that $[R] = [R]_0$ ($[P] = 0$, $\beta A \to \infty$) or $[P] = [R]_0$ ($[R] = 0$, $\beta A \to -\infty$), which corresponds to an early stage of the reaction starting from the pure R or P, respectively, the reaction flux (7.18) takes the asymptotic values J_+ or $-J_-$. The hyperbolic dependence of the reaction fluxes J_+ or J_- on the substrate concentrations $[R]_0 = [R]$ or $[P]$ is called the *Michaelis–Menten equation* (Fig. 7.7a). For low concentrations $[R]_0$, the fluxes $J_\pm$ increase linearly with $[R]_0$:

$$J_\pm = \frac{k_\pm}{K_\pm} [R]_0 . \tag{7.20}$$

On the other hand, for high concentrations $[R]_0$, they reach the saturation values:

$$J_\pm = k_\pm . \tag{7.21}$$

The parameters k_+ or k_- in the Michaelis–Menten equation have a meaning of *turnover number* for the enzyme, i.e., the number of product molecules P or R, respectively, produced per unit time. The parameters K_+ or K_- are called the *Michaelis–Menten constants* and also the *apparent dissociation constants*. Indeed, if $k_+'' \ll k_-'$ (or, alternatively, $k_-' \ll k_+''$), then $K_\pm$ become identical with the actual dissociation constants of R or P from the complex M. The conditions $[P] = 0$, $[R] = [R]_0 = \text{const.}$ or $[R] = 0$, $[P] = [R]_0 = \text{const.}$ can also be satisfied in an open stationary reactor as a result of a constant removal of reaction products. This is a typical situation *in vivo* and in most experimental systems *in vitro*. The experimental results of steady state kinetics are more conveniently represented in the form of the so-called

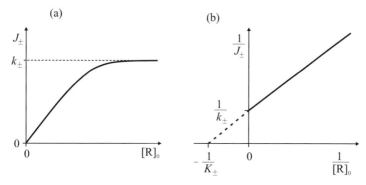

Fig. 7.7. (a) Dependence of the asymptotic enzymatic reaction flux $J_\pm$ on the substrate concentration $[R]_0 = [R]$ or $[P]$ in the absence of the product ($[P] = 0$ or $[R] = 0$, respectively). (b) The Lineweaver–Burk plot for an enzyme with Michaelis–Menten kinetics

Lineweaver–Burk plot, which gives the dependence of the reciprocal reaction flux on the reciprocal concentration of the substrate:

$$\frac{1}{J_\pm} = \frac{1}{k_\pm} + \frac{K_\pm}{k_\pm [R]_0} \ . \tag{7.22}$$

It is then easy to determine both the enzyme turnover number and the value of the apparent dissociation constant (Fig. 7.7b).

Equations of the type (7.11) and (7.19) for the steady state enzymatic kinetics can be derived for a much larger class of kinetic schemes than (7.4), by taking into account more than one intermediate state of the enzyme and even a quasi-continuum of such states (Chap. 9). These types of equation usually describe experimental situations quite well (Fersht, 1999). Therefore, the turnover numbers and the Michaelis constants, in both forward and reverse directions, constitute universal phenomenological parameters for enzymatic reactions and do not have to be linked with specific kinetic schemes. The acceleration of the reaction rate by an enzyme is due to the fact that the turnover numbers $k_\pm$ are much larger than the reaction rates for the noncatalyzed reaction (7.3). According to the Haldane equation (7.14), the presence of an enzyme does not affect the chemical equilibrium between R and P, and the acceleration of the reverse reaction is realized to the same degree as that of the forward reaction.

A good enzyme in the case of the kinetic scheme (7.4) is characterized by high values of the constants k''_+ and k'_-. However, one physical limitation cannot be directly circumvented. Bimolecular reaction rates k'_+ and k''_- are controlled by spatial diffusion and their values cannot exceed 10^9 $M^{-1}s^{-1}$. Consequently, in the case of good enzymes, not only are $k'_+[E]$ and $k''_-[E]$ small compared to the sum $k''_+ + k'_-$, which is a precondition for the applicability of the steady state approximation, but also $k'_+[R]$ and $k''_-[P]$ are small compared to this sum. According to (7.12) and (7.13), this implies that the inequality

$$K_\pm \gg [R]_0 \tag{7.23}$$

is satisfied, and the approximation (7.20) is valid. Indeed, most enzymes under physiological conditions do not reach saturation conditions. The condition of having the larger of $k_+/K_+ \approx k'_+$ or $k_-/K_- \approx k''_-$ close to 10^9 $M^{-1}s^{-1}$ means that the enzyme has reached *kinetic perfection* (Stryer et al., 2002, Chap. 8). For typical substrate concentrations $[R]_0 = 10^{-6}$ M, reactions then take place at rates of the order of 10^3 s^{-1}.

Limitations due to long diffusion times of substrate molecules disappear in the case of enzymes that catalyze subsequent reactions of a metabolic chain into supramolecular multienzymatic complexes. In the process of evolution, nature must have found this solution early on, since this approach is very common in real biological systems (Sect. 5.1).

If the inequality (7.8) is satisfied, the enzymatic reaction (7.4) proceeding in a closed reactor is completely characterized by one thermodynamic variable:

$$X \equiv [P] - [P]^{\text{eq}} = -[R] + [R]^{\text{eq}} . \tag{7.24}$$

According to the relationship (7.14), the steady state kinetics equation (7.11) can be rewritten in terms of this variable as the *Henri equation*:

$$\frac{d}{dt} X = \frac{-\left(k_+ K_+^{-1} - k_- K_-^{-1}\right)[E]_0 X}{\left(1 + K_-^{-1}[R]^{\text{eq}} + K_-^{-1}[P]^{\text{eq}}\right) + \left(K_-^{-1} - K_+^{-1}\right) X} . \tag{7.25}$$

Hitherto we studied the solution to this equation at the beginning of the steady state stage of the reaction, when the conditions $[P] \approx 0$ or $[R] \approx 0$ could be assumed, identical to those realized in the open reactor. Let us now consider the solution to this equation at the end of the steady state stage. We now have conditions close to total thermodynamic equilibrium, when the value of X is low and (7.25) can be linearized to the equation

$$\dot{X} = -\tau^{-1} X , \tag{7.26}$$

which describes an exponential decay of X to zero with the relaxation time

$$\tau = \frac{1 + K_+^{-1}[R]^{eq} + K_-^{-1}[P]^{eq}}{\left(k_+ K_+^{-1} + k_- K_-^{-1}\right)[E]_0} .$$

(7.27)

For an enzyme close to kinetic perfection, the inequalities (7.23) hold, and only unity is left in the numerator of the expression (7.27). It follows that the relaxation time τ is independent of the substrate concentration and inversely proportional to the enzyme concentration:

$$\tau^{-1} \approx \left(k_+ K_+^{-1} + k_- K_-^{-1}\right)[E]_0 .$$

(7.28)

7.3 Competitive and Noncompetitive Inhibition

As already pointed out in Sects. 4.6 and 5.6, the role of enzymes is not only to accelerate biochemical reactions, but also to control them. Hence, there must exist mechanisms both for switching off active enzymes (*inhibition*) and switching on nonactive enzymes (*activation*).

The simplest method for slowing an enzymatic reaction is to reduce the effective number of enzyme molecules by binding them to a molecule that resembles the substrate (*competitive inhibition*). The majority of modern medicines act as competitive inhibitors. Various blockers that bind to the target receptors eliminate or significantly reduce their action thus change the signal transduction pathways. Antibiotics, on the other hand, disturb the metabolism of pathogenic microbes. For example, penicillin can be irreversibly bound to an enzyme that catalyzes the synthesis of the cell wall of Gram-positive bacteria.

In the presence of a competitive inhibitor I, in addition to the reaction (7.4), the reaction

$$E + I \underset{k_{-I}}{\overset{k_{+I}}{\rightleftharpoons}} E'$$

(7.29)

takes place. E' represents an inactive form of the enzyme. Figure 7.8a shows the cumulative kinetic scheme from the point of view of the enzyme. The condition for conservation of the number of enzyme molecules takes the form

$$[E] + [M] + [E'] = [E]_0 = \text{const.} .$$

(7.30)

We are interested only in the steady-state kinetics in an open reactor to which a reagent is supplied at a constant rate and a product removed

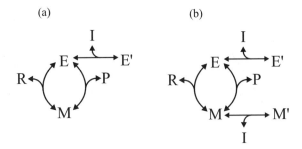

Fig. 7.8. Competitive inhibition (**a**) and noncompetitive inhibition (**b**) of an enzymatic cycle with one intermediate. I denotes the inhibitor, whereas E′ and M′ denote inactive states of the enzyme and enzyme–substrate complex, respectively

in such a way that [R] = const. and [P] = 0. From the relationship (7.30) and the two independent steady-state conditions

$$\frac{\mathrm{d}}{\mathrm{d}t}[\mathrm{M}] = -(k'_- + k''_+)[\mathrm{M}] + k'_+[\mathrm{E}][\mathrm{R}] = 0 \qquad (7.31)$$

and

$$\frac{\mathrm{d}}{\mathrm{d}t}[\mathrm{E}'] = -k_{-\mathrm{I}}[\mathrm{E}'] + k_{+\mathrm{I}}[\mathrm{E}][\mathrm{I}] = 0 , \qquad (7.32)$$

we find the steady-state concentration of the enzyme–substrate complex M as

$$[\mathrm{M}] = \frac{[\mathrm{R}][\mathrm{E}]_0}{K_+(1 + K_{\mathrm{I}}^{-1}[\mathrm{I}]) + [\mathrm{R}]} , \qquad (7.33)$$

where K_{I} is the equilibrium constant for inhibitor binding,

$$K_{\mathrm{I}} \equiv \frac{k_{-\mathrm{I}}}{k_{+\mathrm{I}}} = \frac{[\mathrm{I}]^{\mathrm{eq}}[\mathrm{E}]^{\mathrm{eq}}}{[\mathrm{E}']^{\mathrm{eq}}} , \qquad (7.34)$$

and the constant K_+ is determined by the second equation of (7.12).

The value of the concentration [M] determines the steady-state production rate of P:

$$\frac{\mathrm{d}}{\mathrm{d}t}[\mathrm{P}] = k''_+[\mathrm{M}]. \qquad (7.35)$$

It is still of the Michaelis–Menten form (7.19) but it has a changed apparent dissociation constant. Indeed, the reciprocal flux of the reaction as a function of the reciprocal concentration of the reagent is given by

$$\frac{1}{J_+} = \frac{1}{k_+} + \left(1 + K_{\mathrm{I}}^{-1}[\mathrm{I}]\right) \frac{K_+}{k_+} \frac{1}{[\mathrm{R}]} . \qquad (7.36)$$

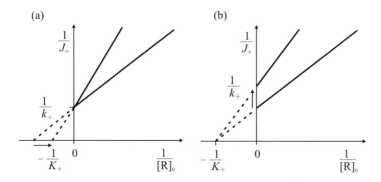

Fig. 7.9. Lineweaver–Burk plot for an inhibited enzyme described by Michaelis–Menten kinetics (see Fig. 7.7b). (**a**) A competitive inhibitor changes the value of the effective Michaelis constant. (**b**) A noncompetitive inhibitor alters the turnover number

Comparing (7.36) with (7.22), we find that a competitive inhibitor does not alter the turnover number but increases the value of the Michaelis constant. On the Lineweaver–Burk plot, the crossing point of the straight line representing the reciprocal flux with the horizontal axis moves toward zero (Fig. 7.9a).

In contrast to a competitive inhibitor, a *noncompetitive inhibitor* binds to both states E and M of the enzyme. Hence its binding site must be different from the binding site of the substrate. In addition to the reaction (7.4), we have two reactions:

$$\text{E} + \text{I} \xrightarrow{K_\text{I}} \text{E}' , \qquad \text{M} + \text{I} \xrightarrow{K_\text{I}} \text{M}' . \qquad (7.37)$$

In the schemes above we gave only the equilibrium constants and assumed them to be equal. The kinetic scheme of noncompetitive inhibition shown from the viewpoint of the enzyme is illustrated in Fig. 7.8b. The condition for the conservation of the number of enzyme molecules takes the form

$$[\text{E}] + [\text{M}] + [\text{E}'] + [\text{M}'] = [\text{E}]_0 = \text{const.} , \qquad (7.38)$$

and three independent steady-state conditions are given by

$$[\text{M}] = K_+^{-1}[\text{E}][\text{R}] , \quad [\text{E}'] = K_\text{I}^{-1}[\text{E}][\text{I}] , \quad [\text{M}'] = K_\text{I}^{-1}[\text{M}][\text{I}] . \quad (7.39)$$

The first condition (7.39) follows from (7.31), valid in the absence of product P, $[\text{P}] = 0$, whereas the other two conditions represent the equilibrium conditions for the reactions (7.37).

Using the four equations (7.38) and (7.39), we find the value of the stationary concentration of the Michaelis complex as

$$[M] = \frac{[R][E]_0}{(1 + K_I^{-1}[I])(K_+ + [R])} \; . \tag{7.40}$$

Hence, the reciprocal reaction rate is given by

$$\frac{1}{J_+} = \left(1 + K_I^{-1}[I]\right)\left(\frac{1}{k_+} + \frac{K_+}{k_+}\frac{1}{[R]}\right) \; . \tag{7.41}$$

A noncompetitive inhibitor, in contrast to the competitive one, reduces the turnover number without altering the value of the Michaelis constant (see Fig. 7.9b).

In Sects. 5.2 and 6.2, it has been argued that no hydrophilic molecule, even water itself, can cross the phospholipid membrane without the intermediacy of protein channels. Channels play an obvious role as enzymes and the steady-state kinetics of transport processes across membranes is in fact an enzymatic kinetics. In particular, all the mechanisms of competitive and noncompetitive inhibition also apply to transport across biological membranes.

7.4 Two-Substrate Enzyme

The scheme shown in Fig. 7.8b can be generalized to the one with two equivalent reagents shown in Fig. 7.10. In the new notation, the conservation law for the number of enzyme molecules takes the form

$$[E] + [M_1] + [M_2] + [M] = [E]_0 = \text{const.} \; . \tag{7.42}$$

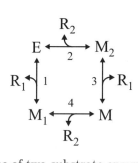

Fig. 7.10. General scheme of two-substrate enzymatic reactions. Not shown are the component reactions detaching products

Assuming that the reactions detaching the products (not shown in Fig. 7.10) are much slower than the reactions binding the reagents, we can approximate the values of the steady-state concentrations of the particular enzyme states by the values of their equilibrium concentrations, and replace the apparent dissociation constants K_- by the actual dissociation constants:

$$K_1 = \frac{[\mathrm{E}][\mathrm{R_1}]}{[\mathrm{M_1}]} , \qquad K_2 = \frac{[\mathrm{E}][\mathrm{R_2}]}{[\mathrm{M_2}]} ,$$
$$K_3 = \frac{[\mathrm{M_2}][\mathrm{R_1}]}{[\mathrm{M}]} , \qquad K_4 = \frac{[\mathrm{M_1}][\mathrm{R_2}]}{[\mathrm{M}]} . \tag{7.43}$$

Not all of them are independent, as the equality

$$\frac{K_1}{K_2}\frac{K_4}{K_3} = 1 \tag{7.44}$$

is satisfied.

From the relationships (7.43) and the conservation law (7.42), equilibrium values of the three intermediate enzyme state concentrations can be found that approximate the corresponding steady-state values:

$$[\mathrm{M}] = \frac{[\mathrm{R_1}][\mathrm{R_2}][\mathrm{E}]_0}{K_3\,(K_2 + [\mathrm{R_2}]) + (K_4 + [\mathrm{R_2}])\,[\mathrm{R_1}]} , \tag{7.45}$$

$$[\mathrm{M_1}] = \frac{K_4[\mathrm{R_1}][\mathrm{E}]_0}{K_3\,(K_2 + [\mathrm{R_2}]) + (K_4 + [\mathrm{R_2}])\,[\mathrm{R_1}]} , \tag{7.46}$$

$$[\mathrm{M_2}] = \frac{K_3[\mathrm{R_2}][\mathrm{E}]_0}{K_3\,(K_2 + [\mathrm{R_2}]) + (K_4 + [\mathrm{R_2}])\,[\mathrm{R_1}]} . \tag{7.47}$$

Many bimolecular reactions can be effectively treated as unimolecular reactions. A good example are hydrolysis reactions catalyzed by hydrolases (e.g., proteases considered in Sect. 7.1), in which one reagent is water, always occurring in excess. However, the reverse reactions of the generalized ester bond synthesis are catalyzed by other enzymes, the synthases, since they are endoergic and proceed along other pathways. In the latter reactions, two reagents of comparable concentrations occur, one of which is phosphorylated (see Appendix C.2). Such reactions are of the form

$$\mathrm{R_1} + \mathrm{R_2} \longrightarrow \text{products} . \tag{7.48}$$

The order of reagent binding can be arbitrary (Fig. 7.10).

The rate of product formation (the enzymatic reaction flux) is proportional to the steady-state concentration of the enzyme–two reagent complex [M]. From (7.45) it follows that the dependence of this rate both on $[\mathrm{R_1}]$ with $[\mathrm{R_2}]$ fixed, and on $[\mathrm{R_2}]$ with $[\mathrm{R_1}]$ fixed, obeys the Michaelis–Menten law.

7.5 Allosteric Control of Enzymatic Activity

The molecule R_2 does not have to be treated as one of the substrates. Instead, as in Sect. 7.3, it can be treated as a regulatory molecule, or *effector*. The effector can act both as a noncompetitive inhibitor or as an *activator*. In the latter case, the active form of the enzyme is that bound to the effector, whence the steady-state rate of the enzymatic reaction is proportional to the concentration [M]. Assuming for simplicity that $K_2 = K_4$ and denoting $R_1 = R$ (the reagent molecule) and $R_2 = A$ (the activator), we obtain from (7.45):

$$[M] = \frac{[R][E]_0}{(1 + K_2[A]^{-1})(K_3 + [R])} , \qquad (7.49)$$

and the reciprocal of the reaction flux

$$\frac{1}{J_+} = \left(1 + K_2[A]^{-1}\right) \left(\frac{1}{k_+} + \frac{K_3}{k_+}\frac{1}{[R]}\right) . \qquad (7.50)$$

An increase in the concentration [A] causes an increase in the effective turnover number, with an unchanged value of the Michaelis constant which is equal, under the approximation used here and in the present notation, to the binding constant K_3.

The effector is a molecule which binds to the enzyme at a site different from the active center of the substrate binding and affects the latter process over a certain distance. Without discussing its physical nature, this type of interaction is referred to in molecular biology as *allostery* (from the Greek *allos*, meaning 'other', and *stereos*, meaning 'space', 'location'), see, e.g., Cantor and Schimmel (1980, Chap. 17). The term was introduced in 1965 by Monod, Wyman and Changeaux. In the case of allosteric control, the enzyme can be formed by two separate entities: one *catalytic* and the other *regulatory*. We then refer to an allosteric *heterotropic* effect (see Fig. 7.11a). Also possible is an allosteric *homotropic* effect, which involves the cooperative action of two or more identical catalytic entities (Fig. 7.11b), as described in the original paper by the above-mentioned authors.

In simple terms, Monod, Wyman and Changeaux relate the allosteric interaction to a simultaneous cooperative change of all component entities from a less active tense (T) conformational state to a more active relaxed (R) conformational state. This change is affected by binding the first ligand (Fig. 7.11c). An alternative model was proposed in 1966 by Koshland and coworkers. According to this model,

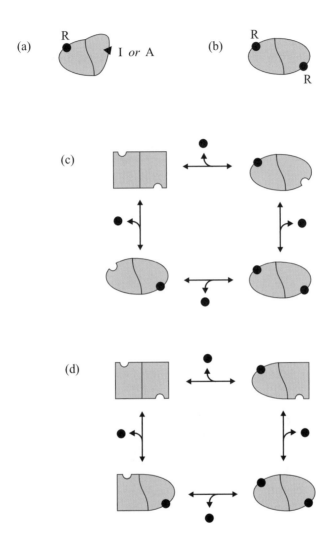

Fig. 7.11. Allosteric interaction. (**a**) Allosteric heterotropic effect. A substrate and an effector (noncompetitive inhibitor I or activator A) bind at different locations on the enzyme. The enzyme consists of two separate entities: a catalytic and a regulatory one. (**b**) Allosteric homotropic effect. An enzyme is composed of two (in general more than two) identical or nearly identical catalytic sub-units. (**c**) The Monod–Wyman–Changeaux concerted-transition interpretation of the scheme in Fig. 7.10. (**d**) The Koshland and coworkers sequential-transition interpretation of that scheme. The *rectangle* and *half-oval* represent the 'tense' and the 'relaxed' conformational states of the enzyme, respectively

a conformational change takes place separately in each entity, and co-operativity consists in facilitating the subsequent ligand binding after the preceding one (Fig. 7.11d).

Substituting $R_1 = R_2 = R$ into (7.45) and assuming $K_1 = K_2$ and $K_3 = K_4$, we obtain

$$[M] = \frac{[R]^2[E]_0}{K_1 K_3 + 2K_3[R] + [R]^2} \cdot \qquad (7.51)$$

For low concentrations, the dependence of [M] on [R] is quadratic,

$$[M] = \frac{[R]^2[E]_0}{K_1 K_3} , \qquad (7.52)$$

whereas for high concentrations, saturation is reached,

$$[M] = [E]_0 . \qquad (7.53)$$

Consequently, the dependence of the reaction flux on [R] is no longer hyperbolic, of the Michaelis–Menten type, but *sigmoidal* (see Fig. 7.12).

The allosteric homotropic effect occurs commonly in proteins in which a reduction of enzymatic functions has taken place, probably as a result of evolution. These are *transport* and *regulatory* proteins (Stryer et al., 2002, Chap. 8). A classic example of the allosteric protein of the first type is hemoglobin – a protein that transports molecular oxygen between regions with high concentrations and regions with low concentrations of oxygen. An example of the allosteric protein of the second type is calmodulin, a regulatory protein for many kinases (Sect. 5.5). It cooperatively binds ions of Ca^{2+}.

The reason for the organization of enzymes into supramolecular allosteric structures is that it signifies an ability to make a large change in the active enzyme form concentration by changing the activator concentration [A] only slightly. In general, the process is described by an effective kinetic scheme

$$E' + m\,A \; \underset{k'''_-}{\overset{k'''_+}{\rightleftarrows}} \; E , \qquad (7.54)$$

in which E' and E denote the enzyme molecule in nonactive and active states, respectively. Binding the first activator molecule A makes binding the next one easier, and so on until the last (mth) molecule. There is a conservation law,

$$[E'] + [E] = [E]_0 = \text{const.} , \qquad (7.55)$$

and the equilibrium condition for the reaction (7.54) is

$$k''_+[E'][A]^m = k'''_-[E] ,$$ (7.56)

from which we obtain the dependence of the active enzyme form concentration $[E]$ on the activator concentration $[A]$:

$$[E] = \frac{[A]^m}{[A]^m + K'''}[E]_0 ,$$ (7.57)

where

$$K''' = k'''_-/k'''_+ .$$ (7.58)

For high values of the concentration $[A]$, the dependence (7.57) reaches saturation,

$$[E] = [E]_0 ,$$ (7.59)

whereas for low values of $[A]$, it can be approximated by a function

$$[E] = (K''')^{-1}[A]^m[E]_0 .$$ (7.60)

From these asymptotic types of behavior, it follows that for $m = 1$ the dependence (7.57) is hyperbolic (the *Langmuir law*, Fig. 7.12a), whereas for $m > 1$, when the function (7.60) becomes concave, the dependence (7.57) is *sigmoidal* (*Hill's law*, Fig. 7.12.b).

For m much larger than unity the dependence of the active enzyme form concentration $[E]$ on the activator concentration $[A]$ becomes close to a step function. One consequence is the possibility of switching an enzymatic reaction on or off via small changes in the activator concentration. The activators can be either protein macromolecules or inorganic ions like Ca^{2+}.

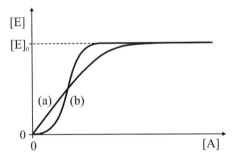

Fig. 7.12. Dependence of the active enzyme form concentration $[E]$ on the activator concentration $[A]$, given by (7.57), for $m = 1$ (the Langmuir hyperbola) and $m > 1$ (the sigmoidal curve of Hill). In the latter case, the value of the concentration $[E]$ changes significantly for only slight changes in the concentration $[A]$ in the vicinity of the inclination point

7.6 Oscillations in Enzymatic Reactions

If the role of activator is played by the very product of an enzymatic reaction, i.e., if a *positive feedback* is realized feedback activation (see Sect. 5.6), stable temporal oscillations can appear in the system (Goldbeter, 1996). A steady state becomes a dissipative structure (Sect. 3.6).

Presumably, the best examined are the glycolytic oscillations resulting from the positive feedback on phosphofructokinase, the key control enzyme of glycolysis (Stryer et al., 2002, Chap. 16). It can be assumed that one reagent of this enzyme always occurs in excess. The second reagent is ATP. One of the two products is ADP. ADP molecules are the essential activators of phosphofructokinase which usually occur as a tetramer.

Here we present the historical model of glycolytic oscillations proposed by Selkov (1968). It is based on the schemes (7.4) and (7.54) already considered. We identify R with ATP and P with ADP. However, the product molecules P are not simply removed from the system, but also play the role of enzyme activators. Hence, besides replacing A with P in the scheme (7.54), we have to consider the possibility of different rates of reagent influx and product outflow from the system:

$$\text{environment} \xrightarrow{v} \text{R} , \qquad \text{P} \xrightarrow{w} \text{environment} . \tag{7.61}$$

Following the reactions (7.61), concentrations [R] and [P] become independent variables in contrast to the scheme (7.4) alone.

The values of the concentrations [E'], [E] and [M] of the three enzyme forms E', E and M are defined by the conservation condition

$$[\text{E}'] + [\text{E}] + [\text{M}] = [\text{E}]_0 = \text{const.} , \tag{7.62}$$

the equilibrium condition (7.56), and the steady-state condition for the enzyme–substrate complex:

$$\frac{\mathrm{d}}{\mathrm{d}t}[\text{M}] = -(k'_- + k''_+)[\text{M}] + k'_+[\text{R}][\text{E}] = 0 . \tag{7.63}$$

From those three conditions, there result three relationships between the concentrations [E'], [E] and [M] and the concentrations [R] and [P]:

$$[\text{E}'] = \frac{[\text{E}]_0}{1 + (1 + x)y^m} , \qquad [\text{E}] = \frac{y^m [\text{E}]_0}{1 + (1 + x)y^m} ,$$

$$[\text{M}] = \frac{xy^m [\text{E}]_0}{1 + (1 + x)y^m} . \tag{7.64}$$

Above, we have introduced the two dimensionless variables

$$x \equiv \frac{[R]}{K_+} , \qquad y^m \equiv \frac{[P]^m}{K'''} . \tag{7.65}$$

The apparent dissociation constant for the reagent, viz.,

$$K_+ = \frac{k'_- + k''_+}{k'_+} , \tag{7.66}$$

has the dimensions of a molar concentration, whereas the dissociation constant for the activator K''' in (7.58) has the dimensions of a molar concentration raised to the power of m.

The concentrations $[R]$ and $[P]$ vary in time according to the system of differential kinetic equations:

$$\frac{d}{dt}[R] = v - k'_+[E][R] + k'_-[M] ,$$
$$\frac{d}{dt}[P] = k''_+[M] - k'''_+[E'][P]^m + k_-[E] - w[P] . \tag{7.67}$$

Taking into account the relationships (7.64) and the definitions (7.65), and then introducing a dimensionless time expressed in units of $(k''_+[E]_0/K_+)^{-1}$,

$$(k''_+[E]_0/K_+)\,t \to t , \tag{7.68}$$

the system (7.67) takes the form

$$\dot{x} = c - \frac{xy^m}{1 + (1 + x)y^m} ,$$
$$\dot{y} = \frac{axy^m}{1 + (1 + x)y^m} - by , \tag{7.69}$$

with three parameters a, b and c.

Unfortunately, a more detailed discussion of this system is rather complex so we simplify the model by assuming that $c \ll 1$, which allows us to neglect the denominators in (7.69). The resulting system, still nonlinear, reads:

$$\dot{x} = c - xy^m ,$$
$$\dot{y} = axy^m - by . \tag{7.70}$$

For $m = 2$, this has a stationary (steady-state) solution

$$x^{st} = \frac{b^2}{a^2 c} , \qquad y^{st} = \frac{ca}{b} . \tag{7.71}$$

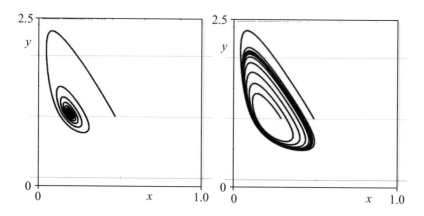

Fig. 7.13. Time course of the enzymatic reaction described by (7.70). The dimensionless variable x determines the reagent concentration, and the dimensionless variable y, the concentration of the product which is simultaneously the enzyme activator. In both diagrams, it is assumed that $a = 5$ and $b = 1$. In the *left-hand diagram* (oscillations decaying to a stable steady-state point), the value $c = 0.21$ is higher and in the *right-hand diagram* (oscillations tending to a stable limit cycle), the value $c = 0.19$ is lower than the critical value $c = 0.2$. Drawings were made using the program DiGraph written by Tomasz Jarus

Oscillations can appear when this solution becomes unstable. Linearizing the right-hand sides of the equations (7.70) in the vicinity of the point (7.71) by expansion in Taylor series, we obtain the instability condition

$$\frac{c^2 a^2}{b^3} = \frac{y^{\text{st}}}{a x^{\text{st}}} < 1 . \tag{7.72}$$

Figure 7.13 presents the numerical solutions to (7.70) for $a = 5$, $b = 1$ and two values of c, one above and one below the critical value $c = 0.2$.

A still simpler positive feedback mechanism is realized when the product of a reaction acts, not as catalyst activator, but as the catalyst itself (*autocatalysis*). This happens frequently in the chemistry of biological processes. In fact, the autocatalysis phenomenon is the basis for replication of genetic material. Many cyclic biological phenomena (so-called biological clocks) are related to processes of replication, or more generally, reproduction. As a second example, let us consider a simple model, but with a long history, associated with the names of Lotka and Volterra (Nicolis and Prigogine, 1977; Haken, 1990). The model consists of three irreversible reactions:

$$\text{A} + \text{X} \xrightarrow{k_1} 2\text{X} , \quad \text{X} + \text{Y} \xrightarrow{k_2} 2\text{Y} , \quad \text{Y} \xrightarrow{k_3} \text{products} . \tag{7.73}$$

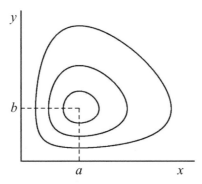

Fig. 7.14. The Lotka–Volterra system of equations (7.75) possesses (besides the limiting cases $x = 0$ and $y = 0$) only periodic trajectories that encircle the steady-state point with coordinates (b, a). The variable x determines the concentrations of molecules that autoreplicate using a material which occurs in unlimited amounts ('hares'), whereas the variable y determines the concentration of molecules that autoreplicate using a material consisting of molecules of the first kind ('lynxes'). Drawing made using the program Di-Graph written by Tomasz Jarus

The first reaction describes the autoreproduction of particles X from a material composed of particles A. The second reaction describes the autoreproduction of particles Y from a material composed of particles X. And the third reaction describes the decay of particles Y. The kinetic equations for molar concentrations of particles X and Y, respectively, are

$$\frac{\mathrm{d}}{\mathrm{d}t}[\mathrm{X}] = k_1[\mathrm{A}]x - k_2[\mathrm{X}][\mathrm{Y}] \ ,$$
$$\frac{\mathrm{d}}{\mathrm{d}t}[\mathrm{Y}] = k_2[\mathrm{X}][\mathrm{Y}] - k_3[\mathrm{Y}] \ . \tag{7.74}$$

Introducing the dimensionless concentrations $x \equiv [\mathrm{X}]\mathrm{M}^{-1}$ and $y \equiv [\mathrm{Y}]\mathrm{M}^{-1}$, where M is the concentration 1 mol/dm^3, and the unit of time $(k_2\mathrm{M})^{-1}$, we obtain the set of two equations

$$\dot{x} = ax - xy \ ,$$
$$\dot{y} = xy - by \ , \tag{7.75}$$

with two parameters a and b.

The set of equations (7.75) has a constant of motion

$$F(x, y) = x - b\ln x + y - a\ln y \ , \tag{7.76}$$

as is easily proved by differentiation:

$$\frac{\mathrm{d}}{\mathrm{d}t}F = \frac{\partial F}{\partial x}\dot{x} + \frac{\partial F}{\partial y}\dot{y} = 0 \ . \tag{7.77}$$

The function (7.76) is a simple solution to the equation

$$\left(1 - \frac{b}{x}\right)\mathrm{d}x = \left(\frac{a}{y} - 1\right)\mathrm{d}y \ , \tag{7.78}$$

obtained by dividing one equation of (7.75) by the other. The trajectories found numerically for the set (7.75), shown in Fig. 7.14, are indeed identical to the contour lines of the surface (7.76).

Originally, the Lotka–Volterra model was not formulated in terms of the kinetics of autocatalytic reactions, but in terms of the population dynamics of two competing animal species: predator and prey (see the caption to Fig. 7.14). In one of the relevant papers, the time course of a solution to the system (7.75) was compared with the number of lynx and hare skins being bought yearly in Canada in the period from 1845 to 1935. Since the time of that paper, the Lotka–Volterra model has been commonly termed the lynxes-and-hares model.

8 Biological Free Energy Transduction

8.1 Isothermal Machines

For many historical reasons, the word 'machine' has had several different meanings in most European languages. In our context, a machine is understood to be a physical system that enables two other systems to perform work on one another. Undoubtedly the oldest machine used by man is the lever (see Fig. 8.1a). If we place weights A_1 and A_2 at the ends, the work to be performed will involve transfer of gravitational energy from one weight to the other. Hence,

$$A_1 \Delta X_1 = -A_2 \Delta X_2 . \tag{8.1}$$

The changes in height ΔX_1 and ΔX_2 are inversely proportional to the corresponding weights. Their values can differ greatly, which attests to the usefulness of this simple machine. Equation (8.1) can be interpreted as a condition for the conservation of gravitational potential energy, namely

$$A_1 \Delta X_1 + A_2 \Delta X_2 = 0 . \tag{8.2}$$

With the help of the lever, work can be done only once. The state of the lever after performing work is different form the state before performing work. However, other machines can perform work cyclically. An example of such a machine acting on the same basis as the lever is the winch (Fig. 8.1b).

So far in our discussion the effects of friction (energy dissipation) have not been considered. Friction occurs at the fulcrum of the lever and on the axis of the winch. The possibility of an irreversible bending of the lever at the fulcrum corresponds to slippage at the axis of the winch. Friction is accounted for by replacing (8.2) by the inequality

$$A_1 \Delta X_1 + A_2 \Delta X_2 = D \geq 0 , \tag{8.3}$$

where D represents irreversibly lost gravitational energy. In the presence of friction, a cyclical machine can work in a stationary manner

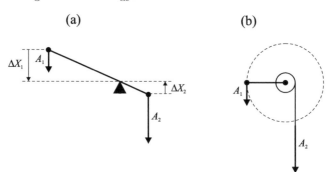

Fig. 8.1. Lever (**a**) and winch (**b**)

at a constant rate (see Fig. 8.2). The inequality (8.3) divided by a short interval of time Δt and after proceeding to the limit can then be rewritten in the form of the inequality [see (3.66)]

$$A_1\dot{X}_1 + A_2\dot{X}_2 = \Phi \geq 0 . \tag{8.4}$$

Work done per unit time is called *power*. The first term in (8.4) is the *input power* and the negative of the second term expresses the *output power*, while the dissipation function Φ represents the *dissipated power* of the machine. The ratio of the output power to the input power,

$$\eta = \frac{-A_2\dot{X}_2}{A_1\dot{X}_1} = 1 - \frac{\Phi}{A_1\dot{X}_1} , \tag{8.5}$$

is called the *efficiency* of the machine.

The winch and the car shown in Fig. 8.2 (regardless of the type of engine used) are two extreme examples of *mechanomechanical* machines since both variables X_i in (8.3) represent displacements, i.e., *mechanical* variables. Every machine that works cyclically and in which the output variable X_2 is a mechanical variable is called an *engine* or a *motor*. The variable X_1 does not have to be mechanical. In addition to the *mechanical* engines discussed here, we also distinguish *electrical* engines for which the variable X_1 represents electrical charge. Its time derivative $\dot{X}_1$ is an electrical current and the force A_1 conjugate to it is a voltage. Furthermore, we often deal with *thermal* or *heat* engines, in which X_1 corresponds to entropy and its time derivative $\dot{X}_1$, when multiplied by the conjugate force A_1 denoting a temperature difference, expresses heat flux intensity. Finally, still in the realm of speculation, is an efficient *chemical* engine that would directly convert chemical energy into mechanical work.

(a)   (b)   (c)

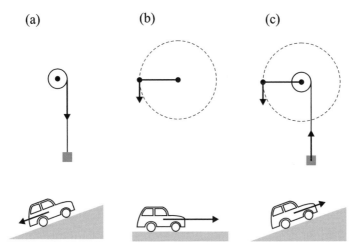

Fig. 8.2. Simple winch and a more complicated machine, viz., a car. (**a**) A weight attached to a freely unwinding rope attains a constant velocity in its downward motion when the force of friction proportional to velocity balances the force due to gravity. A car left unbraked on an incline will sooner or later attain a constant velocity in its backward rolling motion. In both cases the total gravitational energy is being dissipated when this happens. (**b**) Rotational motion of the crank can achieve a stationary velocity with complete dissipation of the energy put into it. This corresponds to the motion of a car at a constant velocity on a horizontal plane. (**c**) Coupling of the rotation of the crank and the tumbling-wheel can lead to upward motion of the weight at a constant velocity and only in this case is work performed against gravity. Analogously, a car engine performs work when it drives at a constant velocity up the hill

Note that, in a car engine, the chemical energy of gasoline is first converted via its internal combustion into a heat flux and then into the motion of pistons in the cylinders. This is in turn converted via axles into the rotational energy of the wheels. In the general case, both variables X_1 and X_2 can have an arbitrary character. For example, an alternator is a *mechanoelectrical* machine while a battery is a *chemoelectrical* machine.

The general theory of machine action is a subfield of thermodynamics (Kondepudi and Prigogine, 1998). The origin of thermodynamics can be traced back to the early 19th century when physical foundations of the action of heat engines were searched for. The French engineer Carnot was credited at the time with the development of a theoretical model of a heat engine that works under the condition of an alternating contact with two thermostats kept at different temperatures. The assumption was that the changes were sufficiently slow to allow the state of complete thermodynamic equilibrium to be maintained and to ignore energy dissipation. In fact, 19th century thermodynamics was first and foremost equilibrium thermodynamics.

However, the Carnot engine is a bad model for most modern machines that work under *isothermal* conditions of constant temperature and *far from thermodynamic equilibrium*. The inclusion of dissipation in such cases is very important. Furthermore, the molecular biological machines that will be discussed in this chapter, such as chemomechanical motors, chemoosmotic pumps and chemochemical receptors work under isothermal conditions far from thermodynamic equilibrium.

Under isothermal conditions, $T = $ const., the total energy of an arbitrary thermodynamic system, regardless of whether the thermodynamic variables X_i do or do not reach their total equilibrium values, can be unambiguously divided into free energy G and bound energy TS (see Sect. 3.5). The former changes as a result of work being done either by the system or on the system. The latter changes due to a transfer of heat. According to the second law of thermodynamics, free energy can be transformed into bound energy in a dissipative process, but not the other way around. In other words, bound energy can never be used under isothermal conditions for the purpose of performing work. In our context, gravitational, chemical and electrical energies are special cases of free energy. In the most interesting case for us, where chemical reactions take place under both isothermal and isobaric ($P = $ const.) conditions, the role of free energy is played by free enthalpy (also called the Gibbs potential, see Sect. 6.1 and Appendix A.2). Hence G is used throughout the present chapter to denote free energy.

Machines that operate under the condition $T = $ const. are *free energy transducers* (Hill, 1989). This means that work done by the environment on one of the machine's subsystems is transformed into its free energy and this, in turn, is passed on to the second subsystem which subsequently uses it to perform work on the environment (Fig. 8.3). The free energy transfer between the two subsystems is reduced by dissipation, which increases the bound energy of the two subsystems. The heat transferred to the thermostat in the course of the process can have an arbitrary sign, as can the entropy transfer between the two subsystems. At a steady state, the values of the bound free energy (i.e., entropy) of the two subsystems remains constant, $G_i = $ const. and $S_i = $ const.. Therefore, balancing the transformations shown in Fig. 8.3, the total work performed on and by the system per unit time must be equal to the rate of dissipation, and this cannot be

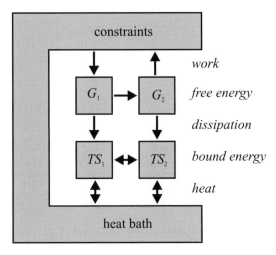

Fig. 8.3. Pathways of energy transformations in a machine working under isothermal conditions

negative according to the second law of thermodynamics. The latter is equal to the heat flux transferred to the environment. This is precisely the meaning of (8.4) which, originally derived for a winch, is generally correct for an arbitrary isothermal machine. A transformation of free energy takes place when, despite a positive value of the entire sum in (8.4), one term has a negative sign.

8.2 Chemochemical Machines.
The Necessity of Enzyme Intermediacy

Consider two chemical reactions: ATP hydrolysis,

$$\mathrm{ATP} \underset{\longleftarrow}{\overset{K_1}{\longrightarrow}} \mathrm{ADP} + \mathrm{P_i} \, ,$$

and phosphorylation of a certain substrate Sub,

$$\mathrm{Sub} + \mathrm{P_i} \underset{\longleftarrow}{\overset{K_2}{\longrightarrow}} \mathrm{SubP} \, .$$

Both reactions ignore the participation of a water molecule which is in excess. The equilibrium constants K_1 and K_2 are determined by the quotients

$$K_1 = \frac{[\mathrm{ADP}]^{\mathrm{eq}} \, [\mathrm{P_i}]^{\mathrm{eq}}}{[\mathrm{ATP}]^{\mathrm{eq}}} \, , \qquad K_2 = \frac{[\mathrm{SubP}]^{\mathrm{eq}}}{[\mathrm{Sub}]^{\mathrm{eq}} \, [\mathrm{P_i}]^{\mathrm{eq}}} \, . \tag{8.6}$$

(a)

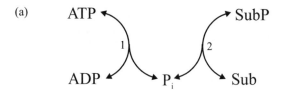

(b)

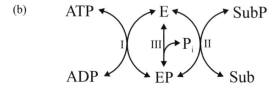

Fig. 8.4. Coupling of two phosphorylation reactions through a common reagent P_i (**a**) and through a kinase enzyme E (**b**)

The first reaction is exoergic from left to right ($K_1 M^{-1} > 1$, $\Delta G^\circ < 0$). The second is assumed to be endoergic in this direction ($K_2 M < 1$, $\Delta G^\circ > 0$). However, the second reaction can take place from left to right if both reactions take place in the same reactor and if the first reaction ensures a sufficiently high concentration of the common reagent P_i (Fig. 8.4a). It appears at first sight that this system acts as a chemochemical machine, where the first reaction transfers free energy to the second. But is this really so?

The thermodynamic forces acting in the two reactions are, respectively

$$A_1 = k_B T \ln K_1 \frac{[\text{ATP}]}{[\text{ADP}]\,[\text{P}_i]}\,,$$
$$A_2 = k_B T \ln K_2 \frac{[\text{Sub}]\,[\text{P}_i]}{[\text{SubP}]}\,.$$

(8.7)

Under steady-state conditions, the corresponding fluxes are

$$J_1 = \dot{X}_1\,, \qquad J_2 = \dot{X}_2\,,$$

(8.8)

where

$$X_1 \equiv [\text{ADP}]\,, \qquad X_2 \equiv [\text{SubP}]\,,$$

(8.9)

and they are equal. The forces (8.7) are not, however, independent of these fluxes. In Sect. 6.1 we demonstrated that the flux and force for each reaction taken separately are always of the same sign and hence

$$A_1 J_1 \geq 0\,, \qquad A_2 J_2 \geq 0\,,$$

(8.10)

in the sum (8.4). Therefore, no free energy transduction can occur. At most, an entropy transfer is possible (Blumenfeld, 1974; see Fig. 8.3) as a result of a simultaneous influence of both reactions on the probability of finding the molecule P_i in the reactor.

Many *in vivo* biochemical reactions take place simultaneously in the same volume of a part or the whole of a cell. However, under physiological conditions, the coupling of two reactions via a common reagent is not easily implemented. As an example, consider the first reaction in the glycolysis chain, i.e., glucose phosphorylation (Sub = Glu) into Glu6P, which is coupled to the ATP hydrolysis reaction. For this reaction,

$$K_2 = \frac{[\text{Glu6P}]^{\text{eq}}}{[\text{Glu}]^{\text{eq}} [P_i]^{\text{eq}}} = 6.7 \times 10^{-3} \text{M}^{-1} \, .$$

Under steady-state (nonequilibrium) physiological conditions $[P_i] = 10^{-2}$ M and $[\text{Glu6P}] = 10^{-4}$ M. As a consequence, for the reaction to proceed in the forward direction, i.e., for the force A_2 to be positive, the concentration of glucose [Glu] must exceed 1.6 M = 300 g/dm^3, which is an unrealistic value.

Nature has a different way to achieve the coupling of ATP hydrolysis and phosphorylation of the substrate, namely through an enzyme that catalyzes both reactions simultaneously (Fig. 8.4b). According to the terminology presented in Sect. 4.6, enzymes that catalyze the transfer of a phosphate group are called kinases. Figure 8.4b shows three reactions labeled I, II and III. In addition to the transfer of the phosphate group P_i from ATP to Sub, it is also possible to detach this group from the enzyme unproductively. This is characterized by the following equilibrium constants:

$$K_{\text{I}} = \frac{[\text{ADP}]^{\text{eq}} [\text{EP}]^{\text{eq}}}{[\text{ATP}]^{\text{eq}} [\text{E}]^{\text{eq}}} \, , \quad K_{\text{II}} = \frac{[\text{E}]^{\text{eq}} [\text{SubP}]^{\text{eq}}}{[\text{EP}]^{\text{eq}} [\text{Sub}]^{\text{eq}}} \, , \quad K_{\text{III}} = \frac{[\text{E}]^{\text{eq}} [P_i]^{\text{eq}}}{[\text{EP}]^{\text{eq}}} \, .$$
$$(8.11)$$

Since an enzyme cannot affect chemical equilibrium conditions, the constants (8.11) must depend on the constants (8.6), and the following relations are satisfied:

$$K_{\text{I}} K_{\text{III}} = K_1 \, , \quad K_{\text{II}}/K_{\text{III}} = K_2 \, . \quad (8.12)$$

As with relationships between equilibrium constants, we find corresponding relationships between thermodynamic forces:

$$A_1 = A_{\text{I}} + A_{\text{III}} \, , \quad A_2 = A_{\text{II}} - A_{\text{III}} \, , \quad (8.13)$$

where the forces acting on reactions I, II, and III are defined analogously to those in (8.7). Using the definitions (8.8) and (8.9), one can identify the reaction fluxes shown in Fig. 8.4:

$$J_1 = J_I , \qquad J_2 = J_{II} , \tag{8.14}$$

and from the steady-state conditions [E] = const. and [EP] = const.,

$$J_{III} = J_I - J_{II} . \tag{8.15}$$

It follows from the relationships (8.15) and (8.13) that the dissipation function for the system of the three reactions illustrated in Fig. 8.4b can be rewritten in two forms:

$$\begin{aligned} \Phi &= A_I J_I + A_{II} J_{II} + A_{III} J_{III} \\ &= (A_I + A_{III}) J_I + (A_{II} - A_{III}) J_{II} \\ &= A_1 J_1 + A_2 J_2 . \end{aligned} \tag{8.16}$$

The three terms in the first equation are nonnegative, but this does not mean that the two terms in the third equation must also be non-negative. It is sufficient that, for $J_1, J_2 > 0$ when $A_I, A_{II} > 0$, we have $A_{III} > A_{II}$. The other condition $A_{III} < -A_I$ is impossible to satisfy for P_i concentrations under physiological conditions. However, for $J_1, J_2 < 0$ when $A_I, A_{II} < 0$, the inequality $A_{III} > -A_I$ can hold. In both cases we are dealing with real transduction of the free energy. The first case, namely the transfer of free energy from subsystem 1 to subsystem 2, takes place in the already discussed process of glucose phosphorylation at the expense of ATP hydrolysis. The second case, namely the transfer of free energy from subsystem 2 to subsystem 1 occurs in the next two stages of the glycolysis chain, in which ADP phosphorylation to ATP proceeds at the expense of even higher energy substrates.

8.3 Universality of the Enzymatic Mechanism of Free Energy Transduction

The kinetic scheme shown in Fig. 8.4b can be generalized to the case of two arbitrary coupled chemical reactions:

$$R_1 \longleftrightarrow P_1 \quad \text{and} \quad R_2 \longleftrightarrow P_2 .$$

We assume for simplicity that both reactions are unimolecular. One reaction is a *donor* of free energy and the other is a free energy *acceptor*.

Each biochemical reaction must be catalyzed by a protein enzyme. *Separately*, each reaction takes place in the direction determined by the second law of thermodynamics, i.e., the condition that the amount of chemical energy dissipated is positive (Figs. 8.5a and b). Only when both reactions occur *simultaneously* using the same enzyme, with one reaction taking place according to the second law of thermodynamics, can the second reaction be forced to take place against the second law. In this case, the first reaction transfers a part of its free energy recovered from dissipation performing work on the second reaction (Fig. 8.5c). The mechanism of energy transfer is very simple: if both reactions occur in a common cycle, they must proceed in the same direction.

Figure 8.5 is intended to resemble Fig. 8.2 in order to emphasize the similarity between a chemochemical machine and a winch device. In the same way, cranking a winch requires the input of work acting against the force of gravity exerted by the weight attached through a rope to the spool: the first reaction performs work against the chemical force acting on the second reaction, forcing it to proceed in the

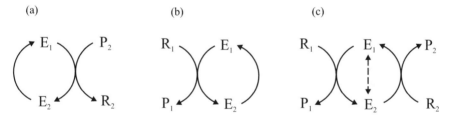

(a)   (b)   (c)

Fig. 8.5. (**a**) and (**b**): Two different chemical reactions proceeding independently of each other. We assume that they are catalyzed by the same enzyme, but the reagent R_1 for the first reaction binds to the state E_1 of the enzyme, and the reagent R_2 for the second reaction binds to the state E_2. Reactions take place under stationary conditions as a result of keeping the reagent and product concentrations fixed but different from the equilibrium values. These concentrations are selected so that the first reaction proceeds from R_1 to P_1 and the second from P_2 to R_2. (**c**) If both reactions take place simultaneously using the same enzyme, the direction of the first reaction can force a change of direction of the second. The new direction would be opposite to the one dictated by the stationary values of the respective concentrations. The *broken line* denotes a possible unproductive transition between states E_1 and E_2 of the enzyme. It is instructive to compare this diagram with Fig. 8.2

opposite direction. Friction associated with the motion of the winch corresponds to energy dissipation in the common reaction cycle. Slippage of the crank with respect to the spool axis is mirrored by the possible direct reaction between the states E_1 and E_2 of the enzyme. The only difference between the winch and a chemochemical machine is in their organization. The winch is a device characterized by macroscopic spatial organization, while the enzymes enabling the operation of a chemochemical machine are microscopic, or at worst mesoscopic entities. The chemochemical machine is a more or less spatially homogeneous solution of enzymes with a typical concentration of 10^{-6} M, i.e., close to 10^{15} molecules per cubic centimeter, or 10^3 molecules per cubic micrometer (the typical size of a bacterial cell or an organelle of a eukaryotic cell).

Historically, the scheme in Fig. 8.5c is identical to a general scheme of *chemical induction* proposed by the Russian chemist Shilov in 1905 (Blumenfeld, 1974). He called the substrate of the first reaction the *inducer*, the substrate of the second reaction the *acceptor*, and the counterpart of the enzyme, not consumed during the reaction, the *actor* (Fig. 8.6a). If the conditions of free energy transduction are not satisfied, the inducer and the acceptor are not uniquely defined. They need each other and the process is referred to as *mutual chemical induction*.

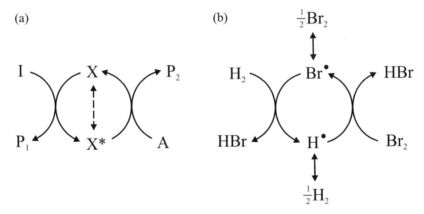

Fig. 8.6. (a) Shilov's scheme of chemical induction. X denotes the actor molecule, a reagent common to the two coupled reactions, I is the inducer molecule which reacts spontaneously with the actor X, and A is the acceptor molecule which reacts with X only in the presence of the inducer. (b) Example of a chain reaction

Particular examples of mutual chemical induction are *chain reactions* (Pauling and Pauling, 1975, Chap. 10), for instance, an association of H_2 and Br_2 into hydrobromide HBr (Fig. 8.6b). The reaction consists of an infinite number of steps in which intermediates, the free radicals $H^\bullet$ and $Br^\bullet$, are alternately created and consumed. Note that the atomic hydrogen and bromide are highly unstable and recombine directly into molecular hydrogen and bromide (Fig. 8.6b). After an accidental initiation of the reaction, a steady-state concentration of the free radicals is very rapidly established and this concentration determines the effective reaction rate. The mechanism of some chain reactions makes the concentration of the intermediates increase exponentially without reaching a saturation level. The chain reaction then becomes an *avalanche reaction*. Examples are the explosive reaction of H_2O synthesis in the gaseous mixture of two parts of hydrogen H_2 and one of oxygen O_2, or the reaction of uranium ^{235}U fission under conditions of an exponentially increasing number of neutrons.

Assuming as in Sect. 7.2 that each component reaction proceeds through one substrate–enzyme intermediate, we replace the three schemes in Fig. 8.5 by three more complex schemes shown in Fig. 8.7. In such a case, however, besides the coupling of the two reactions through a free enzyme E (Fig. 8.7c), three other possible schemes of coupling can be devised (Fig. 8.8). Thus, both reactions are also to be coupled through an intermediate complex M (see Fig. 8.8b), the intermediate complex for one reaction appears to be the free enzyme for the second reaction (see Fig. 8.8c), and both reactions proceed as alternating half-reactions (see Fig. 8.8d). Figure 8.8a repeats the scheme of coupling the reactions through a free enzyme E.

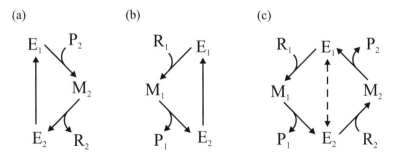

(a)   (b)   (c)

Fig. 8.7. Counterparts of three schemes from Fig. 8.5, assuming that each component reaction proceeds through one intermediate enzyme–substrate complex

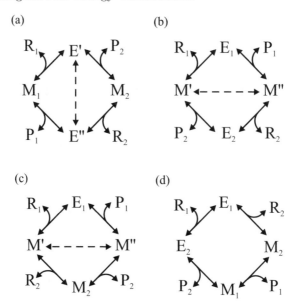

Fig. 8.8. Four possible kinetic schemes for the system of two coupled reactions, assuming that each proceeds through one intermediate state of the enzyme. The *broken lines* represent unproductive (without binding to or unbinding from substrates) direct transitions between different states of the enzyme. They cause a possible mutual slippage of the two component cycles

In general, the chemochemical machine can be viewed as a 'black box', entered and exited by molecules that take part in both coupled chemical reactions (Fig. 8.9). The kinetic scheme of the reactions can be arbitrary, as long as it involves at least one cycle so that the enzyme cannot be exhausted in the course of the reaction.

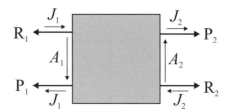

Fig. 8.9. General scheme of a chemochemical machine coupling two reactions: $R_1 \leftrightarrow P_1$ which produces free energy, and $R_2 \leftrightarrow P_2$ which consumes free energy. Both reactions can take place in either direction, as determined by the sign of the flux J_i. The forces A_i are determined by the values of the reagent and product concentrations which are kept stationary

8.4 Molecular Pumps and Motors

Besides simple phosphorylation reactions, the reaction of ATP hydrolysis is coupled to many other biological processes. It can force the transport of ions across membranes in the direction of increasing ion concentration, not directly allowed by the second law of thermodynamics. The chemoosmotic or chemoelectric molecular machines performing such a function are called *molecular pumps*. ATP hydrolysis can also result in a mechanical motion along molecular tracks (microfilaments, microtubules or nucleic acid chains). The corresponding chemomechanical machines are called *molecular motors*. Transport across membranes can also be coupled to a rotational mechanical motion. We then speak about *molecular turbines*.

From a theoretical point of view, it would be convenient to treat all molecular biological machines as chemochemical machines. In fact, the scheme in Fig. 8.8b applies to molecular motors, while those in Figs. 8.8c and d apply to molecular pumps. Treating molecular pumps as chemochemical machines poses no great problem. The molecules present on either side of a biological membrane can be considered to occupy different chemical states, while the transport process across the membrane can be regarded as an ordinary chemical reaction (see Sect. 6.2). Figures 8.10a and b present simplified kinetic cycles of the calcium and the sodium–potassium pumps, respectively (see Sect. 5.2 and Stryer et. al., 2002, Chap. 13). Assuming the concentration of ATP to be much higher than the equilibrium concentration determined by the actual concentration of ADP, we can consider the ATP hydrolysis reaction as an irreversible unimolecular reaction ATP $\rightarrow$ P_i. Hence, both schemes in Fig. 8.10 are identical to the one presented in Fig. 8.8d. A possible slippage is realized, not at the pump itself, but due to the *leakage* of Ca^{2+}, Na^+ or K^+ ions through the corresponding channels (see Sect. 5.2).

Proton pumps taking part in membrane phosphorylation (Sect. 5.3) and representing molecular chemoelectrical machines [fuel cells, see Sect. 6.5)], in particular the quinol : cytochrome c-oxidoreductase (see Fig. 5.14), act according to the scheme presented in Fig. 8.8c. This is clear from Fig. 8.11a, which shows a simplification of the scheme given in Fig. 5.15. From the kinetic point of view, the action of molecular chemoelectrical machines differs only slightly from the action of macroscopic chemoelectrical machines. Figure 8.11b shows the working cycle of the macroscopic Daniell electrochemical cell (see Sect. 6.5).

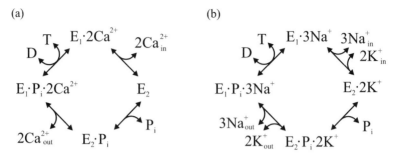

Fig. 8.10. Simplified kinetic cycles of the calcium pump, Ca^{2+}-ATPase (**a**) and the sodium–potassium pump, Na^{+},K^{+}-ATPase (**b**). E_1 and E_2 denote two states of an enzyme that represents a pump with a reaction center oriented inside and outside a compartment, respectively. T, D, and P_i stand for ATP, ADP and inorganic phosphate, respectively

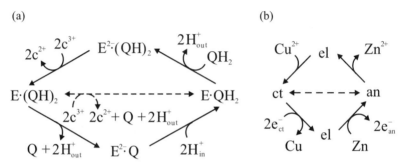

Fig. 8.11. Simplified kinetic cycles of two chemoelectrical machines, one molecular and one macroscopic. Scheme (**a**) represents the action of the quinol:cytochrome c-oxidoreductase (see Fig. 5.15). Scheme (**b**) represents the action of the Daniell electrochemical cell [see Fig. 6.6 and (6.90) or (6.93)]. The main difference is that the vertices of scheme (**a**) correspond to various charge states of a single macromolecular enzymatic complex, whereas those of scheme (**b**) correspond to various parts of a macroscopic device: cathode (ct), anode (an) and electrolyte (el)

Molecular motors represent a more complicated case. Figure 8.12 depicts a simplified version of the Lymn–Taylor–Eisenberg kinetic scheme of the chemomechanical cycle of the actomyosin motor (Howard, 2001; Kurzyński and Chełminiak, 2004; see also Sects. 5.4 and 9.4). The scheme indicates how the ATPase cycle of the myosin head is related to states that are detached, weakly attached and strongly attached to the actin filament. Both the substrate and products of the catalyzed reaction bind to and rebind from the myosin in its strongly attached state, whereas the reaction takes place either in

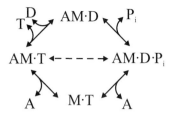

Fig. 8.12. Simplified Lymn–Taylor–Eisenberg kinetic scheme of the chemo-mechanical cycle of the actomyosin motor. M denotes the myosin head and A the actin filament, while T, D and P_i stand for ATP, ADP and inorganic phosphate, respectively

the weakly attached or detached state. Only completion of the whole cycle with ATP hydrolysis achieved in the detached state results in the directed motion of the myosin head along the actin track. ATP hydrolysis in the weakly bound state alone is ineffective and corresponds to slippage.

Assuming a low fixed value of the ADP concentration, the scheme shown in Fig. 8.12 is identical to the one presented in Fig. 8.8b. The question remains as to how to represent a load acting on a motor in terms of concentrations. In Fig. 8.12, A denotes the actin filament before or after translation of the myosin head by a unit step. Experimental evidence and a more deeper reasoning (see Sect. 9.4) indicate that an external load attached to the statistical ensemble of myosin heads (organized, in the case of myofibrils, into a system of thick filaments) influences the free energy involved in binding the myosin heads to thin actin filaments. The associated changes in the binding free energy can be expressed as changes of effective rather than actual concentrations of the actin filament [A] before and after translation. As a consequence, the actomyosin motor can indeed be effectively treated as a typical chemochemical machine. The output flux J_2 is related to the mean velocity of the myosin head along the actin filament and the force A_2 is proportional to the load (Kurzyński and Chełminiak, 2004).

At a macroscopic level, the action of molecular pumps and motors is manifested by *directed transport* of a substance. The possible functioning of mesoscopic machines on a macroscopic scale is due to appropriate organization of the statistical ensemble. Molecular pumps are embedded in the two-dimensional structure of the membrane (Fig. 8.13a), while molecular motors move along a structurally organized system of tracks: microfilaments or microtubules (Fig. 8.13b). However, not all biological molecular machines perform work on a macroscopic scale.

(a)   (b)

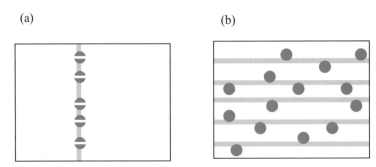

Fig. 8.13. Functioning of biological molecular machines on a thermodynamic (macroscopic) scale is only possible due to appropriate organization of the system. Molecular pumps are embedded into the two-dimensional structure of the membrane (**a**), while molecular motors move along an organized system of tracks, microfilaments or microtubules (**b**)

Examples of such behavior are molecular turbines, e.g., the F_o portion of ATP synthase (Sect. 5.3). Since no mechanism coordinates the rotational motion of individual turbines, no macroscopic thermodynamic variable characterizes this motion. The two portions F_o and F_1 must therefore be treated jointly from the macroscopic point of view, together giving rise to a reversible molecular pump, the H^+ ATPase.

The organization of a molecular pump system in the membrane does not require the individual pumps to be single macromolecules. Thus, the leakage for the calcium or the sodium–potassium pumps is realized through separate Ca^{2+} or Na^+ and K^+ ionic channels. Transport of ions through channels in the direction of increasing concentration is forced by the electric membrane potential created by transport of other ions through other channels (Sect. 6.2). Similarly, osmosis in the direction of increasing solute concentration can be forced by an appropriate hydrostatic pressure.

8.5 Flux–Force Dependence

To provide a better analysis of the character of flux–force relations in biological processes of free energy transduction, we now return to the simplest kinetic scheme presented in Fig. 8.5c for the enzymatic coupling of a reaction supplying free energy with one that absorbs it. Figure 8.14 shows this scheme in more detail and defines reaction rate constants for the individual reactions involved. The thermodynamic forces for the three component reactions are given by the equations

$$A_\mathrm{I} = k_\mathrm{B} T \ln \frac{[E_2]^\mathrm{eq}[P_1]^\mathrm{eq}}{[E_1]^\mathrm{eq}[R_1]^\mathrm{eq}} \frac{[E_1]\,[R_1]}{[E_2]\,[P_1]} \; ,$$

$$A_\mathrm{II} = k_\mathrm{B} T \ln \frac{[E_1]^\mathrm{eq}[P_2]^\mathrm{eq}}{[E_2]^\mathrm{eq}[R_2]^\mathrm{eq}} \frac{[E_2]\,[R_2]}{[E_1]\,[P_2]} \; , \qquad (8.17)$$

$$A_\mathrm{III} \equiv k_\mathrm{B} T \ln \frac{[E_1]^\mathrm{eq}}{[E_2]^\mathrm{eq}} \frac{[E_2]}{[E_1]} \; ,$$

and the corresponding reaction fluxes [see (6.34)] are

$$J_\mathrm{I} = \frac{1 - \mathrm{e}^{-\beta A_\mathrm{I}}}{(k_{+1}[R_1])^{-1} + (k_{-1}[P_1])^{-1}\,\mathrm{e}^{-\beta A_\mathrm{I}}} [E]_0 \; ,$$

$$J_\mathrm{II} = \frac{1 - \mathrm{e}^{-\beta A_\mathrm{II}}}{(k_{+2}[R_2])^{-1} + (k_{-2}[P_2])^{-1}\,\mathrm{e}^{-\beta A_\mathrm{II}}} [E]_0 \; , \qquad (8.18)$$

$$J_\mathrm{III} = \frac{1 - \mathrm{e}^{-\beta A_\mathrm{III}}}{k_{02}^{-1} + k_{01}^{-1}\,\mathrm{e}^{-\beta A_\mathrm{III}}} [E]_0 \; ,$$

where $[E]_0$ is the total enzyme concentration:

$$[E]_0 = [E_1] + [E_2] = [E_1]^\mathrm{eq} + [E_2]^\mathrm{eq} \; . \qquad (8.19)$$

Operational forces and fluxes, respectively (Hill, 1989), are given for noncatalyzed reactions as

$$A_1 \equiv k_\mathrm{B} T \ln \frac{[P_1]^\mathrm{eq}}{[R_1]^\mathrm{eq}} \frac{[R_1]}{[P_1]} \; , \qquad J_1 = \frac{\mathrm{d}}{\mathrm{d}t}[P_1] = -\frac{\mathrm{d}}{\mathrm{d}t}[R_1] \; ,$$

$$\qquad\qquad\qquad\qquad\qquad\qquad\qquad\qquad\qquad\qquad (8.20)$$

$$A_2 \equiv k_\mathrm{B} T \ln \frac{[P_2]^\mathrm{eq}}{[R_2]^\mathrm{eq}} \frac{[R_2]}{[P_2]} \; , \qquad J_1 = \frac{\mathrm{d}}{\mathrm{d}t}[P_2] = -\frac{\mathrm{d}}{\mathrm{d}t}[R_2] \; .$$

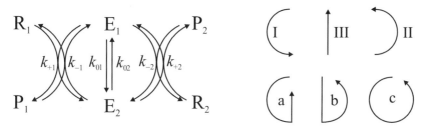

Fig. 8.14. Enzymatic coupling of a reaction providing free energy to one absorbing free energy. The notations for the reaction rate constants, conventional reaction flux directions and cyclical fluxes are shown

Reaction fluxes J_I and J_II directly determine operational fluxes J_1 and J_2:

$$J_1 = J_\mathrm{I} , \qquad J_2 = J_\mathrm{II} . \tag{8.21}$$

Comparing the forces (8.17) and (8.20), we obtain

$$A_1 = A_\mathrm{I} + A_\mathrm{III} , \qquad A_2 = A_\mathrm{II} - A_\mathrm{III} , \tag{8.22}$$

and from the steady-state conditions

$$\frac{\mathrm{d}}{\mathrm{d}t}[\mathrm{E}_1] = \frac{\mathrm{d}}{\mathrm{d}t}[\mathrm{E}_2] = 0 , \tag{8.23}$$

we find the relationship

$$J_1 - J_2 = J_\mathrm{III} . \tag{8.24}$$

The dissipation function (8.4) (the total dissipated power) can be written in three different ways:

$$
\begin{aligned}
\Phi &= A_\mathrm{I} J_\mathrm{I} + A_\mathrm{II} J_\mathrm{II} + A_\mathrm{III} J_\mathrm{III} &&\text{(transition fluxes)} \\
&= A_\mathrm{a} J_\mathrm{a} + A_\mathrm{b} J_\mathrm{b} + A_\mathrm{c} J_\mathrm{c} &&\text{(cycle fluxes)} \\
&= A_1 J_1 + A_2 J_2 &&\text{(operational fluxes) ,}
\end{aligned}
\tag{8.25}
$$

where (see Fig. 8.14)

$$J_\mathrm{a} + J_\mathrm{c} = J_\mathrm{I} , \quad J_\mathrm{b} + J_\mathrm{c} = J_\mathrm{II} , \quad J_\mathrm{c} - J_\mathrm{b} = J_\mathrm{III} , \tag{8.26}$$

and hence,

$$A_\mathrm{a} = A_\mathrm{I} + A_\mathrm{III} , \quad A_\mathrm{b} = A_\mathrm{II} - A_\mathrm{III} , \quad A_\mathrm{c} = A_\mathrm{I} + A_\mathrm{II} . \tag{8.27}$$

All terms in the first and second sum are non-negative (Hill, 1989). Only the two terms in the third sum can have different signs, and when this happens we are dealing with the process of free energy transduction.

As mentioned earlier, $J_1 A_1$ is called the *input power*, whereas $-J_2 A_2$, is the *output power*. The *efficiency* of the process is defined as the ratio of output power to input power:

$$\eta = \frac{-J_2 A_2}{J_1 A_1} = \frac{J_1 A_1 - \Phi}{J_1 A_1} = 1 - \frac{\Phi}{J_1 A_1} . \tag{8.28}$$

This is often written as the product

$$\eta = \epsilon \frac{-A_2}{A_1} , \tag{8.29}$$

in which the *degree of coupling* ϵ of the two subsystems determines the ratio of the operational fluxes:

$$\epsilon \equiv \frac{J_2}{J_1} . \tag{8.30}$$

Using the relations (8.21), (8.22) and (8.24), we can eliminate the force A_{III} from (8.18). Introducing the equilibrium constants

$$K \equiv \frac{[E_1]^{eq}}{[E_2]^{eq}} , \quad K_1 \equiv \frac{[P_1]^{eq}}{[R_1]^{eq}} , \quad K_2 \equiv \frac{[P_2]^{eq}}{[R_2]^{eq}} , \tag{8.31}$$

and taking the fixed values of the total substrate concentrations

$$[R_1] + [P_1] = [R_1]_0 , \qquad [R_2] + [P_2] = [R_2]_0 , \tag{8.32}$$

we can obtain, after a rather tedious calculation making use of the detailed balance conditions for reactions I through III, the equations that link operational fluxes with forces:

$$J_1 = \frac{\left[1 - e^{-\beta(A_1+A_2)} + \left(1 - e^{-\beta A_1}\right) k_{02}\, \tau_2\right] [E]_0}{(1 + K^{-1} e^{-\beta A_2})\, \tau_1 + (1 + K e^{-\beta A_1})\, \tau_2 + (k_{01} + k_{02})\tau_1\tau_2} ,$$

$$J_2 = \frac{\left[1 - e^{-\beta(A_1+A_2)} + \left(1 - e^{-\beta A_2}\right) k_{01}\, \tau_1\right] [E]_0}{(1 + K^{-1} e^{-\beta A_2})\, \tau_1 + (1 + K e^{-\beta A_1})\, \tau_2 + (k_{01} + k_{02})\tau_1\tau_2} , \tag{8.33}$$

where the quantities

$$\tau_i \equiv (k_{+i}[R_i])^{-1} = (k_{+i}[R_i]_0)^{-1} \left(1 + K_i e^{-\beta A_i}\right) , \tag{8.34}$$

for $i = 1, 2$, have the dimensions of time. In (8.34), we used the equation of state (6.18). In the case of no slippage, $k_{01}, k_{02} \to 0$, we are dealing with perfect coupling ($\epsilon = 1$) and the two fluxes are identical:

$$J_1 = J_2 = \frac{\left[1 - e^{-\beta(A_1+A_2)}\right] [E]_0}{(1 + K^{-1} e^{-\beta A_2})\, \tau_1 + (1 + K e^{-\beta A_1})\, \tau_2} . \tag{8.35}$$

The flux–force relations (8.33) for two coupled reactions are of the same functional form as the flux–force relation for the separate reaction (6.18):

$$J_i = \frac{1 - e^{-\beta(A_i - A_i^{st})}}{J_{+i}^{-1} + J_{-i}^{-1} e^{-\beta(A_i - A_i^{st})}} [E]_0 , \tag{8.36}$$

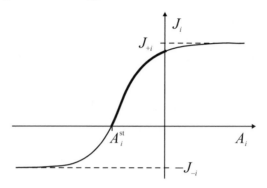

Fig. 8.15. Character of the functional dependence of the output flux J_i versus force A_i. Only when the stalling force A_i^{st} is negative does free energy transduction take place. The $J_i(A_i)$ dependence in this range is marked with a *bold line*

for $i = 1, 2$. However, the parameters J_{+i} and J_{-i} now depend on another force, and their dependence on the reaction rate constants and substrate concentrations is much more complex, so we shall not present them here. Moreover, there are additional parameters A_i^{st}, determining the non-zero values of the *stalling forces*, for which only the fluxes J_i vanish.

The dependence $J_i(A_i)$ given in (8.36) is strictly increasing with a point of inflection and two asymptotes (Fig. 8.15). As noted earlier, free energy transduction takes place if one of the fluxes is of the opposite sign to its conjugate force. From (8.36), it follows that this condition holds when the corresponding stalling force A_i^s is negative (Fig. 8.15). The dependence $J_i(A_i)$ in the range $A_i^{st} \leq A_i \leq 0$ can be convex or concave, or it can involve an inflection point as well.

We do not discuss here the conditions for maximum efficiency of free energy transduction since, even in the linear approximation of the flux–force relations (which is usually a poor approximation as can be seen from Fig. 8.15), the formulas for the values of forces maximizing the efficiency are very complex (Westerhoff and van Dam, 1987). Anyhow, the conditions for maximum efficiency and maximum power output contradict each other. A machine is the more efficient the lower its free energy dissipation, i.e., the more slowly it works. But the more slowly it works, the lower its output power.

However, maximum efficiency and maximum output power are not always at their optimum values from the point of view of living organisms. Very often the output power of biological machines is simply equal to zero, i.e., the output forces stall the machines. This can be the case with molecular motors and molecular pumps as well. The muscles of a man sustaining a big load do not perform any work but, of course, ATP is consumed. The intracellular concentration of Ca^{2+} is kept at a very low level of 100 nM to avoid association with phosphate ions P_i present in the cytosol to form insoluble calcium phosphate. Conversely, potassium K^+ remains at a very high level to secure a constant value, say -60 mV, of the cytoplasmic membrane potential. Because of ATP hydrolysis by the corresponding pumps, there is no resultant flow of the ions into or out of the cell despite the concentration differences. All cases considered are indeed similar to what happens in a car that remains at the same spot on an inclined road with its wheels constantly rotating and slipping (Fig. 8.2).

The steady state of zero output power has several advantages for biological organisms related to their reaction to environmental changes. The first and simplest advantage is the possibility of regulating the degree of a passive reaction to external changes. By linearizing the dependence (8.36) for the output flux J_2 in the vicinity of the stalling force A_2^{st}, we obtain the relation

$$J_2 = \left(J_{+2}^{-1} + J_{-2}^{-1} \right)^{-1} \beta (A_2 - A_2^{st}) . \qquad (8.37)$$

If one of the asymptotic fluxes $J_{\pm 2}$ is small and the other large, the proportionality coefficient in (8.37) is small. Large deviations from the stalling force A_2^{st} result in a small reaction. If both asymptotic fluxes are large, the proportionality coefficient in (8.37) is large. Small deviations from the stalling force, which can even be zero in such a case, result in (almost arbitrarily) large changes in the output flux (see Fig. 8.15).

Active reactions of the organism to external perturbations follow from either a *negative* or a *positive feedback*. The first manifests itself by maintaining the original state regardless of environmental changes (*homeostasis*), whereas the second does so in an on/off-switch-type reaction to arbitrary changes above a certain threshold. The reaction of the cell to environmental changes will be discussed in more detail in Sect. 8.6 devoted to biological *signal transduction*.

The attainable range of variability of the force stalling the machine is nevertheless limited. Comparing (8.33) and (8.36), it follows that

$$-\beta A_1^{st} = \ln \frac{1 + C_1 + K_1 e^{-\beta A_2}}{1 + (C_1 + K_1)e^{-\beta A_2}} ,$$

$$-\beta A_2^{st} = \ln \frac{1 + C_2 + K_2 e^{-\beta A_1}}{1 + (C_2 + K_2)e^{-\beta A_1}} , \qquad (8.38)$$

where the constants

$$C_i = \frac{k_{+i}[R_i]_0}{k_{0i}} , \quad i = 1,2 , \qquad (8.39)$$

represent the ratios of the corresponding productive and nonproductive reaction rates (see Fig. 8.14). The dependences found for the negative stalling forces $-A_1^{\mathrm{st}}$ and $-A_2^{\mathrm{st}}$ on A_2 and A_1, respectively, are strictly increasing but they saturate both for very large positive and for very large negative values of the determining forces. For values outside the variability range, the forces can no longer stall the machine. Figure 8.16 shows both dependences (8.38) and domains in the (A_1, A_2) plane where free energy transduction takes place.

For small values of the forces, (8.38) can be linearized to yield

$$-\beta A_1^{\mathrm{st}} = \frac{C_1}{1 + C_1 + K_1}\beta A_2 , \qquad -\beta A_2^{\mathrm{st}} = \frac{C_2}{1 + C_2 + K_2}\beta A_1 . \quad (8.40)$$

The proportionality coefficients are always less than unity.

In terms of constants C_i, the degree of coupling (8.30) reads

$$\epsilon = \frac{J_2}{J_1} = \frac{1 - e^{-\beta(A_1+A_2)} + C_1^{-1}(1 - e^{-\beta A_2})(1 + K_1 e^{-\beta A_1})}{1 - e^{-\beta(A_1+A_2)} + C_2^{-1}(1 - e^{-\beta A_1})(1 + K_2 e^{-\beta A_2})} . \quad (8.41)$$

It can be readily seen that in the case when $\beta A_1 \geq 0$ and $\beta A_2 \leq 0$, the coupling coefficient is less than or equal to unity. Conversely, when

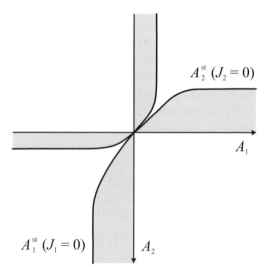

Fig. 8.16. Regions in the (A_1, A_2) plane for which free energy transduction takes place. The boundaries of the vanishing fluxes are determined by (8.38)

$\beta A_1 \leq 0$ and $\beta A_2 \geq 0$, the coupling coefficient is greater than or equal to unity. Therefore, in a similar way to macroscopic machines (see Fig. 8.2), slippage lowers the efficiency of a chemochemical machine. This statement, however, is not entirely general since certain mechanisms for coupling between the two reactions that are more complex than the one defined by the kinetic scheme in Fig. 8.14 can in principle lead to any value of the parameter ϵ, e.g., much greater than unity for $\beta A_1 \geq 0$ and $\beta A_2 \leq 0$. This kind of coupling will be discussed in Sect. 9.5 in the context of the action of the actomyosin motor.

8.6 Biological Signal Transduction

In order to survive, a living organism has to react even to very weak external signals. The sensitivity of biological receptors is extraordinarily high. Textbook examples are the reaction of the human eye to a single photon of light or the reaction of a butterfly male to a single ferromone molecule coming from a female at a distance of several kilometers. Receptors of internal cells of biological organisms react to hormones, cytokines or antigens at very low concentrations (Sect. 5.6). Axons of neural cells generate a complete action potential if their membrane potential only slightly exceeds a threshold value.

Strong reaction to a weak impulse needs an amplification which is performed by virtue of special processes of free energy transduction. In biological cells, the direct source of free energy is ATP or GTP hydrolysis.

As a well-studied example, let us consider the mechanism of action potential generation in the cytoplasmic membrane of an axon (Solomon et al., 2004; Darnell et al., 1999, Chap. 21; Hille, 2001). The Na^+–K^+ pump and at least five different ionic channels take part in this process (Fig. 8.17a). Three channels of Na^+, K^+ and Cl^- ions are non-gated, and two channels of Na^+ and K^+ ions are voltage-gated (see Sect. 5.2). The non-gated Na^+ channel has an extraordinarily low conductance.

We study a patch of the membrane small enough for the distributions of ions on either side to be spatially homogeneous. Electrical properties of this patch can be described in terms of the equivalent circuit presented in Fig. 8.17b. The voltage applied to the circuit equals the axon membrane potential (see Sect. 6.2)

$$u = \Delta\phi \equiv \phi_{\text{in}} - \phi_{\text{out}} , \qquad (8.42)$$

and the pump is considered as a source of constant current i_P. i_C denotes a current that flows through the membrane capacitance c,

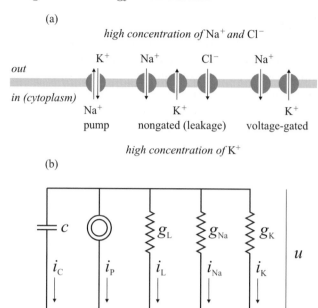

Fig. 8.17. (a) In the process of action potential generation, the Na^+–K^+ pump, the three non-gated channels of Na^+, K^+ and Cl^+ ions, and the two voltage-gated channels of Na^+ and K^+ ions all take part. Permanent pumping of ions across the cytoplasmic membrane increases the Na^+ and Cl^- ion concentration outside the axon and the K^+ ion concentration inside. **(b)** Equivalent electrical circuit of a patch of the axon membrane. The voltage u equals the membrane potential and i_P is a constant current generated by the pump. Currents i_C, i_L, i_{Na} and i_K flow through the membrane capacitance c and the conductances g_L, g_{Na} and g_K, respectively. g_L represents leakage through the non-gated channels, whereas g_{Na} and g_K represent leakage through the voltage-gated channels

and i_L a current that flows through the membrane conductance g_L characterizing a leakage through non-gated channels of Na^+, K^+ and Cl^- ions. i_{Na} and i_K are currents through the conductances g_{Na} and g_K of the voltage-gated Na^+ and K^+ channels, respectively. All currents obey the conservation law

$$i_C + i_P + i_L + i_{Na} + i_K = 0 \,. \tag{8.43}$$

The steady-state value of the membrane potential, referred to as a *resting potential*, is fixed. Three factors maintain this value: a constant intracellular concentration of large organic anions unable to pass through the membrane, a permanent action of the Na^+–K^+ pumps, and leakage through the non-gated Na^+, K^+ and Cl^+ ion channels. The result is an increase in the Na^+ and Cl^- ion concentration outside the axon and the K^+ ion concentration inside (Fig. 8.17a). For typical axons, the resting potential $u^{rest} = -70$ mV. Under the constant resting potential, $i_C = i_{Na} = i_K = 0$, whereupon the pump current compensates the leakage current:

$$i_P + i_L = 0 . \tag{8.44}$$

Current generated by the pump does not vary with a change in the value of the membrane potential. Hence, in the general case, (8.44) should be replaced by

$$i_P + i_L = g_L(u - u^{rest}) . \tag{8.45}$$

The capacity current i_C is determined by the time derivative of the membrane potential:

$$i_C = c\frac{d}{dt}u . \tag{8.46}$$

Moreover, the gated currents i_{Na} and i_K depend on the time variation of the potential:

$$i_{Na} = g_{Na}(t, u)u , \qquad i_K = g_K(t, u)u . \tag{8.47}$$

(We noted an explicit dependence of the corresponding conductances on time and voltage.) Substituting (8.45), (8.46), and (8.47) into the conservation law (8.43), we obtain the *Hodgkin–Huxley equation* (Hille, 2001):

$$\frac{d}{dt}u = -c^{-1}\left[g_{Na}(t, u)u + g_K(t, u)u + g_L(u - u^{rest})\right] . \tag{8.48}$$

Equation (8.48) is an ordinary differential equation but its right-hand side is highly nonlinear and has time-dependent coefficients. The form of the time and voltage dependence of the conductances g_{Na} and i_K is not simple (Hille, 2001). We shall therefore discuss solutions to (8.48) qualitatively rather than quantitatively.

Under the resting potential, both Na^+ and K^+ voltage-gated channels are closed, but a local concentration of cytoplasmic Na^+ ions can increase as a result of opening the neighboring voltage-gated Na^+ channels or ligand-gated channels controlled by neurotransmitters (see Sect. 5.2 and, e.g., Solomon et al., 2004; Darnell et al., 1999, Chap. 21). Due to the low conductance of the non-gated Na^+ channels, a local increase in the Na^+ ion concentration that occurs rapidly enough causes

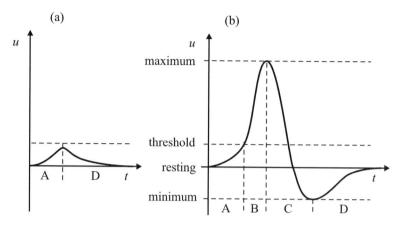

Fig. 8.18. Character of the time variation of the local membrane potential of the axon. (**a**) If the stimulating potential does not reach a threshold value, no action potential develops and the initial resting potential is slowly recovered. (**b**) The reaction to a stimulating potential of a value exceeding the threshold is a spike of action potential. In a depolarization stage, it is mainly the voltage-gated Na^+ channels that are active, and in a repolarization state, it is the voltage-gated K^+ channels. A long-lasting refraction stage is determined mainly by the activity of the slow, non-gated Na^+ channels. Successive stages are indicated as follows: A stimulation, B depolarization, C repolarization, and D refraction to the resting state

a local *depolarization* of the axon membrane. If the membrane potential does not reach a *threshold* value (-55 mV for typical axons), the voltage-gated Na^+ channels remain closed and the initial resting potential is slowly recovered due to leakage of excessive Na^+ ions through the non-gated channels (Fig. 8.18a). This is an example of the *homeostasis* phenomenon mentioned in the last section.

If the membrane potential exceeds the threshold, both Na^+ and K^+ voltage-gated channels open. Initially the Na^+ channels conduct more strongly and depolarization continues until the activity of the K^+ channels prevails (at a *maximum* potential $u = +35$ mV for typical axons) and *repolarization* begins. Finally, after reaching a *minimum* value (-100 mV for typical axons), the membrane potential returns to the resting value in a process called *refraction*. The latter stage is determined mainly by the activity of the slow, non-gated Na^+ channels.

Reaction to the membrane potential stimulation of a value exceeding the threshold is called an *action potential* (Fig. 8.18b). Its spike-like time duration of the order of 1 ms remains the same regardless of how much the level of stimulation exceeds the threshold value. Positive

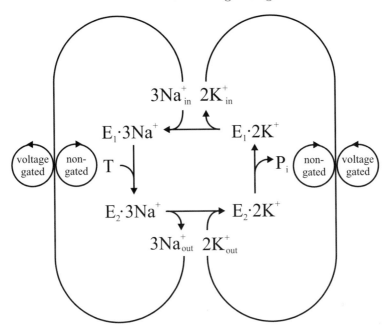

Fig. 8.19. Generation of action potential considered as a complement of the free energy transduction process taking part at the Na^+-K^+ pump. For notation, see Fig. 8.10b

feedback realized by the voltage-gated channels causes the membrane patch to behave as an *on/off switch*. A spike of action potential generated at one site of the axon membrane stimulates a spike of action potential at neighboring sites, giving a reason for neural signal propagation (Solomon et al., 2004; Darnell et al., 1999, Chap. 21).

The process of action potential generation can be seen as a complement to the free energy transduction process realized by the Na^+-K^+ pump (Fig. 8.19). In fact, a similar approach can be applied to any process of signal transduction starting with an arbitrary receptor (see Sect. 5.7). Figure 8.20 presents a simplified kinetic scheme that can be attributed to the signaling pathway shown in Fig. 5.36. Here, the signal transducer molecule tr, e.g., G or Ras protein, is simultaneously, as a GTPase, the free energy transducer. In the signaling pathways omitting the signal transducer molecules, e.g., JAK-STAT pathways of tyrosine kinase-linking receptors (Sect. 5.7), the free energy transduction takes place at each kinase molecule involved in the process.

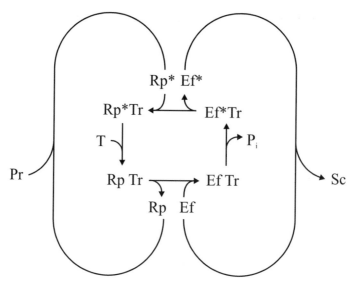

Fig. 8.20. Simplified kinetic cycle of a receptor signal transduction considered as a free energy transduction process. The notation used is as follows: Pr primary messenger, Rp receptor, Tr transducer, Ef effector and finally, Sd second messenger (see Fig. 5.36). Here, T denotes GTP and P_i inorganic phosphate. An *asterisk* distinguishes excited states of the receptor and the effector molecules

9 Lack of Partial Thermodynamic Equilibrium

9.1 Two Classes of Experiments

In Sect. 6.8, devoted to the phenomenological theory of reaction rates (and in Appendix B.3, devoted to the stochastic theory of these rates), we showed that the reciprocal rate of any effectively unimolecular chemical reaction consists in general of three time components:

$$k^{-1} = (k^{\mathrm{eq}})^{-1} + \tau' + K^{-1}\tau'' . \tag{9.1}$$

The term $(k^{\mathrm{eq}})^{-1}$ defines the time needed for the reaction to happen under the assumption that the initial and transition states of a molecule involved in it are in partial thermodynamic equilibrium, both internally and with one another. τ' and τ'' the define the times needed to achieve partial equilibrium in the initial and final states, respectively. K is the chemical equilibrium constant.

In the transition state theory (Atkins, 1998, Chap. 27) which is still commonly used for interpreting enzymatic reactions (Fersht, 1999; Stryer et al., 2002, Chap. 8), times τ' and τ'' are considered as negligibly short. To determine $(k^{\mathrm{eq}})^{-1}$, one needs only to know the average structure of an enzyme–substrate complex in the initial and transition states. No knowledge of dynamics is required. The original purpose of the transition state theory was to describe the reaction rates of small molecules in the gas phase, where a high frequency of collisions and fast vibrational relaxation really do achieve partial thermodynamic equilibrium in a short time. However, this is not the case for biochemical reactions involving protein macromolecules. Here, the second and third contributions in (9.1) dominate over the first. In this, the last chapter of the book, we present the main evidence for the slow intramolecular dynamics of protein enzymes and its possible consequences.

That biochemical reactions are controlled by the intramolecular dynamics of proteins follows directly from two classes of experiments. The first includes observations of the non-exponential initial stages of reactions during which the internal degrees of freedom of the molecules involved only reach a partial thermodynamic equilibrium. The first, already historic experiment of this type was performed thirty years ago by Frauenfelder and coworkers (Austin et al., 1975; Frauenfelder et al., 1991). It concerned the kinetics of ligand binding to myoglobin. Myoglobin, like the more complex hemoglobin, is a protein that stores molecular oxygen. It is well known that the replacement of oxygen by carbon monoxide poisons the organism. This is related to the fact that the CO binding process, as opposed to the O_2 binding process, is irreversible. There are two steps: a reversible bimolecular reaction of ligand adsorption from the solution and an irreversible unimolecular reaction of ligand covalent binding to heme from the protein interior:

$$Mb + CO \rightleftharpoons Mb \cdot CO \longrightarrow MbCO \ .$$

The experimentalists broke up the heme–CO bond in a nonthermal way using a laser flash and observed the process of ligand rebinding to heme in various conditions after the photolysis (Fig. 9.1). At 300 K only the bimolecular reaction of binding from the solution was observed, with its usual exponential time course. The essential novelty of the experiment was to study the process at low non-physiological temperatures. In such conditions the time curve of the bimolecular reaction reveals the clearly non-exponential time course of the unimolecular reaction of ligand binding from the protein matrix.

In standard kinetic experiments with an *ensemble* of molecules, the initial distribution of microstates is not specially prepared and not usually much different from the local equilibrium distribution which results practically in the absence of the preexponential stage of the reaction, even if the reaction rate is controlled by the intramolecular dynamics. In Frauenfelder's experiment the laser photodissociation prepares the ensemble of myoglobin molecules complexing CO so that they are initially in the transition state of the irreversible CO rebinding reaction. In such a case, the short initial stage of the reaction can even dominate the main, exponential stage. The initial stage kinetics, when the partial equilibrium state is still being achieved, cannot be described in terms of a conventional reaction rate constant. The more sophisticated notion of first-passage time has to be used. Quite generally, the complete time course of any irreversible unimolecular reaction is described by the equation (Kurzyński et al., 1998; see also Appendix B.4)

$$\dot{P}(t) = -f(t) \ , \tag{9.2}$$

where $P(t)$ is the fraction of molecules remaining in the initial state, the dot denotes the time derivative, and $f(t)$ is the distribution func-

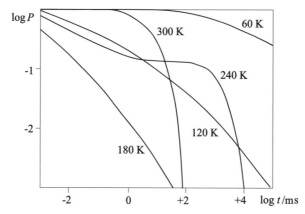

Fig. 9.1. Sketch of the time dependence of the rebinding of CO molecules after the photodissociation of CO-bound sperm wale myoglobin at various temperatures, after Frauenfelder and coworkers (Austin et al., 1975; Frauenfelder et al., 1991). $P(t)$ represents the fraction of the myoglobin molecules that have not rebound CO at time t after the laser flash. At low temperatures, only the unimolecular reaction of CO rebinding from the protein interior is observed. Its time course is clearly non-exponential. The exponential stage observed at 240 K and higher temperatures is attributed to the bimolecular reaction of CO rebinding from the solution. The latter process masks the exponential stage of the unimolecular reaction of CO rebinding from the protein interior (incomplete masking has been observed for horse myoglobin by Post et al., 1993)

tion of first-passage times from the transition to the final state. If it differs from a simple exponential, this function points to the existence of a whole spectrum of relaxation times not well separated from the complete chemical equilibration time k^{-1} in (9.1), whence the predominance of the second and third components over the first.

The first-passage time distribution $f(t)$ between two, in general, transition states (separately, in the forward and reverse reaction) can also be determined in the second class of experiments. These concern *single molecules* that occur in two states effectively treated as chemical states. After each reactive transition, the molecule starts its further microscopic motion from the transition state of the reverse reaction. Observation reveals a dichotomous *telegraphic noise* showing successive dwell times in the alternating chemical states (Fig. 9.2). From the statistics of these dwell times, one is able to determine the first-passage time distribution for the reaction in both directions. If the distribution densities are exponential, the telegraphic noise is a Markov stochastic

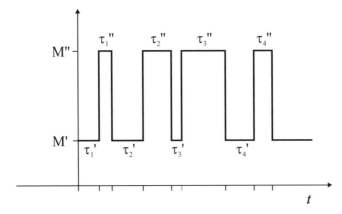

Fig. 9.2. Schematic 'telegraphic noise' recorded for a single molecule experiencing transitions between two states M' and M''. It may be an ionic current flowing through a single protein channel, that fluctuates between two values corresponding to two 'chemical' states: open (O) and closed (C). Alternatively, it may also be the fluorescence intensity of a single protein enzyme, that fluctuates between two levels corresponding to the fluorophore-containing on state and the fluorophore-lacking off state

process (see Appendix B.2), with no memory of the intramolecular dynamics in the preceding chemical state. Non-exponential distributions allow one to determine some features of this intramolecular dynamics (Quin et al., 1996; Edman and Rigler, 2000; Flomenbom et al., 2005; see also Appendixes B.5 and D.4).

The first single-molecule detection technique was the *patch-clamp* technique developed at the turn of the 1970s (Sackmann and Naher, 1995). It enabled observation of the fluctuations of ionic current flowing through single protein channels. It soon appeared that most of the channels occur in two discrete states named 'open' and 'closed', and that the statistics of open and closed times very often show a non-exponential distribution, i.e., the noise is non-Markovian (Fig. 9.3).

A more recent technique is *single fluorophore detection* using confocal fluorescence microscopy (Eigen and Rigler, 1994) or total internal reflection fluorescence microscopy (Funatsu at al., 1995). This enables direct observation of the kinetics of a single protein enzyme immobilized in a femtoliter (10^{-15} dm^3 = 1   m^3) viewing volume. The enzyme takes part in a reaction

$$\mathrm{E} + \mathrm{R} \rightleftarrows \mathrm{ER} \rightleftarrows \mathrm{EP} \rightleftarrows \mathrm{E} + \mathrm{P} \ .$$

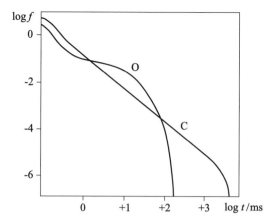

Fig. 9.3. Sketch of the time dependence of the closed time (C) and open time (O) distribution $f(t)$ observed with the help of the patch-clamp technique for the protein K^+ channel of NG 109-15 cells, after Sansom et al. (1989). Both curves show a short-time non-exponential behavior

The R molecule binds a fluorophore that is activated in the EP and the P state. As the product P molecules quickly diffuse out of the viewing region, a blinking fluorescence signal is observed with the form of dichotomous telegraphic noise, as shown in Fig. 9.2. This originates from the single enzyme molecule jumping between the non-fluorescent E or ER states and the fluorescent EP state.

The dichotomous noise overlaps with noises of other origins so that the simplest way of extracting the information included in it is to determine its autocorrelation function $G(t)$ (for the definition see, e.g., Appendix B.2). Figure 9.4 sketches the time dependence of the correlation function of the fluorescence signal emitted by a single horseradish peroxidase molecule (Edman et al., 1999). This enzyme produces the fluorescent product by oxidation, after the decomposition of hydrogen peroxide (H_2O_2), of the non-fluorescent substrate dihydrorhodamine 6G. Note that the initial time course over the first 10 ms is corrected by a non-exponential contribution. It has a stretched exponential form (see Appendix B.4)

$$\exp[-(t/\tau)^\alpha] \, , \tag{9.3}$$

where the exponent has value $\alpha = 0.2$.

The autocorrelation function $C(t)$ is related to the time distribution density functions for the lifetimes of both the fluorescent and non-fluorescent states, but the latter can also be determined more directly by digitalizing the blinking fluorescence signal. This procedure, applied

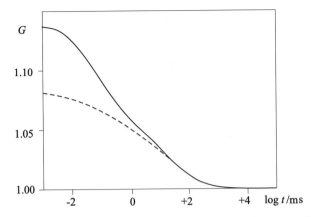

Fig. 9.4. Sketched time dependence of the correlation function of the fluorescence signal emitted by a single horseradish peroxidase molecule immobilized in the light cavity of a confocal microscope. The substrate concentration is [R] = 65 nM. The short-time course does not follow the curve (*broken line*) that fits the long-time exponential behavior. After Edman et al. (1999)

to a lipase B enzyme, also indicates the presence of a 300-ms long initial course of the autocorrelation function with the stretched exponential form (9.3) and exponent $\alpha = 0.15$ (Flomenbom et al., 2005).

9.2 Intramolecular Dynamics of Biomolecules

Because each atom can move in three directions, a molecule consisting of N atoms has $3N$ *degrees of freedom*, numbers that are independent functions of all atomic positions and completely characterize the momentary spatial organization of the molecule. Three of these numbers (coordinates of the mass center) define the *translational* motion of the molecule as a whole, three angles define the resultant *rotational* motion of the molecule, and the remaining $3N - 6$ are *internal* degrees of freedom. They can be identified with covalent bond lengths and angles, as well as dihedral angles of rotations about the bonds (Fig. 9.5a). It is the ability to perform such rotations (limited only to some degree by steric hindrance), combined with the possibility of hydrogen bond break-up and reformation, that makes the landscape of the configurational potential energy of biomolecules (proteins, nucleic acids or polysaccharides) extremely complicated. A general feature of this landscape is the presence of an astronomical number of local minima separated by higher or lower energy barriers of non-covalent nature (McCammon and Harvey, 1987; Brooks et al., 1988; Frauenfelder et

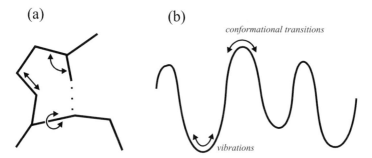

Fig. 9.5. (**a**) Intramolecular dynamics of biomolecules consists in changing the values of covalent bond lengths and angles as well as dihedral angles of rotations about the bonds. Break-up and reformation of the weak hydrogen bonds is also important. (**b**) In the many-dimensional landscape of the configurational potential energy of a biomolecule, one can distinguish conformational substates – the local minima separated by higher or lower energy barriers. On assuming interconformational barriers to be high enough, the intramolecular dynamics can be reasonably decomposed into vibrations within particular conformational substates and conformational transitions

al., 1991; 1999; Kurzyński, 1998). As in the stereochemistry of low-molecular weight organic compounds (see Appendix C.3), regions of the configurational space surrounding the local minima can be referred to as *conformational states* (*substates* in particular contexts).

In a reasonable approximation, assuming interconformational barriers to be high enough, the internal dynamics can be decomposed into more or less damped *vibrations* within particular conformational substates and purely stochastic *conformational transitions* (Fig. 9.5b). As a lower bound of the interconformational barrier heights, one can assume a few units of $k_B T$, say 10 to 20 kJ/mol, which is a typical energy barrier height for a local rotation about a single covalent bond in the absence of any steric constraints or a more collective transition in small cyclic chains of sugar 'puckering' type. At the same time, it is the typical energy needed to break up and reform a hydrogen bond. The vibrational dynamics is characterized by a spectrum of periods of vibrational normal modes whose number equals the number of internal degrees of freedom. Vibrational periods range from 10^{-14} s (weakly damped localized N–H or C–H stretching modes) to 10^{-11} s (overdamped collective modes involving whole macromolecular domains). The conformational transition dynamics is characterized by a spectrum of relaxation times whose number equals the number of conformational substates. In physiological conditions, this spectrum begins at 10^{-11} s (overcoming the just-assumed lowest energy barrier of the order of 10 kJ/mol), and its character depends on the kind of biomolecule.

In proteins, which usually play the role of enzymes and thus take part in almost any biochemical process, the relaxation time spectrum of conformational transition dynamics seems to be practically quasi-continuous, at least in the range from 10^{-11} to 10^{-7} s (Kurzyński, 1998; see also Appendix D.2). Two classes of mathematical models can be proposed for the stochastic dynamics which display such a property. In the first, 'protein-machine' class of models (Appendix D.4), the dynamics of conformational transitions is represented by quasi-continuous diffusion in a certain effective potential along a few 'mechanical' coordinates, e.g., angles or distances describing mutual orientation of approximately rigid fragments of protein secondary structure (α-helices, β-pleated sheets) or larger structural elements. The spectrum of reciprocal relaxation times for dynamics of this type is more or less homogeneous. Otherwise, in the second class of models, the spectrum of reciprocal relaxation times is assumed to have a hierarchical organization. The latter is considered to be a generic property of glassy materials, and we refer to this second, more extensive class of models as protein-glass models (Appendix D.3). Time hierarchies, often observed in the case of proteins (Frauenfelder et al., 1991; 1999), can originate either from a hierarchy of barrier heights in the conformational potential energy landscape or from a hierarchy of bottlenecks (the entropy barrier heights) in the network joining conformations between which direct transitions take place.

Conformational transitions do not take part in the entire body of the protein macromolecules. Figure 9.6 presents the structure of the presumably universal statistically independent unit of biochemical processes, a supramolecular multienzyme protein complex. From the dynamical point of view it is essential to distinguish within its body between solid-like fragments of secondary structure (α-helices or β-pleated sheets) and liquid-like surrounding regions, either nonpolar (domain interiors, lipid membrane environment), or polar ('channels' between domains, water environment). A complex with size exceeding 20 nm (see Fig. 5.4) is too large to be described in terms of *microscopic* mechanics of individual atoms, and too inhomogeneous to be described in terms of *macroscopic* thermodynamics. The proper language is the *mesoscopic* theory of stochastic processes, a short introduction to which can be found in Appendix B.

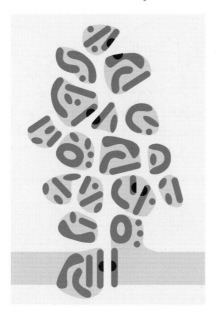

Fig. 9.6. Schematic cross-section of a universal unit of biochemical processes, a supramolecular multienzyme protein complex. *Darkly shaded areas* are solid-like fragments of secondary structures. *Medium shaded areas* are non-polar and *lightly shaded areas* are polar liquid-like regions. *Black marks* indicate individual catalytic centers, usually localized at two neighboring solid-like elements. After Kurzyński (1998)

Both mentioned classes of conformational dynamics models use this language. Models of protein-glass type treat the dynamics of conformational transitions as a quasi-continuous diffusion of structural defects through the liquid-like medium. Alternatively, models of protein-machine type treat this dynamics as a relative motion of solid-like elements, also with the nature of quasi-continuous diffusion. Intermediate metabolites are channeled to internal liquid-like regions and the corresponding concentrations are also mesoscopic variables characterizing the state of the complex rather than the thermodynamic variables. In principle, slow diffusion dynamics controls all chemical reactions that take place in localized catalytic centers.

Besides very fast processes of non-adiabatic charge or energy transfer (see Appendix D.6), typical time scales of biochemical processes range from microseconds to seconds. Hence the vibrational dynamics is too fast to appreciably influence chemical reactions involving proteins. Only much slower conformational transition dynamics can effect the majority of biochemical processes, and therefore any ade-

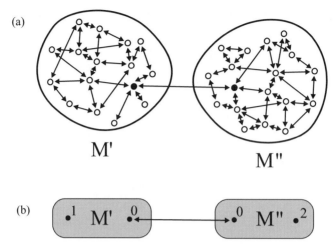

Fig. 9.7. (**a**) Exemplifying realization of model intramolecular dynamics underlying a unimolecular reaction M′ ↔ M″. Chemical states M′ and M″ of a macromolecule are composed of many conformational substates (*white* and *black circles*) and the intramolecular dynamics involves purely stochastic conformational transitions (*arrows*). In fact, a much larger number of conformational substates are expected. The chemical reaction is realized through transitions between a distinguished conformational substate (the gate) in M′, forming the transition state $(M')^{\ddagger}$, and a distinguished conformational substate in M″, forming the transition state $(M'')^{\ddagger}$. (**b**) The representation assumed in subsequent sections. *Shaded boxes* represent diagrams of an arbitrary number of sites, whereas *black dots* represent gates

quate statistical theory of these processes has to be a development of the stochastic theory of reaction rates (Kurzyński, 1993; 1998; see also Appendix B.3). Following that theory, the intramolecular dynamics in both the initial and the final chemical states comprises only purely stochastic transitions between a multitude of substates, some of which form the forward and reverse reaction transition states. Figure 9.7 shows the situation of a *gated* reaction when the transition states are reduced to single conformational substates, the gates. A more general case is presented in Fig. B.1 of Appendix B.3. The reason why we restrict our considerations to the gated reactions will be explained in the next section.

Let us consider the hierarchy of relaxation times that occur in the problem considered. The fastest is vibrational relaxation, which results in an equilibrium of microstates within individual *conformational substates*. On a timescale longer than a few picoseconds, these substates can be described in thermodynamic terms including values of the conformational free energy and the mean lifetime. However, the concentration of a conformational substate is not a macroscopic quantity since, apart from a few exceptional cases, one cannot prepare a macroscopic ensemble of biomacromolecules in a given conformational substate. In fact, only a time average is realized, not the actual ensemble average (Sect. 2.7).

The macroscopic quantities are the concentrations (molar fractions) of *chemical states* of a biomacromolecule.[1] To attribute a value of the free energy to a chemical state, a partial equilibrium must be achieved between conformational substates composing this state. However, we argued that the time evolution of individual conformational substates can be described only in probabilistic and not in thermodynamic terms, i.e., only the probability of a given substate occupation and not the corresponding concentration has a physical meaning. As a consequence, the initial time course of a given chemical reaction, described by a function like (9.3), has a physical meaning on a timescale longer than the vibrational relaxation time, but until the conformational relaxation time has passed, it remains beyond the scope of nonequilibrium thermodynamics as we understand it in the present book. It is an open question as to whether it can be described by a generalized, nonadditive thermodynamics (Tsallis, 1999; 2001).

Direct observation of non-exponential initial stages of reactions in experiments and simulations is only possible for a special preparation of the initial conformational substate of the protein confined to the reaction transition state. Usually, the initial distribution of conformational substates is not much different from partial equilibrium and no initial-condition-dependent stages are observed in the time course of biochemical reactions proceeding in standard conditions. But the specially prepared initial substates of protein macromolecules also occur in standard conditions, if several coupled reactions gated by conformational transition dynamics proceed in the steady state. Because of the slow character of the intramolecular dynamics, the succeeding reactions proceed before the partial equilibria in the preceding chemical species have been reached. As a consequence, the steady-state kinetics, like the initial stage kinetics, cannot be described in terms of the usual rate constants. This possibility was already suggested thirty years ago by Blumenfeld (1974). More adequate physical quantities that should

---

[1]We treat the notion of 'chemical state' generally to mean any thermodynamically distinguishable state. It thus also comprises, e.g., the occurrence of a small molecule on one or other side of a membrane, or the occurrence of a membrane channel macromolecule in the open or the closed state (see the last section).

be used are the mean first-passage times (Kurzyński and Chełminiak, 2003; see also Appendix D.5). In the three last sections of this book, we consider how this statement changes the conventional view of the enzymatic catalysis process itself and biological free energy transduction.

9.3 Enzyme in a Multitude of Conformational States

In Sect. 7.2, we derived (7.18) and (7.19) which determine the steady-state kinetics of the enzymatic reaction involving one intermediate (see Fig. 9.8a). Identical equations can be derived for the enzymatic reaction involving two intermediate enzyme–substrate states (Haldane's kinetics, Fig. 9.8b) (Cantor and Schimmel, 1980, Chap. 16). For the enzymatic reaction involving two discriminated free-enzyme states (Fig. 9.8c), simple but rather tedious calculations result in a slight generalization of (7.18) expressing the reaction flux per enzyme molecule J as a function of the chemical force A:

$$
J = \frac{1 - e^{-\beta A}}{J_+^{-1} + J_-^{-1} e^{-\beta A} + J_0^{-1} \left(K + e^{\beta A} \right)^{-1}} \, .
\tag{9.4}
$$

As in (7.18), the quantities $J_\pm$ have the meaning of asymptotic flux periods and their dependence on the substrate concentration $[R]_0$ is of the conventional Michaelis–Menten form [see (7.19)]:

$$
J_\pm = \frac{k_\pm [R]_0}{K_\pm + [R]_0} \, ,
\tag{9.5}
$$

where $k_\pm$ is the enzyme turnover number and $K_\pm$ the apparent dissociation constant. An additional quantity J_0 determines the position of an inflection point on the flux–force functional dependence (9.4).

As a matter of fact, we argued in the last section that a still more complicated kinetic scheme for the enzymatic reaction is appropriate, involving a whole quasi-continuum of intermediate conformational substates and stochastic transitions between them (Fig. 9.8d). The slow character of this dynamics results in a steady-state occupation distribution of the conformational substates, rather than a partial equilibrium distribution. As a consequence, the steady-state kinetics of the enzymatic reaction considered cannot generally be described in terms of conventional chemical kinetics, i.e., reaction rate constants that need partial equilibrium to be achieved in the chemically discriminated states. A more sophisticated language of mean first-passage

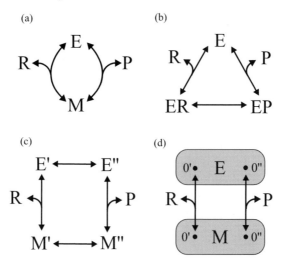

Fig. 9.8. (**a**) to (**c**) Single enzymatic reaction involving one, two, and three intermediates, respectively. (**d**) Single enzymatic reaction with a quasi-continuum of intermediate conformational substates considered in the present section. As in Fig. 9.7b multitudes of conformational transitions within E (the free enzyme) and M (the enzyme–substrate complex) are represented by *shaded boxes*. The reactant and product binding–rebinding bimolecular reactions are assumed to be gated, i.e., they take place only in certain conformational substates represented by *black dots*

times has to be used (Kurzyński and Chełminiak, 2003). It is worth noting here that, in contrast to the reaction rate constants, the mean first-passage times do not generally obey the detailed balance condition.

A technique has been developed to enable calculation of steady-state fluxes for systems of enzymatic reactions controlled and gated by arbitrary-type stochastic dynamics of the enzyme molecule (Kurzyński and Chełminiak, 2003). The corresponding basic ideas are discussed in Appendix D.5. Here we merely note that, using this technique, we found that the reaction flux per enzyme molecule for the scheme in Fig. 9.8d is also determined by the conventional (9.4) and (9.5). However, the expressions for the phenomenological parameters $k_{\pm}$ and $K_{\pm}$ we obtained were unconventional. In the notation explained in Fig. 9.8d, the reciprocal turnover numbers are

$$k_{+}^{-1} = (k_{+}''^{\,\mathrm{eq}})^{-1} + \tau_{\mathrm{E}}(0'' \to 0') + \tau_{\mathrm{M}}(0' \to 0'') \,, \qquad (9.6)$$

$$k_{-}^{-1} = (k_{-}'^{\,\mathrm{eq}})^{-1} + \tau_{\mathrm{E}}(0' \to 0'') + \tau_{\mathrm{M}}(0'' \to 0') \,, \qquad (9.7)$$

and the apparent dissociation constant is

$$K_+ = K'k_+ \left[(k_+''^{\text{eq}})^{-1} + (k_-'^{\text{eq}})^{-1} + \tau_{\text{M}}(0' \leftrightarrow 0'')\right] , \qquad (9.8)$$

where K' is the actual dissociation constant given by

$$K' = [\text{R}]^{\text{eq}}[\text{E}]^{\text{eq}}/[\text{M}]^{\text{eq}} . \qquad (9.9)$$

The apparent dissociation constant K_- is related to K_+ by the Haldane equation (7.14) and the flux J_0 is given by

$$J_0 = \frac{K'[\text{R}]_0^{-1}}{\tau_{\text{E}}(0' \leftrightarrow 0'')} . \qquad (9.10)$$

The quantities $\tau_{\text{E}}(0' \to 0'')$, $\tau_{\text{E}}(0'' \to 0')$, $\tau_{\text{M}}(0' \to 0'')$ and $\tau_{\text{M}}(0'' \to 0')$ are the mean first-passage times between the specified gates within E and M, and $k_+''^{\text{eq}}$ and $k_-'^{\text{eq}}$ are the equilibrium (transition state theory) rate constants for the product and the reactant dissociation reaction, respectively. The mean first-passage times in (9.6) to (9.8) are between the succeeding gates and not between the 'typical' average states and the gates as in the full expression for the rate constant (see Appendix B.4). Therefore, the parameters $k_\pm$ and $K_\pm$ cannot be expressed in terms of the full rate constants $k_\pm'$ and $k_\pm''$ describing the conventional kinetics presented in Sect. 7.2.

If the transition states of the component reactions consisted of many conformational substates with different transition probabilities to the final state (the model with *fluctuating barriers*, see Appendix B.3), the total steady-state forward or reverse reaction flux would be a sum of several terms like (9.5) with different values of the apparent dissociation constant K_+ or K_-. However, this sum could not generally have the Michaelis–Menten form. Consequently, a sufficient condition for the dynamically controlled enzymatic process to follow the Michaelis–Menten steady-state kinetics is gating of the component binding–rebinding reactions. That the vast majority of enzymatic reactions actually obey the Michaelis–Menten law (Fersht, 1999) is, when confronted with the proofs of the slow character of intramolecular protein dynamics presented in the last two sections, a strong argument in favour of the gated mechanism for protein-involving reactions (Fig. 9.7) which we assumed as the basis for all theoretical models considered.

Equations (9.6) to (9.10) describe the steady-state kinetics of enzymatic reactions one level deeper than (9.4) and (9.5). However, this is still phenomenology. A task for theorists is to fill the interiors of the shaded boxes in Fig. 9.8d with simple but adequate models of conformational transition dynamics. It should comprise not only the formation of the catalytic center in a proper conformation (Fersht, 1999), but also even more complicated processes of molecular recognition and formation of a diffusional encounter complex (Verkhivker et al., 2002; Gabdouline and Wade, 2002; Schreiber, 2002).

We would like to conclude this presentation of results with two possibly speculative comments. The first concerns the role of the equilibrium rate constants in enzymatic catalysis. Assuming that billions of years of biological evolution has acted to optimize the rates of enzymatic reactions and that the optimum rate is the fastest possible, one can speculate that present day enzymes have the entrance and exit gates for the reaction very close to each other, so that the corresponding mean first-passage times are negligible. Putting them equal to zero in (9.6) to (9.8) leads to the reconstruction of the simple conventional expressions (7.12) and (7.13), but with the full reaction rate constants replaced by their transition state theory counterparts. This could explain the applicability of the transition state theory for description of enzymatic catalysis, commonly assumed by most enzymologists (Fersht, 1999).

The second comment concerns the role of the mean first-passage times between the gates in the control of enzymatic catalysis. The activity of a protein enzyme can change greatly upon binding an effector molecule. The conventional approach to heterotropic allosteric regulation, in particular noncompetitive inhibition, assumes effector binding to induce long-range *structural* changes (Fersht, 1999). However, there is serious evidence that it can induce some *dynamical* changes as well (Jardetzky, 1996; Stivers et al., 1996; Hodson and Cistola, 1997; Hoofnagle et al., 2001; Kern and Zuiderweg, 2003). The theory presented predicts the enzyme turnover number to depend on both the equilibrium rate constants and the mean first-passage times between the entrance and exit gates. The former are determined by the structure, but the latter by the dynamics. It is physically reasonable to suppose that some inhibitor molecules can act so as to increase the mean first-passage times between the gates rather then to decrease the equilibrium rate constants. The importance of this supposition, if actually true, in particular for pharmacology, can hardly be overestimated.

9.4 Two Coupled Enzymatic Processes: Case of the Actomyosin Motor

Effects due to the lack of partial thermodynamic equilibrium in the case of two coupled enzymatic reactions will be discussed here in the context of free energy transduction. In Chap. 8, the action of biological molecular machines was described in terms of simple chemical kinetics. However, the protein macromolecules shown in many figures in Chap. 5 do not resemble the small molecules of conventional physical chemistry with rapidly equilibrating intramolecular dynamics. Instead, they look more like highly organized assemblies of mechanical elements: levers, hinges, springs (or pistons) and triggers. Moreover, some electrical elements such as conductors, semiconductors and insulators can be distinguished. All these elements seem to cooperate in a similar way to the elements of typical macroscopic machines.

However, molecular machines are not macroscopic but mesoscopic systems, and they are 'soft' in the sense that their elements stick to one another (Jones, 2004). As a matter of fact, molecular machines act due to thermal fluctuations: energy is borrowed from and returned to the surroundings. ATP hydrolysis makes this process unidirectional. Consequently, the action of molecular machines has to be described in the same terms as common chemical reactions, except that a multitude of specially organized conformational substates have to be taken into account, just as in the case of the single enzymatic reaction considered in the last section.

The technique we have used for the single enzymatic reaction applies also for two coupled enzymatic reactions, but the general formulas obtained in this way are complex (Kurzyński and Chełminiak, 2003) and not very perspicuous so we do not quote them here. Instead, we shall restrict our considerations to a case study of the actomyosin motor.

The structure of the actin and myosin filaments organized in the myofibrils, as well as the myosin head itself, were considered in detail in Sect. 5.4. Here, we only recall that the main result from structural studies is the *swinging lever-arm* picture of the myosin head. It refines the classical H.E. Huxley (1969) swinging cross-bridge model and relates the force that the myosin head exerts on the actin filament to rotational motion of its regulatory subunit, the lever arm, relative to the catalytic subunit strongly attached to the actin filament (Rayment et al., 1993; Spudich, 1994; Geeves and Holmes, 1999; Houdusse and Sweeney, 2001; Howard, 2001).

But what drives the movement of one filament relative to another? The question that remains open is how to combine chemistry and mechanics and describe the mechanism of chemomechanical coupling for molecular motors. Presumably, much of the superfluous discussion on this topic results from the fact that authors usually do not clearly define which notion of force they have in mind, i.e., force on the micro-, meso- or macroscopic level. These are formally different quantities. Force in the Newtonian sense can be defined only on the microscopic level of motion of individual atoms and is the subject of molecular dynamics (Hansson et al., 2002), which will not be considered here. Note that this force changes only the velocity of the motion, which itself proceeds due to inertia. The forces exerted by a motor on a track and by a track on a motor have a meaning on the mesoscopic level of the stochastic dynamics of a single motor macromolecule and compete with the fluctuating Brownian forces. They are directly observed only using single-molecule detection techniques (Ishijima and Yanagida, 2001). Otherwise, the external load acts on a statistical ensemble of motor molecules composing, e.g., a myofibril or the whole muscle, and can be directly defined only on the macroscopic level of irreversible thermodynamics (Hill, 1989).

Stochastic translational motion of two macromolecules relative to each other is overdamped, which means that the force F one molecule exerts on another is balanced not only by external forces and fluctuating Brownian forces, but also by friction forces. The friction force is proportional to the velocity and so the one-dimensional translational motion of an actin filament relative to a single myosin head is determined by a stochastic differential *Langevin equation* (van Kampen, 2001; see also Appendix B.2):

$$\zeta \dot{\mathcal{X}} = F_i(\mathcal{X}) + F^{\text{ext}}(\mathcal{X}) + \mathcal{F} . \tag{9.11}$$

The random variable $\mathcal{X}$ describes the relative position of the actin filament with respect to the considered myosin head, and the time derivative $\dot{\mathcal{X}}$ describes the corresponding velocity. ζ denotes a friction coefficient and $\mathcal{F}$ is an infinitely short-time correlated random force ('white noise') with amplitude proportional to temperature. An external force F^{ext} is exerted, e.g., by optical traps or micro-cantilevers in the single-molecule assays or by the remaining myosin heads and any external boundary constraints in the case when the molecules described by (9.11) enter into a statistical ensemble of myosin and actin filaments composing the myofibril.

The force F_i the given myosin head exerts on the actin filament is a strictly molecular property, not dependent on macroscopic external constraints such as the external load (Hill, 1989). Besides the position $\mathcal{X}$, it still depends on the chemical or conformational substate i the myosin molecule is in, and equals the negative derivative of a conformational free energy G_i considered to be a function of the value x assumed[1] by the random variable $\mathcal{X}$:

$$F_i(x) = -\frac{\partial}{\partial x}G_i(x) \ . \tag{9.12}$$

In the simplest model including two states of the myosin head, *attached* to and *detached* from the actin filament, $i = $ att and det, respectively (Huxley, 1957), $G_{\mathrm{att}}(x)$ can be approximated by a parabola and $G_{\mathrm{det}}(x)$ by a constant (no force is exerted in the detached state).

In order to determine the dynamics of the system completely, transition probabilities per unit time between the two substates have yet to be defined. These transitions are limited to more or less localized regions of the variable x values and depend on the concentrations of the ATP hydrolysis substrates (Fig. 9.9a). If we replace the quasi-continua of translational conformational states between successive transitions by two discrete states numbered with a position index l (Fig. 9.9b), we get a chain of sites representing the simple Lymn and Taylor (1971) model of the mechanochemical cycle of the actomyosin motor. The kinetic equations for this model (usually without the index l), with molecular ratios interpreted as occupation probabilities of the corresponding states, can be considered as a discrete counterpart to the Langevin equations (9.11), completed by appropriate interstate transition probabilities. In the theory of stochastic processes, such equations are known as the *master equations* (van Kampen, 2001; see also Appendix B.2).

Generalizations taking into account more conformational substates are possible, both for the continuum model (Hill, 1989; Duke, 1999) and for the discrete model. Among the latter, the Lymn–Taylor–Eisenberg model distinguishing between *strongly-attached* and *weakly-*

---

[1]Following the discussion in Sect. 9.2, a macromolecule that occurs in a multitude of conformational substates quickly reaches a partial equilibrium in the vibrational degrees of freedom, and each of its conformational substates can be attributed by some partial equilibrium free energy G. It corresponds to a *basic free energy level* in the terminology of Hill (1989). The conformational substates can be labeled with both a discrete index i and a quasi-continuous variable x (see the protein-machine model discussed in Appendix D.4).

attached states of the myosin head to the actin filament is commonly used at the present time (Ma and Taylor, 1994; Howard, 2001).

For the equilibrium concentrations of ATP and P_i (the concentration of ADP is assumed to be held constant), the transition probabilities obey the detailed balance condition and the forward and reverse stochastic motions of the myosin head relative to the actin filament are equally probable. Nonequilibrium concentrations of ATP and/or P_i break the detailed balance and result in a systematic decrease or increase in the free energy (Fig. 9.9c). The change in free energy after completion of each ATPase cycle corresponds exactly to the chemical force A that drives the catalytic reaction (see Sect. 7.2). It is clearly from Fig. 9.9c that the presence of a positive chemical force makes stochastic motion more probable to the right than to the left. For more complex models, the translational motion becomes partly independent of the chemical reaction and the introduction of an additional axis representing *reaction progress* is necessary (Keller and Bustamante, 2000; Bustamante, Keller and Oster, 2001).

The reasoning presented above reconciles two seemingly opposite approaches to the motor movement: the so-called power-stroke and thermal-ratchet models (Howard, 2001). The first approach assumes that the motion takes place in the attached state of the myosin head (Fig. 9.10a) during release from the force (strain) that emerged after transition to the distorted high-free-energy conformation with the cocked lever arm (see Fig. 5.25). Following the second approach, the motion is mainly of a diffusive character and takes place in the detached state (Fig. 9.10b). Short-lived nonequilibrium transitions to the attached state with a strongly asymmetric free-energy dependence on the position make this free diffusion unidirectionally biased. The essence of this model does not change if the sequence of attached asymmetric states is replaced by a single detached state with a saw-tooth free energy dependence (see Fig. 9.10c) (Astumian, 1997; Jülicher et. al, 1997). Figures 9.10a, b and c are drawn for total chemical and mechanical equilibrium, when the motions to the right and to the left have to proceed with the same probability. Under such conditions, it can be seen that, as a matter of fact, there is no essential difference between the power-stroke and the thermal-ratchet models. All the motions have a diffusion character, either driven or free. Presumably, the case presented in Fig. 9.9a, combining the two approaches, is the closest to reality. Strong evidence for this statement comes from observations of the long step sizes in unconventional myosins V and VI, far exceeding their lever-arm lengths (Rock et al., 2001; Tanaka et al., 2002).

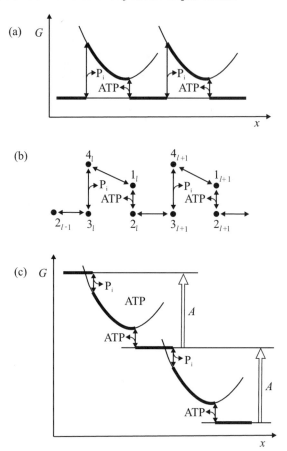

Fig. 9.9. (a) Two-state model of the translational motion of the single myosin head relative to the actin filament. The *vertical axis* represents the conformational free energy G of the system. The *horizontal axis* represents the position x of a fixed point of the myosin head. In the attached states of the myosin head to successive actin filament binding sites, G depends parabolically on x. In the detached state, G is constant. *Vertical arrows* represent transitions between the two states. These transitions can take place at sharply defined values of the position x or they can be delocalized to some extent. Their probabilities depend on the concentrations of ATP and P_i (the concentration of ADP is assumed to be held constant). The picture shows the situation of total chemical equilibrium, when a stochastic motion to the right has the same probability as a stochastic motion to the left. **(b)** On replacing the quasi-continua of translational states in the sequence of alternating attached and detached states by two discrete states, the model from drawing **(a)** transforms into the kinetic model of Lymn and Taylor. **(c)** The change in free energy by A after each ATPase cycle resulting from nonequilibrium concentrations of ATP and P_i in the scheme shown in drawing **(a)**

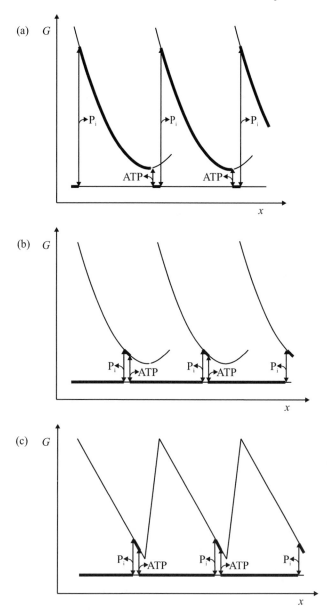

Fig. 9.10. (a) The power-stroke model of myosin motor motion along the actin filament: the movement takes place mainly in the attached state. (b) The thermal-ratchet model of myosin motor motion: the movement takes place mainly in the detached state. (c) Replacement of the sequence of attached asymmetric states by a single detached state with a saw-tooth free-energy dependence on the position

In terms of (9.11), no comparison of the force exerted by the myosin heads with the external load is possible as the later is a macroscopic quantity and acts on the whole statistical ensemble of motor molecules composing a myofibril or a muscle. Only after ensemble averaging, taking into account the fact that the mean value of the Brownian force is zero, do we get a macroscopic balance equation:

$$\zeta v = F^{\text{load}} + F^{\text{mots}} , \qquad (9.13)$$

where $v \equiv \langle \dot{\mathcal{X}} \rangle$ is the mean velocity of the myofibril (muscle) contraction, F^{load} denotes the load, and F^{mots} is the mean force exerted by all the myosin heads.

In physiology, there are two common techniques for studying muscle contraction dynamics (Woledge et al., 1985). In *isometric contraction* experiments, the length of the muscle fiber is held constant, i.e., the velocity v is zero. Then

$$F^{\text{load}} + F^{\text{mots}} = 0 . \qquad (9.14)$$

The force exerted by the myosin heads is proportional to the number of heads strongly attached to the actin filaments, so that the load is also proportional to this number. In *isotonic contraction* experiments, the load is held constant, in particular, at zero. Then

$$\zeta v = F^{\text{mots}} . \qquad (9.15)$$

The force F^{mots} is proportional to the number of myosin heads strongly attached to the actin filaments, whereas the friction coefficient ζ is proportional to the number of weakly attached myosin heads (Stehle and Brenner, 2000). Both numbers are proportional to the overlap between the filaments (see Fig. 5.26). The unloaded shortening velocity should therefore be independent of this overlap and constant (stationary) in time, as observed (Lionne et al., 1996).

As already noted, the external load is not a microscopic quantity and determines only the special organization of the statistical ensemble it is applied to. It changes the number of myosin heads attached to the actin filament but does not directly change their conformation (Hill, 1989; Duke, 1999). In other words, the external load applied to the myofibril influences the free energy of binding of the myosin heads to the filaments and not the free energy of the particular conformational substates. An experimental pendant to this reasoning are the results of EPR studies by Baker et al. (1998; 1999), indicating that the load, like the chemical force, does not change the fraction of myosin heads in any distinguished orientational state.

The idea of attributing changing nonequilibrium transition rates to the load appeared explicitly for the first time in papers by Quian (1997; 2000), Fisher and Kolomeisky (1999), Baker et al. (1999), and Baker and Thomas (2000). However, the reasoning of all these authors was not quite correct. Quian, Fisher and Kolomeisky consider the kinetics of chains of the type presented in Fig. 9.9b, but relate the load directly to the behavior of single motor molecules. Conversely, Baker and Thomas correctly relate the load to a property of the whole ensemble of motor molecules, but try incorrectly to interpret it in terms of a force the myosin heads exert on the actin filaments.

Figure 9.11a shows the commonly accepted Lymn–Taylor–Eisenberg kinetic scheme (Ma and Taylor, 1994; Howard, 2001) indicating how the ATPase cycle of myosin is related to detached, weakly-attached and strongly-attached states of the myosin head to the actin filament. Both the substrate and the products of the catalyzed reaction bind to and rebind from the myosin in its strongly-attached state, whereas the reaction itself takes place either in the weakly-attached or in the detached state. The changes in the binding free energy due to the external load can be expressed as changes in the effective rather than actual concentrations of the actin filament A, which allows the motor to be treated as a standard chemochemical machine. This approach was discussed in Sect. 8.4 (see the simplified scheme in Fig. 8.12).

However, the kinetic scheme in Fig. 9.11a is insufficient for the proper description of the actomyosin mechanical cycle. We have shown that, in order to determine the force exerted by the myosin head on the actin filament in the strongly-attached state, one has to consider a quasi-continuum of conformational substates labeled with the help of a one-dimensional quasi-continuous variable. In the weakly-attached and detached states of the myosin head, the stochastic dynamics of conformational transitions appears to be still more complex than one-dimensional diffusion (Volkman and Hanein, 2000).

In fact, both X-ray crystallography (Houdusse et al., 2000) and the study of cross-linking between various thiols (Konno et al., 2000; Nitao and Reisler, 2000) show that bonding of ATP causes melting of the SH1–SH2 helix, crucial for myosin head rigidity (see Fig. 5.24). Several flexible surface loops, important for the attachment, are not seen at all in X-ray diffraction (Rayment et al., 1993; Dominigues et al., 1998; Houdusse, 2000). Fluorescence polarization data show an essential increase in the dispersion of the lever-arm tilt angle relative to the actin filament axis during the change from rigor to relaxed physiological states of specially prepared muscle (Corrie et al., 1999). A local,

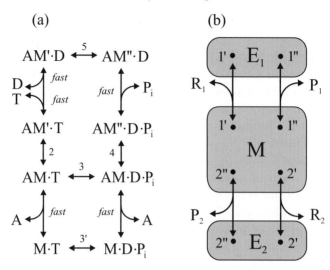

Fig. 9.11. Lymn–Taylor–Eisenberg kinetic model of the chemomechanical cycle of the actomyosin motor. (**a**) The version of Ma and Taylor (1994) with the three distinguished conformational states of the myosin head: open M′, closed M″, and detached M (see Sect. 5.4). A denotes the actin filament, T, D and P_i stand for ATP, ADP and inorganic phosphate, respectively. The original labeling of the reaction steps used by these authors is indicated. The values of particular rate constants are $k_{+2} = 1.8\ k_{-2} \geq 1000\ \text{s}^{-1}$, $k_{+3} \approx k_{-3} \leq 150\ \text{s}^{-1}$, $k_{+3'} = 150\ \text{s}^{-1}$, $k_{-3'} < 15\ \text{s}^{-1}$, $k_{+4} = 2.2\ k_{-4} = 140\ \text{s}^{-1}$, and $k_{+5} = 500\ \text{s}^{-1}$ (see also the compilation by Howard, 2001). (**b**) The version with a quasi-continuum of conformational substates of myosin considered in the present paper. The multitudes of conformational transitions within E_1 (the myosin–ADP complex strongly attached to the actin filament), M (the weakly attached myosin–ATP or ADP·P_i complex) and E_2 (the same complex detached from the actin filament) are represented by *shaded boxes*. R_1 = ATP, P_1 = P_i whereas R_2 and P_2 denote the actin filament before and after translation by a unit step, respectively. All bimolecular reactions are assumed to be gated, i.e., they take place only in certain conformational substates of the myosin head. The distinguished conformational substates composing the gates are labeled 1′, 1″, 2′ and 2″ (see Fig. 5.25)

internal conformational disorder on the nanosecond time scale was observed when analyzing the hyperfine splitting of the EPR signal from a nitro-oxide spin label (Ostap et al., 1995). A global, orientational disorder of the catalytic domain relative to both the actin filament and the lever-arm domain was observed on the microsecond time scale using saturation transfer EPR (Berger and Thomas, 1994; Adhikari et al., 1997) and detection of fluorescence polarization from a single molecule (Warshaw et al., 1998). Orientational disorder was observed on the millisecond time scale by parallel studies of stopped-flow fluorescence and time-resolved electron cryo-microscopy (Walker et al., 1999). According to many investigators (Berger and Thomas, 1994; Baker et al., 1998; Walker et al., 1999), the power stroke proceeds after a conformational disorder-to-order transition.

A consequence of all these observations is the need to replace a few distinguished conformational substates in the scheme of Fig. 9.11a by a quasi-continuum of conformational substates, in the same manner as in the case of a single enzymatic reaction in Sect. 9.3. Figure 9.11b shows an extended version of the Lymn–Taylor–Eisenberg model we consider. Shaded boxes represent the multitudes of conformational substates and transitions within the three main states of the motor: E_1 (the myosin–ADP complex strongly attached to the actin filament), M (the weakly-attached myosin–ATP or ADP·P_i complex), and E_2 (the latter complex detached from the actin filament). All binding–rebinding reactions are assumed to be gated, i.e., they take place only in certain distinguished conformational substates. These substates are supposedly similar to those presented in Fig. 5.25, from which we have taken the notation.

Assuming the ADP concentration to be held fixed, ATP hydrolysis can effectively be treated as a unimolecular reaction $R_1 \leftrightarrow P_1$, where R_1 is ATP and P_1 is the inorganic phosphate P_i. We have already argued that the physical motion itself can be treated as a unimolecular reaction $R_2 \leftrightarrow P_2$, where R_2 and P_2 denote the actin filament non-translated and translated by one step, respectively.

9.5 Flux–Force Dependence for the Actomyosin Motor

As discussed above, the motor is formally considered as a chemochemical machine that enzymatically couples the two unimolecular reactions: the free-energy-donating reaction 1 and the free-energy-accepting reaction 2. The input and output fluxes J_i ($i = 1$ and 2, respectively) and the conjugate thermodynamic forces A_i are defined as in Sects. 7.2 and 9.3:

$$J_i = \frac{d[P_i]/dt}{[E]_0} , \qquad [E]_0 \equiv [E_1] + [M] + [E_2] , \qquad (9.16)$$

and

$$\beta A_i = \ln K_i \frac{[R_i]}{[P_i]} , \qquad K_i \equiv \frac{[P_i]^{eq}}{[R_i]^{eq}} . \qquad (9.17)$$

Free energy transduction is realized if the product $J_2 A_2$ representing the output power is negative. In the absence of reaction 1, reaction 2 proceeds from P_2 to R_2 and can be driven against the conjugate force A_2, provided that reaction 1 occurs.

For the scheme presented in Fig. 9.11b, the flux–force dependence for the two coupled reactions has a general functional form similar to (9.4) (Kurzyński and Chełminiak, 2003):

$$J_i = \frac{1 - e^{-\beta(A_i - A_i^{\mathrm{st}})}}{J_{+i}^{-1} + J_{-i}^{-1} e^{-\beta(A_i - A_i^{\mathrm{st}})} + J_{0i}^{-1}(K_i + e^{\beta A_i})^{-1}} , \qquad (9.18)$$

where $i = 1, 2$. However, the parameters J_{+i}, J_{-i} and J_{0i} now depend on the other force. As in Sect. 8.5, A_i^{st} have the meaning of *stalling forces* for which the fluxes J_i vanish: $J_i(A_i^{\mathrm{st}}) = 0$. The flux J_1 is always positive, i.e., the actomyosin motor cannot work in the opposite direction and reconstruct ATP from ADP and P_i, whereas the force A_2 is always negative, i.e., the myofibril can only be stretched by an external load, as one can pull but not push with a myosin tail (see Figs. 5.23a and b). As noted earlier, energy transduction takes place if the flux J_2 has opposite sign to its conjugate force A_2. From (9.18), it follows that this condition holds when A_2 lies in the range between A_2^{st} and 0.

In the present case, the lack of partial thermodynamic equilibrium between the conformational substates of the enzyme results in the need to replace conventional reaction rate constants by various mean first-passage times between the distinguished conformational substates forming the gates for the component reactions. Unfortunately, the general expressions obtained for the parameters J_{+i}, J_{-i}, J_{0i} and A_i^{st} are complex and not perspicuous (Kurzyński and Chełminiak, 2003). Serious simplifications arise from the assumptions that both the reactions are practically irreversible and values of the corresponding reaction constants differ considerably from unity:

$$1 \le e^{\beta A_1} \ll K_1 \qquad (9.19)$$

and

$$K_2 \ll e^{\beta A_2} \le 1 . \qquad (9.20)$$

These assumptions are well satisfied for the actomyosin motor. Indeed, the equilibrium constant for the complete ATP hydrolysis reaction is

$$\frac{[\mathrm{ADP}]^{\mathrm{eq}}[\mathrm{P_i}]^{\mathrm{eq}}}{[\mathrm{ATP}]^{\mathrm{eq}}} \approx e^{30 \ \mathrm{kJ}/RT} \ \mathrm{M} \approx 2 \times 10^5 \ \mathrm{M} .$$

In physiological and most experimental conditions, in the presence of the creatine phosphate/creatine kinase system, the ADP concentration we assume for $[\text{ADP}]^{\text{eq}}$ is less than 20 M $= 2 \times 10^{-5}$ M (Stryer et al., 2002, Chap. 14), whence

$$K_1 \equiv \frac{[\text{P}_{\text{i}}]^{\text{eq}}}{[\text{ATP}]^{\text{eq}}} \geq 10^{10} \gg 1 \; .$$

Creatinine phosphate acts as a buffer and holds both the ADP and ATP concentrations constant (the latter is of the order of a few mM). The positivity of A_1 is a consequence of the fact that the concentration of P_{i} is never 10 orders of magnitude lower than the concentration of ATP. A low value of K_2 expresses the fact that, in chemical equilibrium, the myosin heads remain strongly attached to the myosin filament and no reasonable external load can stretch the muscle. The second inequality (9.20) corresponds to the already mentioned statement that one can pull but not push with a myosin tail.

Under the assumptions (9.19) and (9.20), the flux–force dependences (9.18) get the simpler form (8.36), but the full expressions for the parameters involved are still rather complicated. However, the additional assumption that the myofibril is able to sustain a high load under the condition of a sufficiently high ATP concentration, i.e.,

$$\beta A_2^{\text{st}} \ll -1 \; , \qquad \text{for} \quad \beta A_1 \gg 1 \; , \tag{9.21}$$

simplifies the expression for the stalling force A_2^{st} sufficiently to be worth quoting:

$$\beta A_2^{\text{st}} = \ln \frac{e^{-\beta A_1} + c_1}{1 + c_1} \; , \tag{9.22}$$

where

$$c_1 \equiv \frac{(k_{+1''}^{\text{eq}})^{-1} + [\text{M}]^{\text{eq}}/[\text{E}_1]^{\text{eq}}\, \tau_{\text{E}_1}(1'' \leftrightarrow 1') + \tau_{\text{M}}(1'' \leftrightarrow \{1', 2'\})}{(k_{-1'}^{\text{eq}})^{-1} + \tau_{\text{M}}(1'' \leftrightarrow 1')} \; . \tag{9.23}$$

As in Equations (9.6) to (9.8), the quantities k^{eq} are the equilibrium (transition state theory) rate constants for the substrate dissociation reactions taking place between the specified gates within M and E_1 (see Fig. 9.11b). The sign $+$ or $-$ is appended depending whether a given reaction proceeds in a counterclockwise or a clockwise direction, respectively. The quantity $\tau_{\text{E}_1}(1'' \leftrightarrow 1')$ denotes the mean first-passage time in E_1 from substate $1''$ to $1'$ *and back*:

$$\tau_{\text{E}_1}(1'' \leftrightarrow 1') \equiv \tau_{\text{E}_1}(1'' \rightarrow 1') + \tau_{\text{E}_1}(1' \rightarrow 1'') \; , \tag{9.24}$$

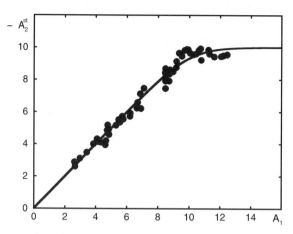

Fig. 9.12. Fit of (9.22) to the data of Paté et al. (1998), Fig. 6. We have assumed that the concentration of inorganic phosphate P_i directly determines the force A_1 in k_BT units (the ratio of ATP to ADP concentrations is held constant). The maximum negative stalling force $-A_2^{st}$ in k_BT units was fitted to be 10.0. After Kurzyński and Chełminiak (2003)

and similarly for the other quantities of this type further on. The quantity $\tau_M(1'' \leftrightarrow \{1', 2'\})$ denotes the mean first-passage time in M from substate $1''$ to $1'$ *or* $2'$ and back, and similarly for the other quantities of this type. Note that relations like the following hold (Kurzyński and Chełminiak, 2003):

$$\tau_M(1'' \leftrightarrow \{1', 2'\}) + \tau_M(1' \leftrightarrow \{1'', 2'\}) = \tau_M(1'' \leftrightarrow 1') . \qquad (9.25)$$

Figure 9.12 shows how a function of the form (9.22) fits the experimental data of Paté et al. (1998). The fitted value of the parameter $c_1 = 0.45 \times 10^{-4}$ corresponds to the maximum negative value of -10.0 for the stalling force A_2^{st} in k_BT units. It is really much smaller than -1, in agreement with the assumption (9.21). Note that, in the linear range, the negative stalling force $-A_2^{st}$ is to high accuracy equal to the force A_1, which implies the highest efficiency of stalling [see (8.40)]. All the experimental data in Fig. 9.12 are for $\beta A_1 \gg 1$, and later we shall therefore assume that reaction 1 is very far from equilibrium.

Very far from chemical equilibrium, for $\beta A_1 \gg 1$, the flux J_1 given by (9.18) for $i = 1$ assumes its asymptotic value

$$J_1 = J_{+1} . \qquad (9.26)$$

To simplify also the flux–force dependence $J_2(A_2)$, we assume it to be convex, which is actually the case for the actomyosin motor (Woledge et al., 1985; Howard, 2001). The convexity condition is

$$J_{+2}^{-1} \ll J_{02}^{-1} + J_{-2}^{-1} e^{\beta A_2^{\text{st}}} , \qquad (9.27)$$

and hence,

$$J_2(A_2) = \frac{e^{\beta A_2} - e^{\beta A_2^{\text{st}}}}{J_{02}^{-1} + J_{-2}^{-1} e^{\beta A_2^{\text{st}}}} = J_2(0) \frac{e^{\beta A_2} - e^{\beta A_2^{\text{st}}}}{1 - e^{\beta A_2^{\text{st}}}} . \qquad (9.28)$$

A similar relation was obtained for a simpler model by Qian (2000). The flux J_2 is proportional to the mean velocity of the myosin head along the actin filament and the force A_2 is proportional to the load. Since our description of the motion is in terms of an effective reaction, we consider only the dimensionless quantities $J_2(A_2)/J_2(0)$ and $-A_2/A_2^{\text{st}}$.

The function (9.28) describes experimental behavior as well as A.V. Hill's conventional hyperbolic dependence (Woledge et al., 1985). Figure 9.13 shows how it fits the data of He et al. (2000). The fitted values of 4.8 and 7.9 for the negative stalling force $-\beta A_2^{\text{st}}$ in $k_B T$ units are comparable to the maximum value of 10.0 determined for another sample from the $A_2^{\text{st}}(A_1)$ dependence (see Fig. 9.11).

Far from chemical equilibrium, for $\beta A_1 \gg 1$, assuming (9.19) and (9.20) and neglecting the parameter c_1, i.e., the exponential $e^{\beta A_2^{\text{st}}}$, the expression for the ratio of the two fluxes (the degree of coupling) also simplifies:

$$\frac{J_2}{J_1} \equiv \epsilon = \frac{c}{1 + c_2(e^{-\beta A_2} - 1)} , \qquad (9.29)$$

where

$$c_2 \equiv \frac{(k_{-2'}^{\text{eq}})^{-1} + \tau_{\text{M}}(2' \leftrightarrow \{1'', 2''\})}{(k_{+2''}^{\text{eq}})^{-1} + [\text{M}]^{\text{eq}}/[\text{E}_2]^{\text{eq}} \, \tau_{\text{E}_2}(2'' \leftrightarrow 2') + (k_{-2'}^{\text{eq}})^{-1} + \tau_{\text{M}}(2' \leftrightarrow 2'')} , \qquad (9.30)$$

and

$$c \equiv \frac{\tau_{\text{M}}(1' \leftrightarrow 1'')}{(k_{+2''}^{\text{eq}})^{-1} + [\text{M}]^{\text{eq}}/[\text{E}_2]^{\text{eq}} \, \tau_{\text{E}_2}(2'' \leftrightarrow 2') + (k_{-2'}^{\text{eq}})^{-1} + \tau_{\text{M}}(2' \leftrightarrow 2'')} . \qquad (9.31)$$

Neglecting $e^{\beta A_2^{\text{st}}}$ in (9.28), the degree of coupling (9.29) states a linear dependence between the rate of ATP utilization and the rate of muscle shortening:

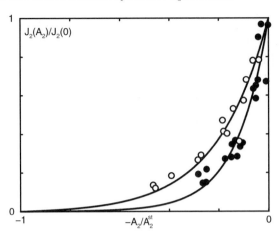

Fig. 9.13. Fit of (9.28) to the data of He et al. (2000), Fig. 3, for the sarcomere shortening velocity. *Black circles* correspond to slow fibers and *white circles* to fast 2A fibers. The fitted values of the negative stalling force $-\beta A_2^{\mathrm{st}}$ in $k_{\mathrm{B}}T$ units were found to be 4.8 and 7.9, respectively. After Kurzyński and Chełminiak (2003)

$$J_1 \propto 1 + \frac{1 - c_2}{c_2} \frac{J_2(A_2)}{J_2(0)} \ . \tag{9.32}$$

Figure 9.14 shows how this dependence fits the data of He et al. (2000). The parameter c_2, with value approximately equal to $1/3$, appears to control the *Fenn effect*, a decrease in the rate of ATP consumption when the muscle shortening rate is reduced (Woledge et al., 1985).

Assuming in addition that the reciprocal equilibrium rate constants $(k^{\mathrm{eq}})^{-1}$ are negligibly small, i.e., that the binding–rebinding reactions are much faster than most conformational transitions (see the classical kinetic model in Fig. 9.11a), the flux (9.28) can be rewritten in terms of a simple Michaelis–Menten formula (Kurzyński and Chełminiak, 2003):

$$J_2 = \frac{e^{\beta A_2}}{\tau_{\mathrm{E}_2}(2' \to 2'')} \frac{[\mathrm{ATP}]}{K + [\mathrm{ATP}]} \ , \tag{9.33}$$

with an apparent dissociation constant

$$K = \frac{\tau_{\mathrm{E}_2}(2'' \leftrightarrow 2')}{\tau_{\mathrm{E}_2}(2' \to 2'')} \frac{[\mathrm{E}_1]^{\mathrm{eq}}}{[\mathrm{E}_2]^{\mathrm{eq}}} [\mathrm{ATP}]^{\mathrm{eq}} \ . \tag{9.34}$$

In the above, we set $[\mathrm{R}_1] = [\mathrm{ATP}]$. Because the ATP hydrolysis process usually proceeds for nonsaturating values of the ATP concentration,

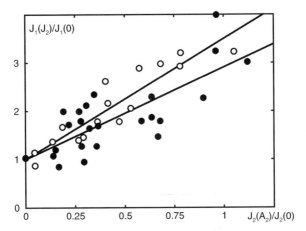

Fig. 9.14. Fit of (9.32) to the data of He et al. (2000), Fig. 4A. *Black circles* correspond to slow fibers and *white circles* to fast 2A fibers. The fitted values of the parameter c_2 were found to equal 0.35 and 0.30 for the slow and the fast fibers, respectively. After Kurzyński and Chełminiak (2003)

[ATP] $\ll K$, the flux (9.33) is in fact inversely proportional to K, and hence to $\tau_{E_2}(2'' \leftrightarrow 2')$ rather than to $\tau_{E_2}(2' \to 2'')$.

As the degree of coupling (9.29) does not depend on the ATP concentration, both J_1 and J_2 should show the same Michaelis–Menten-type dependence on this concentration. Such a dependence has been recorded experimentally (Ma and Taylor, 1994; Lionne et al., 1996; Amitani et al., 2001) with comparable values of the constant K of the order of 10^{-5} M to 10^{-4} M. It is difficult to discuss whether the values of K for the ATPase coincide with those for the motion since, in particular experimental conditions, various values of the parameters occurring in (9.34) were assumed.

The degree of coupling (9.29) is a ratio of two factors. The denominator, tending to unity as the load approaches zero, determines the slippage of the two reaction cycles. The numerator c can be accounted for as a *transmission coefficient*. It represents the mean number of steps the myosin head travels without slippage per ATP molecule hydrolyzed. Until recently, it was more or less commonly accepted that only one step could be made per ATP molecule consumed (the *tight coupling* hypothesis, Howard, 2001). This seemed to be well grounded on the results of single myosin molecule motility assays (Finer et al., 1994). However, new instrumentation with higher resolution convincingly indicates that each movement consists of several shorter regular steps. The step size depends on the geometry of the experiment and can equal either 5.5 nm, the actin molecule diameter (Kitamura et al., 1999), or 2.7 nm, the monomeric repeat along the actin filament (Liu and Pollack, 2004). As the mean distance of a single movement in the motility assays equals about 10 nm (Finer et al., 1994; Kitamura et al., 1999), some 2 to 5 steps are made on average per ATP molecule consumed. A two-step movement per ATPase cycle was observed in the single molecule motility assay for an unconventional myosin I (Veigel et al., 1999).

Presumably, this sliding through several steps per ATP molecule hydrolyzed also takes place in an assembly of actomyosin motors composing a myofibril. Such a *loose coupling* hypothesis has been assumed in order to explain various experimental findings, mainly using non-steady techniques (Brenner, 1991; Cooke et al., 1994; Higuchi and Goldman, 1995; Piazzessi and Lombardi, 1995). It could also explain what are at first sight rather strange experimental findings by Lionne et al. (1996), in which the total ATP consumed during unloaded myofibril contraction is independent of the initial sarcomere length and the myofibrillar ATPase does not vary in time.

Recall from Sect. 9.4 that the contraction velocity v (flux J_2 multiplied by step size d) of the unloaded myofibril does not depend on the initial overlap between the filaments. On the other hand, the ATPase flux J_1 increases with the initial overlap (Lionne et al., 1996). A consequence of these facts and (9.29) is that the product of the transmission coefficient c with the step size d should decrease with the initial overlap. The product cd can be interpreted as the distance traveled by the myosin head per ATP molecule consumed and is referred to as the *sliding* or *interaction distance*. The sliding distance evaluated by Lionne et al. (1996) actually varies from ~ 270 nm for a small initial sarcomere length (large overlap) to ~ 600 nm for a large initial sarcomere length (small overlap). The order of these values agrees with the order of values defined by the ratio of the shortening velocity per half sarcomere of the unloaded filament (a few ms^{-1}) and the turnover number for the ATPase (a few tens s^{-1}), compiled by Howard (2001). It should be noted that the sliding distance value depends sensitively on temperature (Candau et al., 2003).

From (9.33), (9.34) and the constancy of the sliding velocity, it follows that the step size d, if variable, should be proportional to the time $\tau_{E_2}(2'' \leftrightarrow 2')$. In addition, the transmission coefficient (9.31) depends on that time. Let us emphasize that the model presented in Fig. 9.11b and considered here concerns the kinetics of a *single* actomyosin motor. The time $\tau_{E_2}(2'' \leftrightarrow 2')$ is the only parameter in the model that contains any information about the force the remaining motors exert on the common actin filament,[1] and it is natural for this time to depend on the filament overlap. We have also to remember that the step size d should be a multiple of the actin filament period 36 nm, which can be considered as an elementary step distance for the myofibril. Conventionally, assuming the tight-coupling hypothesis, the sliding distance value was interpreted in terms of the *duty ratio*, the fraction of the myosin heads performing the work, i.e., the power stroke (Howard, 2001). Equations (9.29) and (9.31) suggest that a more correct interpretation is in terms of the transmission coefficient.

To conclude, there is stronger and stronger experimental evidence that the actomyosin motor makes several steps per ATP molecule hydrolyzed. Equation (9.31) is able to explain this fact, provided that the mean first-passage time $\tau_M(1'' \leftrightarrow 1')$ is several times longer than the sum of times occurring in the denominator. The long mean first-passage time $\tau_M(1'' \leftrightarrow 1')$ can be explained by the necessity of the already discussed melting and recrystallization of the SH1–SH2 helix during a transition from the substate $1''$ to $1'$ and back within the state M. The relatively short mean first-passage time $\tau_M(2' \leftrightarrow \{1'', 2''\})$, shorter than $\tau_M(2' \leftrightarrow 2'')$, is the reason why, before coming back to the strongly attached state E_1, the myosin head can stochastically undergo several mechanical cycles through the detached state E_2. A slight generalization of the model can be considered in which the complete refolding of the motor molecule is achieved only after several such cycles (Terrada et al., 2002).

This result is very important. It can help to answer a still open question about the origin of the free energy for the directed motor motion (Howard, 2001): one-step conformational changes or a thermal ratchet-type mechanism? The power stroke, being the source of the force the myosin head exerts on the actin filament, takes place in the strongly-attached state E_1. A diffusive motion of the myosin head with respect to the actin filament takes place in the detached state E_2. Equation (9.29) implies that no details of the dynamics in E_1 are important for the action of the actomyosin motor. The dynamics in M and, to a lesser extent, in E_2 are essential. This result is vaguely reminiscent of A.F. Huxley's (1957) idea of a thermal ratchet including the intramolecular dynamics of the myosin head (see also Cordova, Ermentrout and Oster, 1992).

---

[1] This way of describing the properties of an assembly of elements in terms of the single-element property is called the molecular field approximation in theoretical physics.

The stochastic model of the actomyosin motor action presented here is consistent with all presently available experimental data. However, it is only phenomenology. A challenge for theorists is to fill the interiors of the shaded boxes in Fig. 9.11b with simple but adequate models of conformational transition dynamics that afford possibilities for calculating appropriate mean first-passage times (see Appendixes B and D). The dynamics in both E_1 and E_2 can be reasonably approximated by driven one-dimensional diffusion (Hill, 1989; Duke, 1999; Astumian, 1997; Jülicher et al., 1997), described in the continuous limit by a simple Langevin equation like (9.11). However, the dynamics in M is certainly not one-dimensional diffusion. Presumably, it should involve diffusion on fractal lattices (see Appendix D.4 and Chełminiak and Kurzyński, 2004) which has been demonstrated to be a reasonable model of the protein's intramolecular dynamics (Kurzyński et al., 1998). Against such a dynamical background, transitory detachments from the actin filaments resulting in no translational motion, and also the melting of the SH1–SH2 helix, should be described. Here, experience with protein folding dynamics can be helpful (Lee et al., 2003). An open question concerns the mechanism of energy transfer between the binding and catalytic centers, which can proceed either as a sequence of local conformational transitions (Böckmann and Grubmüller, 2002), or as a process analogous to non-adiabatic electron or proton transfer (Cruzeiro-Hansson and Takeno, 1997; see also Appendix D.6).

Models of driven and biased one-dimensional diffusion (translational or rotational) have been proposed for many molecular biological motors. Let us mention RNA polymerase (Jülicher and Bruinsma, 1998; Wang et al., 1998), kinesin (Peskin and Oster, 1995; Duke and Leibler, 1996), ATP synthase (Wang and Oster, 1998; Elston, Wang and Oster, 1998) and, in general, arbitrary F- or V-type ATPase pumps (Grabe, Wang and Oster, 2000). The main problem solved in the cited papers was to describe the coordinated action of several subunits: two heads in the case of the kinesin, three β units in the case of F_1 or V_1 portions, and up to ten protonatable entities in the case of F_o or V_o rotors of F- or V-type ATPase pumps (see Chap. 5). An intriguing question concerns the role of the internal dynamics of the component protein molecules in the action of these motors.

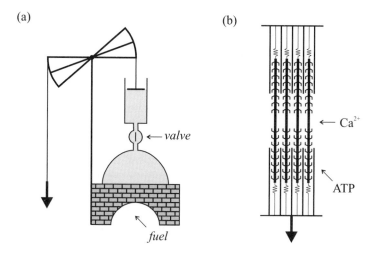

Fig. 9.15. (a) Newcomen's heat engine from 1712. Steam from the boiler moves the piston up the cylinder. Then, after closing the valve, the cylinder is cooled, the steam is precipitated, and atmospheric pressure moves the piston downwards, raising the load. (**b**) A muscle consists of many periodically repeated structures, the sarcomeres. They are composed of the thick myosin filaments along which the thin actin filaments can slide. After activation by calcium ions, the muscle contracts raising the load. This motion, is the result of many Brownian motions of the myosin heads, directed due to ATP hydrolysis

9.6 Biological Molecular Machines as Biased Maxwell Demons

A nontrivial challenge for contemporary statistical physics is to find an adequate conceptual apparatus to describe the action of biological molecular machines. Functionally, there is no essential difference between the muscle and, e.g., the steam engine. They both perform macroscopic work at the expense of certain chemical reactions, either ATP hydrolysis or fuel burning. However, the steam engine has macroscopic structure (Fig. 9.15a), whereas the muscle is organized on a microscopic or, more precisely, a mesoscopic level (Fig. 9.15b).

Viewed macroscopically, any biological machine is an appropriately organized assembly of enzymes. Similarly to the suspension particles in the solution observed by Brown, macromolecular enzymes playing the role of the component molecular machines move about and, in particular, change their chemical state due to thermal fluctuations. On a short time scale, energy is 'borrowed' from and 'returned' to the heat bath. That the stochastic motion of biological machines is not purely random results from their highly organized structure and the constant input of free energy, mainly due to the hydrolysis of ATP. A large number of ingenious techniques have been designed to enable precise observation of the behavior of single molecular machines. For molecular pumps and ion channels, one such a technique is the patch-clamp method (Sackmann and Naher, 1995). In the case of molecular motors, there have been various motility assays (Mehta et al., 1999; Ishijima and Yanagida, 2001). All these observations reveal the stochastic nature of the behavior exhibited by biological molecular machines. We must emphasize the essential difference between the stochastic behavior of biological macromolecules and small molecules that are the subject of conventional physical chemistry. Small molecules fluctuate only between a small number of discrete chemical states (the microscopic dynamics of translational, rotational and internal degrees of freedom is from that perspective purely random), whereas the stochastic dynamics of large biomolecules is much more complex. Biological macromolecules have a *mesoscopic* level of organization which is lacking in the case of small molecules.

In 1871, James Clerk Maxwell, pondering the foundations of thermodynamics, contemplated the functioning of a hypothetical being that could observe the velocities of individual gas molecules moving about in a container. The special feature of the container would be a partition with an opening that could be covered by a latch (Fig. 9.16a). This being, referred to in the literature as Maxwell's demon, would be in charge of closing and opening the hole in the partition, allowing only sufficiently fast particles to move from right to left and only sufficiently slow ones to pass from left to right. Over time, this would of course result in a temperature increase in the left part of the container and a decrease in the right part. The temperature gradient thereby created clearly contradicts the second law of thermodynamics due to the work that can be performed in the process just by thermal fluctuations in a gas at thermodynamic equilibrium.

Fewer than 100 years later, another great physicist, Richard Feynman (1966) presented this problem in the more provocative manner shown in Fig. 9.16b. A mechanical wheel and a ratchet and pawl are mounted on a common axis which has vanes attached to it. The surfaces of the vanes are bombarded with gas molecules on both sides. The presence of the pawl prevents the ratchet from rotating in one of the directions. As a result the kinetic energy of gas fluctuations is transformed into the rotational kinetic energy of the ratchet's unidi-

(a)

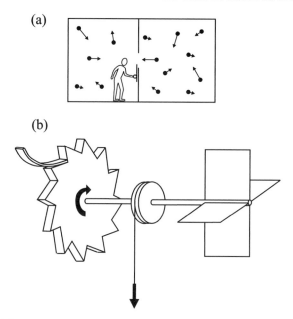

(b)

Fig. 9.16. (a) Maxwell's demon as originally proposed. (b) Feynman's version of the Maxwell demon problem: the ratchet and pawl machine

rectional motion around the axis. This can, in principle, be utilized to raise a weight against the force of gravity.

The logical error in both arguments is made when we consider direct interaction of microscopic systems (gas molecules) with macroscopic ones (hole opened in the partition, pawl in the wheel). However, the pawl device cannot possibly react to the collision of a single molecule unless it, too, is a microscopic object subject to the same types of thermal fluctuations. Random bending of the pawl assists with the rotation of the wheel in the opposite direction. Similarly, with the latch controlled by Maxwell's demon: to measure the speed of individual molecules, it must be in contact via a physical interaction. To react to such an interaction, the observing device must be microscopic but then, of course, it will rapidly be brought to thermal equilibrium as a result of interactions with chaotically moving molecules around it. Hence, it will behave chaotically, in the same way as an average gas molecule does, and consequently, no net macroscopic force will be generated. These unfavorable fluctuations can be reduced by lowering the temperature of the pawl or by freezing the head of Maxwell's demon. In general, this can be done by reducing the entropy or supplying free energy. When this is done the contradiction with the second law of thermodynamics is automatically removed and these mesoscopic machines will work according to the normal rules of behavior discussed earlier. The biological machines that are the centerpiece of this book provide an example of such biased Maxwell demons.

A Thermodynamic Supplement

A.1 Thermodynamics of Ideal Gases

For simple thermodynamic systems, the equations of state (discussed in Sect. 3.3) are of a special form (Callen, 1985; Kondepudi and Prigogine, 1999). The simple systems are spatially homogeneous and can always be divided into an arbitrary large number of identical subsystems. In the continuous limit, the additivity condition (3.3) applied to such a division implies that the entropy of a simple system must be a homogeneous function of its arguments:

$$S(\lambda E, \lambda X_1, \ldots, \lambda X_n) = \lambda S(E, X_1, \ldots, X_n) , \tag{A.1}$$

for arbitrary λ. A similar relation holds for the energy:

$$E(\lambda S, \lambda X_1, \ldots, \lambda X_n) = \lambda E(S, X_1, \ldots, X_n) . \tag{A.2}$$

From the homogeneity (A.2) of the energy and from the definitions of temperature T and thermodynamic forces A_i, it follow that the latter quantities are *intensive*, i.e., independent of the size of the system:

$$
\begin{aligned}
T &= T(S, X_1, \ldots, X_n) = T(\lambda S, \lambda X_1, \ldots, \lambda X_n) , \\
A_i &= A_i(S, X_1, \ldots, X_n) = A_i(\lambda S, \lambda X_1, \ldots, \lambda X_n) ,
\end{aligned}
\tag{A.3}
$$

for $i = 1, \ldots, n$ and arbitrary λ. In particular, setting $\lambda = 1/X_n$, we find from it that $n + 1$ forces (including temperature) are functions of only n variables $S/X_n, X_1/X_n, \ldots, X_{n-1}/X_n$. Hence, all the forces and the temperature cannot be independent. They are linked by a universal relationship called the *Euler equation*:

$$E = TS - \sum_{i=1}^{n} A_i X_i . \tag{A.4}$$

The relationship (A.4) is easily derived by differentiating (A.2) with respect to λ, making use of the definitions (3.17) and (3.24), and finally setting $\lambda = 1$. The variables X_i/X_n like the forces A_i, $i = 1, \ldots, n-1$ are intensive. If X_n denotes the volume V, then we refer to them as *densities*. If X_n denotes the number of molecules (or moles) N, then we speak about *proper* thermodynamic variables.

An ideal gas of classical particles of one kind is a simple thermodynamic system with two thermodynamic degrees of freedom ($X_1 = V$, $X_2 = N$). Besides the temperature T, properties of the environment are determined by two thermodynamic forces: the pressure, $A_1 = P$, and (in open systems) the chemical potential, $A_2 = -\mu$. Energy is related to temperature by the equation

$$E = \frac{3}{2}Nk_\mathrm{B}T \ . \tag{A.5}$$

This is the first equation of state, which says that, for an ideal gas, energy is independent of the volume and the type of particles, and that each translational degree of freedom contributes a quantity $k_\mathrm{B}T/2$ to it, where k_B is the Boltzmann constant. The second equation of state relates the pressure to the volume:

$$PV = Nk_\mathrm{B}T \ . \tag{A.6}$$

This is known as the *Clapeyron equation* (see Fig. A.1a). A more complex equation for real gases which allows for two different phases to occur,[1] is represented by the three-dimensional diagram in Fig. A.1b.

From the Euler equation (A.4), which in the present case takes the form

$$E = TS - PV + \mu N \ , \tag{A.7}$$

and from (A.5) and (A.6), we obtain the third equation of state for the chemical potential:

$$\mu = \left(\frac{5}{2}k_\mathrm{B} - \frac{S}{N}\right)T \ . \tag{A.8}$$

The energy E occurs in the equation of state (A.5), whereas it is replaced by the entropy S in (A.8). The basic equation (A.2) or (A.5) relating the two quantities can be derived from the differential identity

$$\begin{aligned}
\mathrm{d}E(S, V, N) &= \left(\frac{\partial E}{\partial S}\right)_{V,N} \mathrm{d}S + \left(\frac{\partial E}{\partial V}\right)_{S,N} \mathrm{d}V + \left(\frac{\partial E}{\partial N}\right)_{S,V} \mathrm{d}N \\
&= T\mathrm{d}S - P\mathrm{d}V + \mu\mathrm{d}N \ ,
\end{aligned} \tag{A.9}$$

---

[1] Spontaneous division of a system into two or more parts called *phases* follows when the stability condition is broken and equations (A.3) are no longer uniquely reversible (see Appendix A.3).

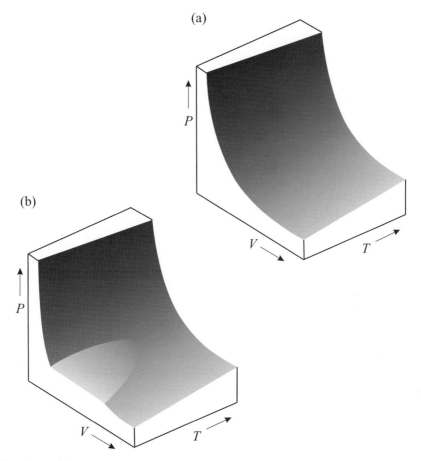

Fig. A.1. (a) Dependence of the pressure P on the volume V and temperature T for an ideal gas. (b) Dependence of P on V and T for a system with the gas–liquid phase change. A region of phase coexistence is shown, where the P–V dependence is not one-to-one. A singular point at which the phases are no longer distinguishable has very interesting physical properties and is referred to as the *critical point*

or, after a transformation,

$$\mathrm{d}S = \frac{1}{T}\mathrm{d}E + \frac{P}{T}\mathrm{d}V - \frac{\mu}{T}\mathrm{d}N \ . \tag{A.10}$$

Substituting (A.5), (A.6) and (A.8), and using the rule for differentiation of a quotient, viz.,

$$\mathrm{d}\left(\frac{S}{N}\right) = \frac{\mathrm{d}S}{N} - \frac{S\mathrm{d}N}{N^2} \ , \tag{A.11}$$

we obtain from (A.10)

$$d\left(\frac{S}{N}\right) = k_B\left(\frac{3}{2}\frac{dE}{E} + \frac{dV}{V} - \frac{5}{2}\frac{dN}{N}\right). \tag{A.12}$$

This relation is easy to integrate and results in the dependence

$$\frac{S}{N} = \frac{S_0}{N_0} + k_B \ln\left[\left(\frac{E}{E_0}\right)^{3/2}\left(\frac{V}{V_0}\right)\left(\frac{N}{N_0}\right)^{-5/2}\right], \tag{A.13}$$

where the subscript 0 characterizes the values of the corresponding variables in a certain reference system. Equation (A.13) can be rewritten in a form that explicitly displays the homogeneity of the system:

$$\frac{S}{N} = \frac{S_0}{N_0} + k_B \ln\left[\left(\frac{E}{N}\frac{N_0}{E_0}\right)^{3/2}\left(\frac{V}{N}\frac{N_0}{V_0}\right)\right]$$

$$= \frac{S_0}{N_0} + k_B \ln\left[\left(\frac{T}{T_0}\right)^{5/2}\left(\frac{P}{P_0}\right)^{-1}\right]. \tag{A.14}$$

Substituting (A.13) or (A.14) into (A.8), we obtain the equation of state for the chemical potential μ depending on the energy or the remaining thermodynamic forces instead of the entropy:

$$\frac{\mu}{T} = \frac{\mu_0}{T_0} - k_B \ln\left[\left(\frac{E}{E_0}\right)^{3/2}\left(\frac{V}{V_0}\right)\left(\frac{N}{N_0}\right)^{-5/2}\right]$$

$$= \frac{\mu_0}{T_0} - k_B \ln\left[\left(\frac{T}{T_0}\right)^{5/2}\left(\frac{P}{P_0}\right)^{-1}\right]. \tag{A.15}$$

As there are many applications in chemical kinetics, let us consider a *two-component* ideal gas, i.e., a homogeneous mixture of two different ideal gases of classical particles. This is a simple system with three thermodynamic degrees of freedom $X_1 = V$, $X_2 = N_1$ and $X_3 = N_2$. The forces conjugate to the numbers N_1 and N_2 of particles of the two different kinds are the chemical potentials μ_1 and μ_2. The equations of state (A.5) and (A.6) do not depend on the type of particle and can be rewritten as

$$E = \frac{3}{2}(N_1 + N_2)k_B T \equiv E_1 + E_2 \tag{A.16}$$

and

$$PV = (N_1 + N_2)k_B T \equiv (P_1 + P_2)V, \tag{A.17}$$

where energies and pressures of particular components are introduced explicitly. The Euler equation (A.4) includes the four terms

$$E = TS - PV + \mu_1 N_1 + \mu_2 N_2 , \qquad (A.18)$$

and can be considered as a sum of two Euler equations, i.e.,

$$E_i = TS_i - P_i V + \mu_i N_i , \qquad (A.19)$$

$i = 1, 2$, provided that S_i is interpreted as the entropy of the ith component when it *alone* occupies the *entire* volume V under the partial pressure

$$P_i = \frac{N_i}{N} P , \qquad N \equiv N_1 + N_2 . \qquad (A.20)$$

The formulas (A.19) are identical in form to (A.4), so that all the equations derived for the one-component ideal gas can be applied to the case of the two-component mixture. In this way, from (A.14) we obtain

$$\frac{S_i}{N_i} = \frac{S_0}{N_0} + k_B \ln \left[\left(\frac{T}{T_0} \right)^{5/2} \left(\frac{P_i}{P_0} \right)^{-1} \right] , \qquad (A.21)$$

and from (A.15),

$$\frac{\mu_i}{T} = \frac{\mu_0}{T_0} - k_B \ln \left[\left(\frac{T}{T_0} \right)^{5/2} \left(\frac{P_i}{P_0} \right)^{-1} \right] . \qquad (A.22)$$

When we use (A.20) to replace the partial pressures by the total pressure (only the latter is recorded in experiment), we obtain the simple relationships

$$\frac{S_i}{N_i} = \frac{S_i^\circ}{N} - k_B \ln \frac{N_i}{N} \qquad (A.23)$$

and

$$\mu_i = \mu_i^\circ + k_B T \ln \frac{N_i}{N} , \qquad (A.24)$$

where S_i° and μ_i° denote the values that the entropy and chemical potential would assume under given conditions (fixed T, P and N) for the pure ith component.

The second term in (A.23) is referred to as the *entropy of mixing*. It has the form of the negative Boltzmann constant multiplied by the logarithm of the probability of a random choice of the ith particle type. This does indeed correspond to the statistical definition of entropy discussed in Sect. 2.6. An important conclusion is that the expressions (A.23) and (A.24) for the entropy and chemical potential of a system of statistically independent molecules each of which occurs in two states also remain valid for inhomogeneous systems.

A.2 Legendre Transformations

Let us allow the temperature T to vary. From (3.52) and (3.34), a general expression results for the free energy change:

$$\Delta F = \Delta E - T\Delta S - S\Delta T = -S\Delta T - \sum_i A_i \Delta X_i \,. \qquad (A.25)$$

The form of this change indicates that the free energy is a function of the temperature T and the parameters X_i:

$$F = F(T, X_1, \ldots, X_n) \,, \qquad (A.26)$$

and that

$$S = -\left(\frac{\partial F}{\partial T}\right)_{X_1,\ldots,X_n} , \qquad A_i = -\left(\frac{\partial F}{\partial X_i}\right)_{T,\ldots} . \qquad (A.27)$$

The second equation (A.27) represents a definition of thermodynamic parameters much more practical than (3.24). It is easier to realize isothermal than isoentropic conditions.

The transition from the energy as a function of entropy $E = E(S)$ to the free energy as a function of temperature $F = F(T)$ can be given a simple geometrical interpretation (Fig. A.2). Since

$$\left(\frac{\partial^2 E}{\partial S^2}\right)_X > 0 \,, \qquad (A.28)$$

a curve that represents the dependence E of S is *concave* for each S. A curve whose second derivative has constant sign (and only such a curve) can be represented equivalently as a set of *points*

$$\bigl(S, E(S)\bigr) \qquad (A.29)$$

or as a set of *straight lines*

$$(T, E - TS) = \bigl(T, F(T)\bigr) \,. \qquad (A.30)$$

The first number in the brackets denotes the tangent of the inclination angle of a straight line tangential to the relevant curve, and the second number, the coordinate of the point at which this line crosses the vertical axis (Fig. A.2).

Formally, the transition from energy to free energy in thermodynamics, i.e.,

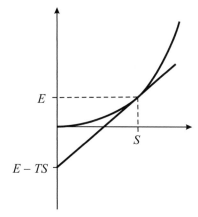

Fig. A.2. Legendre transformation from energy $E = E(S)$ to free energy $F = F(T)$

$$E(X, S) \rightarrow F(X, T) = E - TS , \qquad T \equiv \left(\frac{\partial E}{\partial S} \right)_X , \qquad \text{(A.31)}$$

is identical to the transition from the Lagrangian to the (minus) Hamiltonian function in mechanics, i.e.,

$$\mathcal{L}(q, \dot{q}) \rightarrow -\mathcal{H}(q, p) = \mathcal{L} - p\dot{q} , \qquad p \equiv \left(\frac{\partial \mathcal{L}}{\partial \dot{q}} \right)_q . \qquad \text{(A.32)}$$

Both transitions are examples of the *Legendre transformation*.

Most of the considerations in Chap. 3 concerned isothermal conditions, $T = $ const., since it is the temperature T rather than the entropy S that can be easily controlled during thermodynamic processes. Moreover, a thermodynamic force A often turns out to be more easily controlled than the conjugate thermodynamic variable X (isobaric conditions, $P = $ const., instead of $V = $ const. conditions; constant chemical potential conditions, $\mu = $ const., instead of $N = $ const. conditions; constant magnetic field conditions, $\boldsymbol{H} = $ const., instead of constant magnetization conditions, $\boldsymbol{M} = $ const., etc.). If there is no way of controlling a thermodynamic variable X, we cannot actually use the work related to a change in its value. The really *useful* work is that decreased by a component $-A\Delta X$:

$$W' = W + A\Delta X . \qquad \text{(A.33)}$$

Just as work W is related to changes in energy and free energy [(3.35) and (3.51)], the useful work W' is related to changes in *enthalpy*

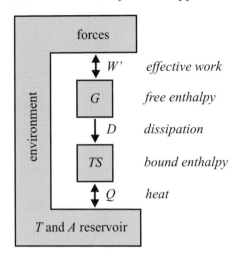

Fig. A.3. Dynamic and thermal subsystems of the thermodynamic system under $T = $ const. and $A = $ const. conditions, and their interaction with each other and with the environment

$$\Delta H = Q + W' = \Delta E + A\Delta X , \qquad (A.34)$$

and *free enthalpy*

$$\Delta G = W' - D = \Delta F + A\Delta X . \qquad (A.35)$$

Under $A = $ const. conditions, when the force A can be inserted under the increment operation Δ, we have

$$H = E + AX \qquad (A.36)$$

and

$$G = F + AX = E - TS + AX , \qquad (A.37)$$

whence

$$H = G + TS . \qquad (A.38)$$

Under $T = $ const. and $A = $ const. conditions, enthalpy transformations should be considered instead of energy transformations (Fig. A.3).

The quantity G is often called the *Gibbs potential* or the *Gibbs free energy* (to be distinguished from the free energy F, which is then referred to as the *Helmholtz free energy*). For simple thermodynamic systems with two thermodynamic degrees of freedom and $X_2 = N$, from the Euler equation (A.4) and from (A.37), a simple interpretation of the Gibbs potential follows as the chemical potential μ multiplied by the number of molecules N:

$$G = \mu N . \qquad (A.39)$$

According to (A.37), the free enthalpy (the Gibbs potential) G is the double Legendre transformation of the energy E. The function G is convenient to use as it depends on two thermodynamic parameters that are easy to control:

$$\Delta G = \Delta E - T\Delta S - S\Delta T + A\Delta X + X\Delta A = -S\Delta T + X\Delta A , \quad \text{(A.40)}$$

whence

$$G = G(T, A) \quad \text{(A.41)}$$

and

$$S = -\left(\frac{\partial G}{\partial T}\right)_A , \qquad X = \left(\frac{\partial G}{\partial A}\right)_T . \quad \text{(A.42)}$$

For simple thermodynamic systems, the third independent argument of the Gibbs potential is the number of molecules N. Only this argument is extensive, so it is no wonder that the Gibbs potential is proportional to the chemical potential:

$$G = N\mu = N\left(\frac{\partial E}{\partial N}\right)_{S,X} = N\left(\frac{\partial G}{\partial N}\right)_{T,A} . \quad \text{(A.43)}$$

Let us summarize all the basic thermodynamic equalities for the simple system with two thermodynamic degrees of freedom under the conditions $N = \text{const.}$:

$$E = E(S, X) , \qquad\qquad \Delta E = T\Delta S - A\Delta X , \quad \text{(A.44)}$$

$$F = E - TS = F(T, X) , \qquad \Delta F = -S\Delta T - A\Delta X , \quad \text{(A.45)}$$

$$H = E + AX = H(S, A) , \qquad \Delta H = T\Delta S + X\Delta A , \quad \text{(A.46)}$$

$$G = E - TS + AX = G(T, A) , \quad \Delta G = -S\Delta T + X\Delta A . \quad \text{(A.47)}$$

Comparing coefficients in the expressions for increments, it follows that:

$$T = \left(\frac{\partial E}{\partial S}\right)_X = \left(\frac{\partial H}{\partial S}\right)_A , \quad \text{(A.48)}$$

$$S = -\left(\frac{\partial F}{\partial T}\right)_X = -\left(\frac{\partial G}{\partial T}\right)_A , \quad \text{(A.49)}$$

$$A = -\left(\frac{\partial E}{\partial X}\right)_S = -\left(\frac{\partial F}{\partial X}\right)_T , \quad \text{(A.50)}$$

$$X = \left(\frac{\partial H}{\partial A}\right)_S = \left(\frac{\partial G}{\partial A}\right)_T . \quad \text{(A.51)}$$

Differentiating (A.48) to (A.51) with respect to the appropriate thermodynamic variables or forces and taking into account the commutativity of second derivatives, we obtain a variety of *Maxwell's relations*. For instance,

$$\left(\frac{\partial X}{\partial T}\right)_A = \frac{\partial^2 G}{\partial T \partial A} = \frac{\partial^2 G}{\partial A \partial T} = -\left(\frac{\partial S}{\partial A}\right)_T . \tag{A.52}$$

This relation gives a simple recipe for an indirect measurement of entropy changes. In particular, substituting $X = V$ and $A = P$, we obtain from it a useful relationship for the *thermal expansion coefficient*, namely,

$$\alpha_P \equiv \frac{1}{V}\left(\frac{\partial V}{\partial T}\right)_P = -\frac{1}{V}\left(\frac{\partial S}{\partial P}\right)_T . \tag{A.53}$$

All Legendre transforms of energy are called *thermodynamic potentials*. The complete $n+1$-tuple Legendre transform of energy is given by a generalization of (A.37):

$$I = F + \sum_{i=1}^{n} A_i X_i = E - TS + \sum_{i=1}^{n} A_i X_i . \tag{A.54}$$

The counterpart of (A.40) is

$$\Delta I = -S\Delta T + \sum_{i=1}^{n} X_i \Delta A_i , \tag{A.55}$$

from which it follows that the thermodynamic potential I is indeed a function of the temperature and n thermodynamic forces:

$$I = I(T, A_1, \ldots, A_n) , \tag{A.56}$$

and that

$$S = -\left(\frac{\partial I}{\partial T}\right)_{A_1,\ldots,A_n} , \qquad X_i = \left(\frac{\partial I}{\partial A_i}\right)_{T,\ldots} . \tag{A.57}$$

The commutativity of the second derivatives of the potential I leads to Maxwell's relations

$$\left(\frac{\partial X_i}{\partial A_j}\right)_{A_i,\ldots} = \left(\frac{\partial X_j}{\partial A_i}\right)_{A_j,\ldots} , \qquad \left(\frac{\partial X_i}{\partial T}\right)_{A_i,\ldots} = -\left(\frac{\partial S}{\partial A_i}\right)_{T,\ldots} , \tag{A.58}$$

just as the commutativity of the second derivatives of the energy E results in Maxwell's relations

$$\left(\frac{\partial A_i}{\partial X_j}\right)_{X_i,\ldots} = \left(\frac{\partial A_j}{\partial X_i}\right)_{X_j,\ldots} , \qquad \left(\frac{\partial A_i}{\partial S}\right)_{X_i,\ldots} = -\left(\frac{\partial T}{\partial X_i}\right)_{S,\ldots} . \tag{A.59}$$

A.3 Capacities and Susceptibilities. Thermodynamic Stability

The negative derivatives of thermodynamic variables with respect to the conjugate thermodynamic forces,

$$C \equiv - \left(\frac{\partial X}{\partial A} \right) \qquad \text{(A.60)}$$

are called *capacities*. An example is the *electric capacity*, which is the derivative of the electric charge $X = Q$ with respect to the electric potential $-A = \phi$ (see Table 3.1). For spatially homogeneous systems, it is convenient to introduce an intensive quantity, the *susceptibility*, equaling the capacity divided by volume:

$$\chi = \frac{C}{V} . \qquad \text{(A.61)}$$

Particular examples are the *compressibility coefficient*

$$\kappa = -\frac{1}{V} \frac{\partial V}{\partial P} , \qquad \text{(A.62)}$$

the *magnetic susceptibility*

$$\chi_m = \frac{\partial \boldsymbol{M}}{\partial \boldsymbol{H}} , \qquad \text{(A.63)}$$

or the *electric susceptibility*

$$\chi_e = \frac{\partial \boldsymbol{P}}{\partial \boldsymbol{E}} . \qquad \text{(A.64)}$$

The magnetic and electric susceptibilities are in general tensor quantities, since the directions of magnetization or polarization need not be the same as the directions of the magnetic or electric fields.

Capacities and susceptibilities are defined under various thermodynamic conditions that relate them to various thermodynamic potentials. For instance, the *isothermal* capacity C^T at constant temperature T is related to the second derivatives of the free energy $G(T, A)$ or $F(T, X)$. Indeed, according to the second relationship of (A.51) and the one-to-one dependence of the thermodynamic variable X on the force A, we obtain from the second relationship of (A.50)

$$C^T = - \left(\frac{\partial X}{\partial A} \right)_T = - \left(\frac{\partial^2 G}{\partial A^2} \right)_T = - \left(\frac{\partial A}{\partial X} \right)_T^{-1} = \left(\frac{\partial^2 F}{\partial X^2} \right)_T^{-1} . \qquad \text{(A.65)}$$

The entropy S is a convex function of X, whereas the energy E is concave. Moreover, because the temperature T is positive, the free energy $F(T, X) = E(S, X) - TS(E, X)$ is a concave function:

$$\frac{\partial^2 F}{\partial X^2} > 0 . \tag{A.66}$$

As a consequence, the isothermal capacity (or susceptibility) is always positive:

$$C^T > 0 . \tag{A.67}$$

A similar inequality can be proved for the capacity under any other thermodynamic condition: constancy of the entropy S, or the constancy of some thermodynamic variable X' or the conjugate force A'.

The counterpart of the capacity in the case when the variable X is replaced by energy E or enthalpy H and the negative force $-A$ by temperature T is the *heat capacity*. In more detail, the heat capacity at constant value of a certain thermodynamic variable X or the conjugate force A is defined as the amount of heat needed to change the temperature by one unit (e.g., 1 K):

$$C^X = \left(\frac{Q}{\Delta T} \right)_X , \qquad C^A = \left(\frac{Q}{\Delta T} \right)_A . \tag{A.68}$$

For a reversible process, $Q = T\Delta S$ and, after the limit transition, the heat capacities (A.68) can be rewritten as derivatives of some function of state:

$$C^X = T \left(\frac{\partial S}{\partial T} \right)_X = T \left(\frac{\partial T}{\partial S} \right)_X^{-1} = T \left(\frac{\partial^2 E}{\partial S^2} \right)^{-1} > 0 , \tag{A.69}$$

$$C^A = T \left(\frac{\partial S}{\partial T} \right)_A = T \left(\frac{\partial T}{\partial S} \right)_A^{-1} = T \left(\frac{\partial^2 H}{\partial S^2} \right)^{-1} > 0 . \tag{A.70}$$

For irreversible processes, these are the expressions (A.69) and (A.70), but not (A.68), which are well-determined. On substituting

$$T = \left(\frac{\partial E}{\partial S} \right)_X = \left(\frac{\partial H}{\partial S} \right)_A , \tag{A.71}$$

we can write

$$C^X = \left(\frac{\partial E}{\partial S} \right)_X \left(\frac{\partial S}{\partial T} \right)_X = \left(\frac{\partial E}{\partial T} \right)_X , \tag{A.72}$$

$$C^A = \left(\frac{\partial H}{\partial S}\right)_A \left(\frac{\partial S}{\partial T}\right)_A = \left(\frac{\partial H}{\partial T}\right)_A . \tag{A.73}$$

The susceptibilities corresponding to the heat capacities

$$c^X = \frac{C^X}{V} , \qquad c^A = \frac{C^A}{V} , \tag{A.74}$$

are referred to as the *specific heats* at constant X or A, respectively.

The positivity of the capacity or susceptibility expresses the *thermal stability* of a system. The state of a system with negative susceptibility is unstable. Such a state can appear when solutions of the equation of state are no longer unique. The system is divided into two or more different *phases* (see Fig. A.1b). At special points of the phase diagram of a system which indicates phase transitions, the susceptibility can become divergent. It will be shown in Appendix A.5 that such divergences are related to infinitely large thermodynamic fluctuations. We than speak about a *critical state*.

Let us consider two thermodynamic variables X_i and X_j and assume that their variations result only in the variation of the force A_i, with the other force remaining fixed: $A_j = \mathrm{const}$. Small changes can be described by general expressions of the form

$$\Delta A_i = \left(\frac{\partial A_i}{\partial X_i}\right)_{X_j} \Delta X_i + \left(\frac{\partial A_i}{\partial X_j}\right)_{X_i} \Delta X_j , \tag{A.75}$$

and

$$\Delta A_j = \left(\frac{\partial A_j}{\partial X_i}\right)_{X_j} \Delta X_i + \left(\frac{\partial A_j}{\partial X_j}\right)_{X_i} \Delta X_j \tag{A.76}$$

$$= \left(\frac{\partial A_i}{\partial X_j}\right)_{X_i} \Delta X_i + \left(\frac{\partial A_j}{\partial X_j}\right)_{X_i} \Delta X_j = 0 .$$

In the latter equation, the Maxwell relations (A.59) were used. Calculating ΔX_j from (A.76) and substituting into (A.75), we obtain

$$\Delta A_i = \left[\left(\frac{\partial A_i}{\partial X_i}\right)_{X_j} - \frac{(\partial A_i/\partial X_j)^2_{X_i}}{(\partial A_j/\partial X_j)_{X_i}}\right] \Delta X_i , \tag{A.77}$$

and after the limit transition as the increment goes to zero:

$$\left(\frac{\partial A_i}{\partial X_i}\right)_{A_j} = \left(\frac{\partial A_i}{\partial X_i}\right)_{X_j} - \frac{(\partial A_i/\partial X_j)^2_{X_i}}{(\partial A_j/\partial X_j)_{X_i}} . \tag{A.78}$$

The second term is a quotient of a square of a certain quantity over a reciprocal capacity. According to the thermodynamic stability condition, it is always positive. In this way we obtain the *Planck inequality*:

$$\left(\frac{\partial A_i}{\partial X_i}\right)_{A_j} > \left(\frac{\partial A_i}{\partial X_i}\right)_{X_j} , \tag{A.79}$$

or, taking into account the fact that the capacity is positive,

$$-\left(\frac{\partial X_i}{\partial A_i}\right)_{A_j} > -\left(\frac{\partial X_i}{\partial A_i}\right)_{X_j} . \tag{A.80}$$

In particular, the relation

$$C^A > C^X \tag{A.81}$$

follows from (A.80). This states that the heat capacity under a fixed value of some thermodynamic force A is larger than the heat capacity under a fixed value of the conjugate variable X. This result has a simple physical meaning: when the system is heated under conditions $A = $ const., it has to perform an additional but not useful work related to a possible change in the value of X.

In general, the inequality (A.79) gives mathematical expression to the *Le Chatelier–Brown principle*: the possibility of a variation in the thermodynamic variable X_j weakens the reaction to a variation in the value of the thermodynamic variable X_i. In other words, the system itself opposes external changes.

A.4 Canonical and Generalized Canonical Probability Distributions

The considerations in Chap. 2 conclude with all the necessary premises to construct an explicit form of the probability distribution $\rho(s)$ which describes a macroscopic system in a state of thermodynamic equilibrium. On the one hand, this distribution must be constant in time, so that the mean values of all the dynamical variables are constant, while on the other hand, it must be a coarse-grained description of a uniform distribution of states of a statistical ensemble over the entire region of the phase space available to motion, so that it must be constant in this region. But we have shown that the notion of a thermodynamic equilibrium is relative. Therefore, the properties of the sought probability distribution must be relativised: it should be constant in time *on a given time scale* and take constant values in the regions available to motion only *on a given time scale*. The distribution would reach a global constant value only on a much longer time scale, when transitions through the bottlenecks separating various available regions become possible (see Fig. 2.9).

The state of thermodynamic equilibrium is related to the maximum value of the entropy of the coarse-grained probability distribution. It follows from Sect. 2.6 that a probability distribution taking a constant value over a certain subset **A**, and equal to zero beyond it, gives an entropy higher than any other probability distribution constant on a set *smaller* than **A**. We now prove that a probability distribution constant on **A** gives the highest value of the entropy from among all distributions different from zero on **A**.

Entropy is a *functional* (a function defined on a set of functions) of the probability distribution:

$$S[\rho] = -k_{\mathrm{B}} \int_{\mathbf{A}} \mathrm{d}s\, \rho(s) \ln \rho(s) \,. \tag{A.82}$$

We require the function ρ to fulfill the following normalization condition on the subset **A**:

$$\int_{\mathbf{A}} \mathrm{d}s\, \rho(s) = 1 \,. \tag{A.83}$$

The task of finding the form of the function ρ for which the functional (A.82) takes the maximum value whilst satisfying the condition (A.83) is equivalent to the task of finding the form of ρ for which an extended functional

$$-k_{\mathrm{B}} \int_{\mathbf{A}} \mathrm{d}s \left[\rho \ln \rho + \alpha \rho \right] \tag{A.84}$$

takes the maximum value without any additional condition. The quantity α, referred to us as a *Lagrange indefinite multiplier* for the condition (A.83), is to be determined by that condition.

Looking for the extrema (minima or maxima) of functionals is the subject of the calculus of variations, well known from, e.g., effective applications to classical mechanics in the Lagrange formulation. According to the rules of this calculus, an integral with fixed boundaries, which depends only on a function ρ but not on its derivatives, takes the extreme value if the derivative of the integrand with respect to the relevant function is zero. In the case considered, this leads to the condition

$$\frac{\mathrm{d}}{\mathrm{d}\rho}\left[\rho \ln \rho + \alpha \rho\right] = \ln \rho + 1 + \alpha = 0 \,, \tag{A.85}$$

or, on introducing a new quantity

$$1 + \alpha = \ln \Omega , \tag{A.86}$$

to the condition

$$\rho(s) = \Omega^{-1} = \text{const.} . \tag{A.87}$$

Hence, we have proved that the probability distribution of the extremal (in our case the maximum) entropy is indeed constant. The corresponding entropy has value

$$S = k_{\mathrm{B}} \ln \Omega . \tag{A.88}$$

From the normalization condition (A.83), it follows that the quantity Ω, related to the Lagrange indefinite multiplier α, satisfies the equality

$$\Omega = \int_{\mathbf{A}} \mathrm{d}s . \tag{A.89}$$

In this way we reconstruct the result (2.27). The probability distribution (A.87) constant on a certain region of the phase space is called the *microcanonical* distribution.

However, a problem appears when we consider how to determine the region in the phase space 'available to motion on a given time scale'. It could be defined as a hypersurface in $\mathbf{S}$ for which a certain dynamical variable $\mathcal{X}$ takes a constant value:

$$\mathcal{X}(s) = X . \tag{A.90}$$

For instance, we certainly know that the energy $\mathcal{H}$ is a constant of motion for the Hamilton equations (2.4). But how do we define a possible variation of that region on a longer time scale? And how do we take into account the case where the probability distribution has constant but *different* values in various subsets composing the entire region accessible for motion (see, e.g., Fig. 2.9)? We can do this by replacing the condition (A.90) by the weaker condition requiring a fixed *mean value* of the dynamical variable $\mathcal{X}$:

$$\langle \mathcal{X} \rangle \equiv \int_{\mathbf{S}} \mathrm{d}s \, \mathcal{X}(s) \, \rho(s) = X . \tag{A.91}$$

A variation of the mean value $\langle \mathcal{X} \rangle$ in time can result from a variation of the region accessible for motion, i.e., from a variation of the very function ρ in (A.91) or, in the case of a dynamical variable $\mathcal{X}$ of the type (2.39) having several different discrete values, from a variation of the values of the function ρ in various regions determining $\mathcal{X}$ (Fig. 2.9). Fluctuations are negligible for macroscopic variables in the form of the sum (2.43), and the condition (A.91), although richer in possibilities, does not in fact differ from the condition (A.90).

Thus, we look for the function ρ that maximizes the functional

$$S[\rho] = -k_{\mathrm{B}} \int_{\mathbf{S}} ds\, \rho(s) \ln \rho(s) , \tag{A.92}$$

the integral being determined on the *entire* space of states $\mathbf{S}$, and satisfying *two* additional conditions:

$$\int_{\mathbf{S}} ds\, \rho(s) = 1 , \tag{A.93}$$

where the normalization now involves the entire space $\mathbf{S}$, and (A.91). On introducing two undefined Lagrange multipliers α and β, we study the necessary condition for maximizing the extended functional

$$-k_{\mathrm{B}} \int_{\mathbf{S}} ds \left[\rho \ln \rho + \alpha \rho + \beta \mathcal{X} \rho \right] . \tag{A.94}$$

This condition is that the derivative of the integrand should be zero:

$$\ln \rho + 1 + \alpha + \beta \mathcal{X} = 0 . \tag{A.95}$$

Hence

$$\rho(s) = Z^{-1} e^{-\beta \mathcal{X}(s)} , \tag{A.96}$$

where

$$\ln Z = 1 + \alpha . \tag{A.97}$$

The entropy that corresponds to the distribution (A.96) is

$$S = k_{\mathrm{B}} \ln Z + k_{\mathrm{B}} \beta X . \tag{A.98}$$

In terms of (A.98), the distribution (A.96) can be rewritten in yet another form:

$$\rho(s) = e^{-S/k_{\mathrm{B}}} e^{-\beta [\mathcal{X}(s) - X]} = e^{-S/k_{\mathrm{B}}} e^{-\beta\, \mathcal{X}(s)} . \tag{A.99}$$

The probability distribution (A.96) is called the *canonical* probability distribution. It was introduced into physics for the first time by Gibbs who considered an appropriately defined statistical ensemble of macroscopic systems. Let us note that when the variable $\mathcal{X}$ has the form of a sum (2.43) than the canonical distribution (A.96), being an exponential of this sum, is factorized according to (2.44). Hence, the distribution (A.96) in fact defines a statistical ensemble of many identical copies of the same subsystem. Following the comment after

(2.44), it is irrelevant whether we describe a single subsystem or the whole ensemble of subsystems.

From the normalization condition, it follows that

$$Z(\beta) = \int_S ds\, e^{-\beta \mathcal{X}(s)} . \tag{A.100}$$

The quantity Z is referred to as the *partition function* or the *sum of states*. Differentiating with respect to β, we get

$$\frac{\partial Z}{\partial \beta} = -\int_S ds\, \mathcal{X} e^{-\beta \mathcal{X}(s)} = -Z\langle \mathcal{X} \rangle , \tag{A.101}$$

whence

$$X = -\frac{\partial}{\partial \beta} \ln Z . \tag{A.102}$$

From this equation the value of the second undefined Lagrange multiplier β can be determined.

Differentiating the function $\ln Z$ once more with respect to β:

$$\frac{\partial^2 \ln Z}{\partial \beta^2} = -\frac{\partial X}{\partial \beta} = -\frac{\partial}{\partial \beta} Z^{-1} \int_S ds\, \mathcal{X}(s) e^{-\beta \mathcal{X}(s)} \tag{A.103}$$
$$= Z^{-2} \frac{\partial Z}{\partial \beta} \int_S ds\, \mathcal{X}(s) e^{-\beta \mathcal{X}(s)} + Z^{-1} \int_S ds\, \mathcal{X}(s)^2 e^{-\beta \mathcal{X}(s)}$$
$$= -\langle \mathcal{X} \rangle^2 + \langle \mathcal{X}^2 \rangle .$$

This is an important result saying that the derivative of the mean value X with respect to the Lagrange multiplier β determines the dispersion (the squared standard deviation) of the corresponding dynamical variable $\mathcal{X}$:

$$\frac{\partial X}{\partial \beta} = -\langle (\ \mathcal{X})^2 \rangle . \tag{A.104}$$

The condition (A.91) can be extended by fixing the mean values of a larger number of dynamical variables $\mathcal{X}_i$, $i = 0, 1, \ldots, n$ ($\mathcal{X}_0 \equiv \mathcal{X}$):

$$\langle \mathcal{X}_i \rangle \equiv \int_S ds\, \mathcal{X}_i(s)\, \rho(s) = X_i . \tag{A.105}$$

The variational procedure then leads to the *generalized canonical* distribution

$$\rho(s) = Z^{-1} e^{-\sum_i \beta_i \mathcal{X}_i(s)} , \tag{A.106}$$

with the entropy

$$S = k_B \ln Z + k_B \sum_i \beta_i X_i , \tag{A.107}$$

and partition function

$$Z(\beta_0, \beta_1, \ldots, \beta_n) = \int_{\mathbf{S}} ds \, e^{-\sum_i \beta_i \mathcal{X}_i(s)} \, . \qquad (A.108)$$

The relationship (A.102) is replaced by $1 + n$ relationships

$$X_i = -\frac{\partial}{\partial \beta_i} \ln Z \, , \qquad (A.109)$$

from which the values of $1 + n$ Lagrange multipliers β_i can be determined, whereas (A.104) is replaced by $1 + n$ more general equations

$$\frac{\partial X_i}{\partial \beta_j} = -\langle \, \mathcal{X}_i \, \mathcal{X}_j \rangle \, . \qquad (A.110)$$

Expression (A.106) describes the most general form of the probability distribution for a system in a state of thermodynamic equilibrium on a certain fixed time scale. This distribution is unambiguously determined by the mean values of a small number of distinguished dynamical variables (small relative to all the microscopic degrees of freedom). In Chap. 3, these mean values were identified with the thermodynamic variables. The mean values of all other dynamical variables either vanish or, in the case of fast variables, are determined by the mean values of the distinguished variables or again, in the case of very slow variables, are frozen. The distinguished variables are considered as slow in the sense that, on a time scale longer than initially assumed, their mean values can change until they reach a constant value characterizing a thermodynamic equilibrium state on a higher level of the time-scale hierarchy. In contrast to some *partial equilibrium* state, we then talk about a *complete equilibrium*. The generalized canonical probability distribution (A.106) therefore describes the state of partial equilibrium and total equilibrium as well, and the only difference is the number of distinguished dynamical variables. A physical interpretation of the Lagrangian multipliers β_i and their relationship with the thermodynamic variables is presented in Appendix A.5.

A.5 Statistical Interpretation of Thermodynamics

Under fixed temperature conditions $T = $ const., it is easy to interpret the macroscopic phenomenological thermodynamics in terms of the microscopic mechanics. The interpretation follows immediately by identifying the Clausius entropy of Chap. 3 with the Boltzmann–Gibbs entropy of Sect. 2.6 for the canonical or generalized canonical probability distributions just derived.

For the canonical probability distribution (A.96) with the dynamical variable $\mathcal{X}$ chosen to be the energy, $X = E$, $\mathcal{X} = \mathcal{H}$, (A.98) reads

$$S = k_B \ln Z + k_B \beta E . \tag{A.111}$$

Comparing it with (3.49), which can be rewritten as

$$S = -T^{-1}F + T^{-1}E , \tag{A.112}$$

we can interpret the undefined Lagrange multipliers β and $\ln Z$, hitherto only formally introduced. Thus, the multiplier β has the physical meaning of reciprocal temperature,

$$\beta = 1/k_B T , \tag{A.113}$$

while the logarithm of the partition function is related to the free energy by

$$F = -k_B T \ln Z = -k_B T \ln \int_{\mathbf{S}} ds \, e^{-\mathcal{H}(s)/k_B T} . \tag{A.114}$$

Comparing (A.72) or (A.73) with (A.104), we obtain the statistical interpretation of heat capacity as a quantity proportional to the mean square fluctuation of the energy:

$$C = \left(\frac{\partial E}{\partial T}\right) = \frac{1}{k_B T^2} \left\langle (\ \mathcal{H})^2 \right\rangle . \tag{A.115}$$

The relationship (A.115) justifies the heat capacity being positive.

Under conditions where not only the temperature T is fixed but so also are n additional thermodynamic forces A_i, the second equation of (A.54) implies

$$S = -T^{-1}I + T^{-1}E + \sum_{i=1}^{n} A_i T^{-1} X_i . \tag{A.116}$$

Comparing this with (A.107), in which X_0 is chosen to be the energy, $\mathcal{X}_0 = \mathcal{H}$, $X_0 = E$, $\beta_0 \equiv \beta$, and hence

$$S = k_B \ln Z + k_B \beta E + k_B \sum_{i=1}^{n} \beta_i X_i , \tag{A.117}$$

we find, besides (A.113), the relationships

$$\beta_i = A_i/k_{\mathrm{B}}T = \beta A_i , \quad (i = 1, \ldots, n) , \tag{A.118}$$

and

$$I = -\beta^{-1} \ln Z = -\beta^{-1} \ln \int_{\mathbf{S}} d s\, e^{-\beta\left[\mathcal{H}(s)+\sum_{i=1}^{n} A_i \mathcal{X}_i(s)\right]} . \tag{A.119}$$

Equation (A.110) implies the following relations between capacities (susceptibilities) and mean square fluctuations or, more generally, the *fluctuation correlation functions* of the dynamical variables $\mathcal{X}_i$ that correspond to the thermodynamic variables X_i:

$$-\left(\frac{\partial X_i}{\partial A_j}\right) = \frac{1}{k_{\mathrm{B}}T}\langle \mathcal{X}_i\, \mathcal{X}_j\rangle . \tag{A.120}$$

Equation (A.114), or more generally (A.119), is the basis for theoretical modeling of the free energy or thermodynamic potential. Knowing these quantities, one can find equations of state that determine all the other thermodynamic quantities [see the relationships (A.109) corresponding to (A.57)].

Statistical interpretation of the kinetic coefficients L_{ij} occurring in the relationships (3.69) as well as the relaxation time (3.78) requires knowledge of the temporal evolution of a partial equilibrium probability distribution function (A.106) toward the complete equilibrium probability distribution function; or equivalently, knowledge of the temporal behavior of fluxes (the time derivatives) $\dot{\mathcal{X}}_i(t)$ of the dynamical variables $\mathcal{X}_i$. In general, all the necessary information can be obtained using thermodynamic perturbation theory which, however, is a rather sophisticated technique. Close to the complete thermodynamic equilibrium, a simple approximation is offered by Onsager's *fluctuation regression hypothesis* (Chandler, 1987), which states that fluctuations decay in time according to the same equation (3.79) as the thermodynamic variables:

$$\dot{\mathcal{X}}(t) = -\tau^{-1}\left[\mathcal{X}(t) - \langle\mathcal{X}\rangle\right] = -\tau^{-1}\, \mathcal{X}(t) . \tag{A.121}$$

Setting $t = 0$, multiplying by $\dot{\mathcal{X}}(t)$, and averaging over the complete thermodynamic equilibrium state, we obtain from (A.121) the expression

$$\langle\dot{\mathcal{X}}(t)\dot{\mathcal{X}}(0)\rangle = -\tau^{-1}\langle\dot{\mathcal{X}}(t)\, \mathcal{X}(0)\rangle . \tag{A.122}$$

Integrating it over time from 0 to ∞ and taking into account the fact that $\mathcal{X}(t)$ tends to the equilibrium mean value $\langle\mathcal{X}\rangle$ as $t \to \infty$, we have

$$\int_0^\infty dt \left\langle \dot{\mathcal{X}}(t)\dot{\mathcal{X}}(0) \right\rangle = -\tau^{-1} \left\langle \left[\langle \mathcal{X} \rangle - \mathcal{X}(0) \right] \mathcal{X}(0) \right\rangle = \tau^{-1} \left\langle \left[\mathcal{X}(0) \right]^2 \right\rangle .$$

$$(A.123)$$

Finally, we obtain from the latter equation a statistical formula for the reciprocal relaxation time (3.78):

$$\tau^{-1} = \int_0^\infty dt \frac{\langle \dot{\mathcal{X}}(t)\dot{\mathcal{X}}(0) \rangle}{\langle (\ \mathcal{X})^2 \rangle} = \frac{L}{C} .$$

$$(A.124)$$

Generalization of (A.122) to the case of time correlation functions of various fluxes and the expression (A.104) for the capacity results in a statistical formula defining the kinetic coefficients:

$$L_{ij} = \frac{1}{k_B T} \int_0^\infty dt \left\langle \dot{\mathcal{X}}_i(t)\dot{\mathcal{X}}_j(0) \right\rangle .$$

$$(A.125)$$

This formula explains both the Onsager symmetry (3.70) and the positivity (3.72) of the kinetic coefficients.

B Stochastic Processes

B.1 From Liouville's Equation to the Diffusion Equation

The time behavior of macroscopic systems is deterministic on the level of both the microscopic and macroscopic descriptions (see Chap. 2). However, the time behavior of individual molecules seems to be purely random. Is this really the case? What kind of randomness could there be in the parts that does not contradict the determinism of the whole?

To answer these questions let us consider a system of N identical molecules of mass m. Their translational motion is described by N vectorial differential Newton equations of the second order:

$$m\ddot{\boldsymbol{r}}_i = \boldsymbol{F}_i(\boldsymbol{r}_1, \ldots, \boldsymbol{r}_N) , \tag{B.1}$$

equivalent to $2N$ vectorial differential Hamilton equations of the first order:

$$\dot{\boldsymbol{r}}_i = \boldsymbol{v}_i , \qquad m\dot{\boldsymbol{v}}_i = \boldsymbol{F}_i . \tag{B.2}$$

(It is convenient here to consider the velocities $\boldsymbol{v}_i$ rather than the momenta $\boldsymbol{p}_i = m\boldsymbol{v}_i$.) With the initial positions and velocities of individual particles unknown, we consider, instead of trajectories, the N-particle probability distribution

$$\rho = \rho(\boldsymbol{r}_1, \boldsymbol{v}_1, \ldots, \boldsymbol{r}_N, \boldsymbol{v}_N, t) . \tag{B.3}$$

According to Liouville's theorem [volume conservation by time transformations of the phase space, Penrose (1979)], the evolution equation of ρ is of the form

$$\frac{\mathrm{d}}{\mathrm{d}t}\rho = \frac{\partial}{\partial t}\rho + \sum_{i=1}^{N}\left(\frac{\partial\rho}{\partial\boldsymbol{r}_i}\cdot\dot{\boldsymbol{r}}_i + \frac{\partial\rho}{\partial\boldsymbol{v}_i}\cdot\dot{\boldsymbol{v}}_i\right) = 0 , \tag{B.4}$$

where the central dots denote the 3-dimensional scalar products, and the partial derivatives with respect to vectors are the 3-dimensional gradient operators. Using the Hamilton equations (B.2), (B.4) leads to the *Liouville equation*:

$$\frac{\partial}{\partial t}\rho + \sum_{i=1}^{N}\left(\boldsymbol{v}_i \cdot \frac{\partial \rho}{\partial \boldsymbol{r}_i} + m^{-1}\boldsymbol{F}_i \cdot \frac{\partial \rho}{\partial \boldsymbol{v}_i}\right) = 0 \; . \tag{B.5}$$

The single particle probability distribution is obtained from ρ by integration over positions and velocities of all particles except for the one considered:

$$p(\boldsymbol{r}_1, \boldsymbol{v}_1, t) = \int d^3\boldsymbol{r}_2\, d^3\boldsymbol{v}_2 \ldots d^3\boldsymbol{r}_N\, d^3\boldsymbol{v}_N\, \rho(\boldsymbol{r}_1, \boldsymbol{v}_1, \boldsymbol{r}_2, \boldsymbol{v}_2, \ldots, \boldsymbol{r}_N, \boldsymbol{v}_N, t) \; . \tag{B.6}$$

This satisfies the equation

$$\left(\frac{\partial}{\partial t} + \boldsymbol{v} \cdot \frac{\partial}{\partial \boldsymbol{r}} + m^{-1}\boldsymbol{F} \cdot \frac{\partial}{\partial \boldsymbol{v}}\right) p(\boldsymbol{r}, \boldsymbol{v}, t) = I \; , \tag{B.7}$$

where $\boldsymbol{F}$ is a resultant force acting on a distinguished particle and I is some functional of ρ referred to as the *collision integral*.

Boltzmann assumed in 1872 that, when the system is so diluted that only binary collisions are possible, the states of the particles in it are not correlated at distances larger than the distance over which interparticle forces act [*hypothesis of molecular chaos*, see (2.44)]:

$$\rho(\boldsymbol{r}_1, \boldsymbol{v}_1, \ldots, \boldsymbol{r}_N, \boldsymbol{v}_N, t) = p(\boldsymbol{r}_1, \boldsymbol{v}_1, t) \ldots p(\boldsymbol{r}_N, \boldsymbol{v}_N, t) \; . \tag{B.8}$$

The collision integral can then be written in the form

$$I = \int d\sigma\, d^3\boldsymbol{v}_1 |\boldsymbol{v} - \boldsymbol{v}_1|\left[p(\boldsymbol{r}, \boldsymbol{v}_1', t)p(\boldsymbol{r}, \boldsymbol{v}', t) - p(\boldsymbol{r}, \boldsymbol{v}_1, t)p(\boldsymbol{r}, \boldsymbol{v}, t)\right] \; , \tag{B.9}$$

where $d\sigma$ is the differential cross-section for scattering and the velocities of two particles before collision, $\boldsymbol{v}$ and $\boldsymbol{v}_1$, and after collision, $\boldsymbol{v}'$ and $\boldsymbol{v}_1'$, satisfy the conditions of momentum and energy conservation. Equation (B.7), with the collision integral I approximated in this way, is called the *Boltzmann equation* (Penrose, 1979; Huang, 1987).

The Boltzmann equation is a nonlinear integro-differential equation and the techniques for solving it are not easy. This equation is the starting point for the derivation of Boltzmann's so-called H-theorem (the law of entropy increase). For the force $\boldsymbol{F} = \boldsymbol{0}$, the equilibrium solution to the Boltzmann equation does not depend on the position and its dependence on the velocity has the form known as the the *Maxwell–Boltzmann distribution*:

$$p^{\mathrm{eq}}(\boldsymbol{v}) \propto e^{-mv^2/2k_{\mathrm{B}}T} \ , \qquad\qquad (\mathrm{B.10})$$

The mean velocity of a molecule calculated for this 3-dimensional distribution in spherical coordinates is

$$\bar{v} = \sqrt{8k_{\mathrm{B}}T/\pi m} \ . \qquad\qquad (\mathrm{B.11})$$

The mean velocity determines the mean free path of a molecule:

$$\bar{l} = \bar{n}\bar{v}/2z \ , \qquad\qquad (\mathrm{B.12})$$

where $\bar{n}$ is the mean value of the spatial density of molecules (the number of molecules per unit volume) and z is the number of collisions per unit volume and unit time (each collision ends the free paths of two molecules).

The Boltzmann equation is also the starting point for derivation of the equations of hydrodynamics in which the transport theory is rooted. In particular, after integrating (B.7) over velocity and taking into account the momentum and energy conservation laws, we get the *continuity equation*:

$$\frac{\partial}{\partial t}p(\boldsymbol{r},t) + \frac{\partial}{\partial \boldsymbol{r}}\cdot \boldsymbol{j}(\boldsymbol{r},t) = 0 \ , \qquad\qquad (\mathrm{B.13})$$

where

$$p(\boldsymbol{r},t) = \int \mathrm{d}^3\boldsymbol{v}\, p(\boldsymbol{r},\boldsymbol{v},t) \qquad\qquad (\mathrm{B.14})$$

is the probability density of individual molecule position and

$$\boldsymbol{j}(\boldsymbol{r},t) = \int \mathrm{d}^3\boldsymbol{v}\, \boldsymbol{v}\, p(\boldsymbol{r},\boldsymbol{v},t) \qquad\qquad (\mathrm{B.15})$$

is the corresponding flux density. When multiplied by the number of molecules N, (B.14) and (B.15) have the meaning of the spatial density of the number of molecules and the corresponding flux.

The driving force for the flux comes from the molecule number density fluctuations. In the linear approximation

$$\boldsymbol{j}(\boldsymbol{r},t) = -D\frac{\partial}{\partial \boldsymbol{r}}p(\boldsymbol{r},t) \ , \qquad\qquad (\mathrm{B.16})$$

from which, after substituting into (B.13), we get the *diffusion equation*:

$$\frac{\partial}{\partial t}p(\boldsymbol{r},t) = D\,\frac{\partial}{\partial \boldsymbol{r}}\cdot\frac{\partial}{\partial \boldsymbol{r}}p(\boldsymbol{r},t) \equiv D\Delta p(\boldsymbol{r},t) \ . \qquad\qquad (\mathrm{B.17})$$

Here Δ denotes the Laplace operator and D is the coefficient of self-diffusion,

$$D = \bar{l}\bar{v}/3 . \tag{B.18}$$

The diffusion equation (B.17) determines the dynamics of the molecule number density fluctuations or the position probability distribution of individual molecules. The position of a given molecule is indeed random, but the position probability distribution behaves in time in a deterministic way.

Instead of considering the *macroscopic hydrodynamic* variable related to underlying microscopic variables, we can consider a *mesoscopic* variable, the position of a massive *Brownian particle* colliding with light (microscopic) particles. On identifying the massive particle with particle number 1 in (B.6), the collision integral can be reinterpreted as (Huang, 1987)

$$I = \int d^3 u \Big[w(v, v+u)\, p(r, v+u, t) - w(v-u, v)\, p(r, v, t) \Big] , \tag{B.19}$$

where $w(v+u, v)$ is the probability of the massive particle velocity changing from v to $v+u$ in a single collision with a light particle per unit time.

If the mass difference between the heavy and light particles is large, u is small and the integrand can be expanded in a Taylor series with the initial velocity v being irrelevant for the result of the collision:

$$w(v, v+u)p(r, v+u, t) - w(v-u, v)p(r, v, t) \tag{B.20}$$

$$= u \cdot \frac{\partial}{\partial v} w(-u, 0)p(r, v, t)$$

$$+ \frac{1}{2}\left(u \cdot \frac{\partial}{\partial v}\right)\left(u \cdot \frac{\partial}{\partial v}\right) w(-u, 0)p(r, v, t) + \cdots .$$

After substituting the collision integral (B.19) approximated in this way into (B.7), we get

$$\frac{\partial}{\partial t}p(r, v, t) = \left[-v \cdot \frac{\partial}{\partial r} + m^{-1}\frac{\partial}{\partial v} \cdot \left(-F + a + b\frac{\partial}{\partial v}\right)\right] p(x, v, t) , \tag{B.21}$$

where m now denotes the mass of a heavy particle, while

$$a = \int d^3 u \, m u \, w(-u, 0) \tag{B.22}$$

and

$$b = \int d^3u \, \frac{mu^2}{2} \, w(-\boldsymbol{u}, \boldsymbol{0}) \tag{B.23}$$

are the mean changes in the momentum and energy per unit time, respectively, experienced by the heavy particle as a result of collisions with many light particles.

The quantities $\boldsymbol{a}$ and b are interrelated. For the external force $\boldsymbol{F} = \boldsymbol{0}$, the equilibrium solution to (B.21) should have the form (B.10), whereupon

$$\left(\boldsymbol{a} + b\frac{\partial}{\partial \boldsymbol{v}}\right) p^{\mathrm{eq}}(\boldsymbol{v}) = 0 \;, \tag{B.24}$$

and we obtain the relation

$$\boldsymbol{a} = b\frac{m\boldsymbol{v}}{k_{\mathrm{B}}T} \;. \tag{B.25}$$

The term $\boldsymbol{a}$, proportional to the velocity, can be interpreted as a *friction* (*dissipation*) force, whereas the term in b corresponds to the *fluctuating* force. Equation (B.25) is a prototype of the *fluctuation–dissipation theorem*. Introducing the *friction coefficient*

$$\zeta = \frac{mb}{k_{\mathrm{B}}T} \;, \tag{B.26}$$

we rewrite (B.21) in the form

$$\frac{\partial}{\partial t}p(\boldsymbol{r}, \boldsymbol{v}, t) \tag{B.27}$$

$$= \left[-\boldsymbol{v}\cdot\frac{\partial}{\partial \boldsymbol{r}} + m^{-1}\frac{\partial}{\partial \boldsymbol{v}}\cdot\left(-\boldsymbol{F} + \zeta\boldsymbol{v} + \zeta k_{\mathrm{B}}Tm^{-1}\frac{\partial}{\partial \boldsymbol{v}}\right)\right]p(x, \boldsymbol{v}, t) \;.$$

This equation, known as the *Fokker–Planck equation*, also leads to the spatial diffusion equation, as will be shown further on. First, however, it may be worth reviewing the basic concepts of the theory of stochastic processes.

B.2 Markov Processes

The variation of probability in time is the subject of the theory of stochastic processes (Gardiner, 1983; van Kampen, 2001). A *stochastic process* can be considered as a family of random variables $\mathcal{X}(t)$ at different moments of time t. Let us recall that the assumption of mechanical determinism allows us to identify the random variables with dynamical variables, i.e., real-valued functions on the space of micro-scopic states s (Sect. 2.6). As a consequence, the stochastic process, being a real-valued function

$$\mathcal{X}(t,s) = x , \qquad (B.28)$$

can alternatively be considered as a family of *realizations* of the process $\mathcal{X}(s)$, functions of time labeled generally by uncontrollable microscopic states s.

For a fixed moment of time t, a random variable $\mathcal{X}(t)$ is character-ized by the *mean value* and the *fluctuation* of the process:

$$X(t) \equiv \langle \mathcal{X}(t) \rangle , \qquad \mathcal{X}(t) \equiv \mathcal{X}(t) - X(t) . \qquad (B.29)$$

For two fixed moments of time t and t' and two random variables $\mathcal{X}(t)$ and $\mathcal{X}(t')$, we define the *fluctuation correlation* function:

$$G(t,t') \equiv \langle\ \mathcal{X}(t)\ \mathcal{X}(t') \rangle = \langle \mathcal{X}(t) \mathcal{X}(t') \rangle - \langle \mathcal{X}(t) \rangle \langle \mathcal{X}(t') \rangle . \qquad (B.30)$$

A stochastic process $\mathcal{X}(t)$ is referred to as *stationary* if its charac-teristics are invariant with respect to translation in time. Thus, for a stationary process the mean does not depend on time:

$$X(t) = X = \text{const.} , \qquad (B.31)$$

and the two-time function of fluctuation correlation depends only on one time argument equal to the time difference:

$$G(t,t') = G(t - t') \equiv \langle\ \mathcal{X}(t - t')\ \mathcal{X}(0) \rangle . \qquad (B.32)$$

The Fourier transform of $G(t)$,

$$S(\omega) \equiv \hat{G}(\omega) \equiv \int_{-\infty}^{\infty} dt\, e^{i\omega t} G(t) , \qquad (B.33)$$

is called the *spectral density* of the stationary process $\mathcal{X}(t)$.

The description of a stochastic process is all the more complete as the higher time correlation functions of the process are established. Unfortunately, the amount of information needed to do this grows rapidly with the increasing number of moments of time. Usually, either no correlation or at most the two-time correlation is considered.

A process for which all random variables $\mathcal{X}(t)$ are uncorrelated is said to be *purely random*. In such a process the two-time function of fluctuation correlation disappears for $t \neq t'$ (no memory), so it should have the form

$$G(t, t') = S\delta(t - t') . \tag{B.34}$$

The *Dirac delta* $\delta(t)$ is a mathematical quantity that vanishes when the argument differs from zero, whilst at zero it behaves in such a way that normalization to unity is possible:

$$\int_{-\infty}^{\infty} dt \, \delta(t) = 1 . \tag{B.35}$$

As a consequence, for any (not necessarily stationary) purely random process, the spectral density does not depend on frequency:

$$S(\omega) = S = \text{const.} . \tag{B.36}$$

All Fourier components are represented with equal weight so another name for the purely random process is *white noise*. Of course, the notion of white noise is meaningful only for continuous time changes. An example of a purely random process with discrete time (and a discrete set of values $\{\, 0, 1\,\}$) is the series of *Bernoulli trials* considered in Sect. 2.7 in the context of coin tossing.

A more complex process, the *Markov process*, is completely characterized by the two-time correlation function. (It has a memory, but a 'short' one.) Markov processes are the only stochastic processes for which the evolution of probability is deterministic. A deeper analysis leads to the conclusion that the evolution equation has to be linear and in general irreversible in time (Gardiner, 1983).

Without going into the deterministic microscopic dynamics underlying a given Markov process, we shall describe it on a phenomenological level in terms of the probability or the probability density being a function of the process values alone. For simplicity, let us start by considering processes involving a discrete (countable) set of values ('states'). In such cases, the probability $p_l(t)$ of being in the state l at time t is described by a system of *master equations*:

$$\dot{p}_l(t) = \sum_{l'} \left[w_{ll'} p_{l'}(t) - w_{l'l} p_l(t) \right] . \tag{B.37}$$

The dot denotes the derivative with respect to time. The *transition probabilities per unit time* $w_{ll'}$ are assumed to satisfy the *detailed balance condition*:

$$w_{l'l} p_l^{\text{eq}} = w_{ll'} p_{l'}^{\text{eq}} , \tag{B.38}$$

where the p^{eq} denote the equilibrium solutions to (B.37).

As a particular example let us consider a linear chain of states $l = \ldots -2, -1, 0, 1, 2, \ldots$ with the only nonzero transition probabilities per unit time being those between nearest neighbors:

$$w_{l+1,l} = w_{l-1,l} = 1/2\tau \ . \tag{B.39}$$

This corresponds to the master equations

$$\dot{p}_l(t) = \frac{1}{2\tau}\left[p_{l+1}(t) + p_{l-1}(t) - 2p_l(t) \right] \ . \tag{B.40}$$

Denoting the change in the process value in one jump by ξ and introducing formally the probability density

$$p(x,t) = \xi^{-1} p_l(t) \ , \qquad x \equiv l\xi \ , \tag{B.41}$$

we rewrite (B.40) as

$$\frac{\partial}{\partial t} p(x,t) = \frac{1}{2\tau}\left[p(x+\xi,t) + p(x-\xi,t) - 2p(x,t) \right] \ . \tag{B.42}$$

Expanding the probability density in the series

$$p(x \pm \xi, t) = p(x,t) \pm \xi \frac{\partial}{\partial x} p(x,t) + \frac{1}{2}\xi^2 \frac{\partial^2}{\partial x^2} p(x,t) + \cdots \ , \tag{B.43}$$

and taking the continuous limit ξ, $\tau \to 0$ with $D \equiv \xi^2/2\tau$ remaining finite, we finally arrive at the equation

$$\frac{\partial}{\partial t} p(x,t) = D \frac{\partial^2}{\partial x^2} p(x,t) \ , \tag{B.44}$$

identical to the diffusion equation considered (in three dimensions) in Appendix B.1. The above line of reasoning implies that diffusion may be identified with the process of a *random walk*. To mathematicians this process is known as the *Wiener process*, whereas physicists stick to the historical name of *Brownian motion*. Of course, the variable x need not necessarily be treated as the position of a randomly walking particle. It can, for example, be the intensity or voltage of an electrical signal or any other physical quantity.

A general solution to (B.44) with an arbitrary initial probability density $p(x,0)$ can be written in the form of the integral

$$p(x,t) = \int_{-\infty}^{\infty} dx' \, p(x,t|x') \, p(x',0) \ , \tag{B.45}$$

where the kernel (the Green function or propagator) $p(x, t|x')$ is the solution to the equation

$$\frac{\partial}{\partial t} p(x, t|x_0) = D \frac{\partial^2}{\partial x^2} p(x, t|x_0) , \qquad (B.46)$$

with initial condition

$$p(x, 0|x_0) = \delta(x - x_0) . \qquad (B.47)$$

It is easy to check that this solution is a Gaussian distribution

$$p(xt|x_0) = (2\pi\sigma^2)^{-1/2} \exp\left[-(x - x_0)^2/2\sigma^2\right] , \qquad (B.48)$$

with a constant mean value x_0 (determined by the initial condition) and dispersion

$$\sigma^2 = \langle \mathcal{X}(t)^2 \rangle = 2Dt . \qquad (B.49)$$

The limit $t \to 0$ of the function (B.48) completed by the relationship (B.49) represents a very intuitive model of the Dirac delta.

The standard deviation σ of the value of a random walk process increases in time as $t^{1/2}$. This implies a specific feature of scaling: to make the effect of such a process look on average the same on a scale of observation reduced by a factor of 2, the time of observation should be reduced, not just by a factor of 2, but by a factor of four! Of course, a continuous curve featuring such a scaling cannot be differentiable and must have an infinite number of 'teeth' so that direct representation of any realization of the diffusion process is technically possible only at discrete moments of time.

In fact, the Gaussian character of the probability distribution for a diffusion process should not be surprising. Let us divide the interval $(0, t)$ into n equal parts: $0 = t_0 < t_1 < \ldots < t_n = t$, $t_i = i\tau$, and let us consider a sequence of increments of the process $\mathcal{X}$:

$$\mathcal{X}(t_1) - \mathcal{X}(t_0) , \quad \mathcal{X}(t_2) - \mathcal{X}(t_1) , \quad \ldots , \quad \mathcal{X}(t_n) - \mathcal{X}(t_{n-1}) . \quad (B.50)$$

According to the definition of a random walk in its discrete form, (B.40), these increments are uncorrelated random variables. According to the law of large numbers and the central limit theorem (Gardiner, 1983) applied to the sum

$$\mathcal{X}(t) - \mathcal{X}(0) = \sum_{i=1}^{n} [\mathcal{X}(t_i) - \mathcal{X}(t_{i-1})] , \qquad (B.51)$$

in the limit $n \to \infty$, the resulting increment $\mathcal{X}(t) - \mathcal{X}(0)$ must be a variable with a Gaussian probability distribution having a dispersion proportional to n, and thus to the time t.

The lack of correlation between the increments (B.50) suggests that the time derivative of the diffusion process should be a purely random process, provided that it is given a definite mathematical meaning. We know that the derivative of a particular realization of the diffusion process does not exist (teeth at each moment of time), but differentiation of the process with respect to time may be defined independently of any realization by resorting to the notion of, e.g., mean square convergence (Gardiner, 1983). Analogously, a time stochastic integral can be defined.

Having defined the meaning of the time derivative of a stochastic process, we can write down the supposed *stochastic differential equation* for the diffusion process $\mathcal{X}(t)$ as

$$\dot{\mathcal{X}}(t) = B\mathcal{Y}(t) , \tag{B.52}$$

where $\mathcal{Y}(t)$ stands for a stationary purely random process (white noise) with vanishing mean:

$$\langle \mathcal{Y}(t) \, \mathcal{Y}(t') \rangle = \delta(t - t') , \qquad \langle \mathcal{Y}(t) \rangle = 0 , \tag{B.53}$$

and B is the amplitude of that process. The solution to (B.52) is a stochastic integral of the process $\mathcal{Y}$:

$$\mathcal{X}(t) - \mathcal{X}(0) = B \int_0^t dt \, \mathcal{Y}(t) . \tag{B.54}$$

As the mean of a stochastic integral is an ordinary Riemann integral of the mean, we find that

$$\langle \mathcal{X}(t) \rangle = \langle \mathcal{X}(0) \rangle = \text{const.} , \tag{B.55}$$

and

$$\langle \, \mathcal{X}(t)^2 \rangle = B^2 \int_0^t dt' \int_0^t dt'' \langle \mathcal{Y}(t') \, \mathcal{Y}(t'') \rangle = B^2 t . \tag{B.56}$$

The dispersion we arrive at here is identical to the dispersion (B.49). Consequently, the ordinary stochastic differential equation (B.52) is indeed equivalent to the partial differential equation (B.44), provided that

$$\frac{1}{2}B^2 = D . \tag{B.57}$$

Diffusion and possible purely deterministic components of the process jointly satisfy the equation

$$\dot{\mathcal{X}} = A(\mathcal{X}) + B\mathcal{Y} , \tag{B.58}$$

where A is some well-determined function of the process $\mathcal{X}$. The stochastic ordinary differential equation (B.58) is equivalent to the usual differential equation with partial derivatives for the probability density (Gardiner, 1983):

$$\frac{\partial}{\partial t} p(x,t) = -\frac{\partial}{\partial x}\left[A(x)p(x,t)\right] + \frac{1}{2}B^2\frac{\partial^2}{\partial x^2}p(x,t) . \tag{B.59}$$

Equations (B.58) and (B.59) can be generalized to the case of a vectorial (many-component) variable x. In the literature, (B.58) is called the *Langevin equation*. If x contains only position coordinates, (B.59) is known as the *Smoluchowski equation*, and if x contains both position and velocity coordinates, the *Fokker–Planck equation*.

The Fokker–Planck equation (B.27) for a massive Brownian particle that moves in one dimension depends on two variables, the position x and the velocity v. As a consequence, there are two coupled Langevin equations for the corresponding stochastic processes $\mathcal{X}(t)$ and $\mathcal{V}(t)$:

$$\begin{aligned}\dot{\mathcal{X}} &= \mathcal{V} , \\ m\dot{\mathcal{V}} &= F - \zeta\mathcal{V} + \zeta B\mathcal{Y} ,\end{aligned} \tag{B.60}$$

where

$$\frac{1}{2}\zeta B^2 = k_{\mathrm{B}}T . \tag{B.61}$$

In the overdamped limit, we assume the velocity $\mathcal{V}$ to reach the stationary stage much faster than the position $\mathcal{X}$, so that after substituting $\dot{\mathcal{V}} = 0$ into (B.60), we get a single Langevin equation:

$$\zeta\dot{\mathcal{X}} = F(\mathcal{X}) + \zeta B\mathcal{Y} . \tag{B.62}$$

This has the physical meaning of a balance equation for the forces: the *viscous (friction) force* balances the *driving force (exerted* by the environment) and the *Brownian fluctuation force*. The partial differential equation corresponding to the ordinary stochastic equation (B.62) is the Smoluchowski equation:

$$\frac{\partial}{\partial t} p(x,t) = -\frac{\partial}{\partial x}\left[\zeta^{-1}F(x)p(x,t)\right] + \frac{1}{2}B^2\frac{\partial^2}{\partial x^2}p(x,t) . \tag{B.63}$$

For $F = 0$ (no external driving force), we reconstruct from (B.63) the diffusion equation (B.44) holding for the Brownian particle. From (B.57) and (B.61), the *Einstein relation* follows:

$$D\zeta = k_B T \, . \tag{B.64}$$

B.3 Stochastic Theory of Reaction Rates

As a rule, the rates of biochemical reactions are interpreted in terms of the *transition state theory* (Stryer et al., 2002, Chap. 8; Fersht, 1999). This theory, known also as the *theory of absolute reaction rates*, was developed by Eyring, Evans and Polanyi in 1935 with the intention of applying it to gas-phase bimolecular reactions. It draws attention to the purely inertial intramolecular dynamics of what is called the activated complex (Atkins, 1998, Chap. 27) and assumes equilibration processes to be negligibly fast.

However, there is a great deal of convincing experimental evidence (see Sect. 9.1) to suggest that, in the case of biochemical reactions involving protein enzymes, processes of both intermolecular and intramolecular equilibration cannot be neglected. These are taken into account in the *stochastic theory of reaction rates*. The basic assumption of this theory is that the molecule under consideration can occur in a number of substates and that the transitions between these states are purely stochastic. The origins of the stochastic theory of reaction rates go back to the Smoluchowski theory of diffusion-controlled coagulation from 1917 and the Kramers one-dimensional theory of reactions in the overdamped limit from 1940. A clear discussion of the concepts involved can be found in papers by Widom (1965, 1971), Northrup and Hynes (1980) and Hänggi et al. (1990). The application to biochemical processes involving macromolecular proteins is considered in papers by Kurzyński (1998) and Kurzyński and Chełminiak (2003).

Let us start with a simple picture presenting an exemplifying realization of the microscopic (or rather mesoscopic) stochastic dynamics underlying a unimolecular reaction

$$R \rightleftarrows P$$

between two chemical species R and P of a given molecule. We assume that the molecule fluctuates among many substates which can be divided into two subsets corresponding to chemical species R and P (Fig. B.1). The chemical reaction is realized through transitions between distinguished substates in R, jointly forming what is called the

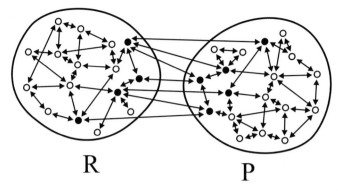

R   P

Fig. B.1. Exemplifying realization of the model intramolecular dynamics underlying the unimolecular reaction R ↔ P. Chemical states R and P of the molecule are composed of many substates (*white* and *black circles*) and the intramolecular dynamics involves purely stochastic transitions between these states (*arrows*). Actually, a much larger number of substates is expected. The chemical reaction is realized through transitions between distinguished substates in R, jointly forming what is called the transition state $R^{\ddagger}$ (*black circles*), and distinguished substates in P, jointly forming the transition state $P^{\ddagger}$. If the transition states comprise all the substates in R and P, we refer to such a situation as a reaction with fluctuating barriers, and if the transition states are reduced to single conformational substates, we talk about a gated reaction

transition state $R^{\ddagger}$, and distinguished substates in P, jointly forming the transition state $P^{\ddagger}$ (see Sect. 6.8). Two limiting cases can be formally distinguished: one where both transition states comprise all the substates in R and P, referred to as a reaction with *fluctuating barriers* (each substate is related to a generally different set of free energy barriers for the reactive transitions); and the opposite one, in which the transition states are reduced to single conformational substates, jointly forming a 'gate', whence we talk about a *gated reaction*.

In the Smoluchowski theory of coagulation, the subset R is a three-dimensional region accessible to translational diffusion of the molecule, the transition state $R^{\ddagger}$ is formed out of uniformly distributed traps, and the subset P is reduced to a single totally absorbing sink (irreversibility). The reaction can be considered as gated when one trap falls, on average, to one diffusing molecule. In the Kramers theory of reaction rates, all the substates lie along a one-dimensional *reaction coordinate* and the transition state is a single substate of the highest free energy.

In formal terms, the model dynamics is generally described by a system of master equations (B.37). The quantity $p_l(t)$ denotes, in our context, the probability of the molecule being in the substate l at time t. In the appropriate linear combinations of probabilities

$$X_k(t) = \sum_l \mathcal{X}_{kl} p_l(t) \equiv \langle \mathcal{X}_k(t) \rangle \,, \tag{B.65}$$

the system of linear equations (B.37) is decoupled into the system of independent equations

$$\dot{X}_k(t) = -\tau_k^{-1} X_k(t) \,. \tag{B.66}$$

If the condition (B.38) is satisfied, the coefficients τ_k^{-1} are real and positive and have the meaning of reciprocal relaxation times. The normal modes of relaxation (B.65) are written in such a way that they can be interpreted as the mean values of some physical quantities $\mathcal{X}_k$ defined on a set of substates labeled with the index l.

The mole fractions of individual species, proportional to the molar concentrations [R] and [P], are the sums of probabilities

$$P_R(t) = \sum_{l \in R} p_l(t) \,, \qquad P_P(t) = \sum_{l \in P} p_l(t) \,. \tag{B.67}$$

These can be rewritten as the mean values

$$P_R(t) = \langle \mathcal{P}_R(t) \rangle \,, \qquad P_P(t) = \langle \mathcal{P}_P(t) \rangle \tag{B.68}$$

of the characteristic functions of the subsets R and P, respectively:

$$\mathcal{P}_{Rl} \equiv \begin{cases} 1 & \text{if } l \in R \,, \\ 0 & \text{if } l \in P \,, \end{cases} \qquad \mathcal{P}_{Pl} \equiv \begin{cases} 1 & \text{if } l \in P \,, \\ 0 & \text{if } l \in R \,. \end{cases} \tag{B.69}$$

Following the normalization of probability to unity, the mole fractions (B.68) are related by

$$P_R + P_P = 1 \,. \tag{B.70}$$

The molar fractions (B.67) satisfy the equation

$$\dot{P}_R(t) = -\dot{P}_P(t) = -\sum_{l \in R^\ddagger, l' \in P^\ddagger} \left[w_{l'l} p_l(t) - w_{ll'} p_{l'}(t) \right] \,, \tag{B.71}$$

where $R^\ddagger$ and $P^\ddagger$ are the appropriate transition states. In general, the solution to (B.71) is non-exponential and depends on the initial values of all the probabilities p_l. The situation simplifies, however, when the reaction is an *activated process*, i.e., as a result of a bottleneck of either

$$\tau_2^{-1}$$

$$\tau_1^{-1}$$
0

Fig. B.2. Schematic spectrum of reciprocal relaxation times characterizing the conformational transition dynamics of the molecule *and* a chemical transformation involving it. The gap in this spectrum between the reciprocals of the longest and the next shorter relaxation times, τ_1^{-1} and τ_2^{-2}, respectively, testifies to the existence of time-scale separation. The ground value of the spectrum equal to 0 (the infinite relaxation time) is related to the sum of all probabilities, which remains constant

energetic or entropic origin, the transitions between the two subsets are not very probable. For such a reaction, equilibration of substates *within* individual chemical species proceeds much faster than equilibration *between* the species. A consequence is a time-scale separation in the system and a gap in the spectrum of relaxation times between the longest and the next shorter relaxation times τ_1 and τ_2, respectively (Fig. B.2).

After the lapse of time τ_2 (*initial stage* of the reaction), equation (B.71) takes the form of the usual kinetic equation

$$\dot{P}_R(t) = -\dot{P}_P(t) = -k_+ P_R(t) + k_- P_P(t)$$
$$= -\tau_1^{-1}(P_R - P_R^{eq}) = \tau_1^{-1}(P_P - P_P^{eq}) \qquad (B.72)$$

of the exponential solution. For given equilibrium values of the mole fractions P_R^{eq} and P_P^{eq}, the longest *chemical relaxation* time τ_1 determines in a unique way the *forward* and *reverse reaction rate constants* k_+ and k_-, respectively, through the equations

$$\tau_1^{-1} = k_+ + k_- \qquad (B.73)$$

and

$$k_+/k_- = P_{\mathrm{P}}^{\mathrm{eq}}/P_{\mathrm{R}}^{\mathrm{eq}} \equiv K \;, \tag{B.74}$$

where K is referred to as the *equilibrium constant*. The quantities P_{R} or P_{P} do not have to coincide exactly (up to some multiplicative and additive constant) with the slowest variable of the system X_1. If it holds, the kinetic equation (B.72) is valid on any time scale, and also at the very beginning stage of the reaction.

Because of the special properties of the characteristic functions

$$\mathcal{P}_{\mathrm{R}}^2 = \mathcal{P}_{\mathrm{R}}\;, \qquad \mathcal{P}_{\mathrm{P}}^2 = \mathcal{P}_{\mathrm{P}}\;, \qquad \mathcal{P}_{\mathrm{R}}\mathcal{P}_{\mathrm{P}} = 0\;, \tag{B.75}$$

the thermodynamic perturbation theory for the problem discussed can be applied exactly, up to infinite order (Kurzyński, 1990). This results in an exact expression (valid arbitrarily far from equilibrium) for the reaction rate constant k_+ in terms of the equilibrium time correlation function of fluxes:

$$k_+ = P_{\mathrm{P}}^{\mathrm{eq}}\tau^{-1} = \int_0^\infty dt'\,\langle\dot{\mathcal{P}}_{\mathrm{R}}(t')\dot{\mathcal{P}}_{\mathrm{R}}(0)\rangle^{\mathrm{eq}}/P_{\mathrm{R}}^{\mathrm{eq}}\;. \tag{B.76}$$

A similar formula determines the reverse reaction rate constant k_-. Equation (B.76) was derived for the first time by Yamamoto (1960), assuming the first order perturbation theory. Chandler (1987) derived it simply by resorting to Onsager's regression hypothesis (see Appendix A.5). After integration over time k_+, (B.76) can be rewritten formally as a limit of the *reactive flux*

$$J_+(t) \equiv \langle\mathcal{P}_{\mathrm{R}}(t)\dot{\mathcal{P}}_{\mathrm{R}}(0)\rangle^{\mathrm{eq}}/P_{\mathrm{R}}^{\mathrm{eq}}\;, \tag{B.77}$$

for a sufficiently long time t. Similarly, k_- can be rewritten as a limit of $J_-(t)$ given by a formula analogous to (B.77). In the derivation of (B.77), we have taken into account the causality principle which says that the reactive fluxes vanish for $t < 0$.

Both time limits $t \to \infty$ and $t \to 0$ of the reactive flux should be treated very carefully (Hänggi et al., 1990). The limit

$$\lim_{t\to 0} J_+(t) = \frac{\langle\mathcal{P}_{\mathrm{R}}\dot{\mathcal{P}}_{\mathrm{R}}\rangle^{\mathrm{eq}}}{P_{\mathrm{R}}^{\mathrm{eq}}} \equiv \nu\frac{P_{\mathrm{R}^{\ddagger}}^{\mathrm{eq}}}{P_{\mathrm{R}}^{\mathrm{eq}}} = \nu\exp(-\Delta G_{\mathrm{R}}^{\ddagger}/k_{\mathrm{B}}T) \equiv k_+^{\mathrm{eq}} \tag{B.78}$$

coincides with the value of the reaction rate constant provided by the transition state theory (see Sect. 6.8). Here, ν is interpreted as the *mean frequency of transitions* from $\mathrm{R}^{\ddagger}$ to $\mathrm{P}^{\ddagger}$, $P_{\mathrm{R}^{\ddagger}}^{\mathrm{eq}}$ is the equilibrium occupation of the transition state and $\Delta G_{\mathrm{R}}^{\ddagger}$ denotes the *free energy of activation*. On the other hand, the exact limit is

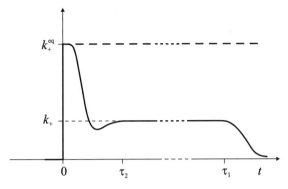

Fig. B.3. Schematic dependence of the reactive flux on time. For the reaction which is an activated process, the plateau value behavior is characteristic. Transition state theory approximates the reactive flux time course by the Heaviside step function. The long-time behavior of the reactive flux is drawn on a much more compressed time scale

$$\lim_{t \to \infty} J_+(t) = \langle \dot{\mathcal{P}}_R \rangle^{eq} = 0 \ . \tag{B.79}$$

Hence, the assumption of the time-scale separation corresponds only to the plateau value behavior of $J_+(t)$ and $J_-(t)$ (Fig. B.3).

Note that the reaction may proceed faster in the initial stage and the long-time exponential decay may need to be cut off by an appropriate regularization factor in the integral (B.76). The jump at $t = 0$ is related to a Dirac delta component of the time correlation function of fluxes, which thus appears to have the form of a sum:

$$\frac{\langle \dot{\mathcal{P}}_R(t) \dot{\mathcal{P}}_R(0) \rangle^{eq}}{\langle \mathcal{P}_R \rangle^{eq}} = k_+^{eq} \delta(t) + S_+(t) \ . \tag{B.80}$$

Equations (B.76) and (B.80) state clearly that the core of the transition state theory is the assumption that the flux $\dot{\mathcal{P}}_R(t)$ is delta-correlated white noise. To determine the transition state theory rate constant (B.78), no knowledge of the intramolecular dynamics is needed. It is the finite correlation time component $S_+(t)$ in the sum (B.80) that results from the intramolecular dynamical processes.

If the transition states $R^{\ddagger}$ and $P^{\ddagger}$ are short-lived intermediates, the exact reciprocal rate constants can be decomposed into three time components (see Sect. 6.8):

$$k_+^{-1} = (k_+^{eq})^{-1} + \tau_R + K^{-1} \tau_P \ , \tag{B.81}$$

and similarly for k_-^{-1}, related to k_+^{-1} by (B.74). The first component in (B.81) determines the time needed to cross the boundary under the assumption, made in the transition state theory, that the transition state $R^{\ddagger}$ is in a partial equilibrium with the rest of the microstates composing the chemical state R. However, as a result of the transition, this equilibrium is disturbed. The second component in (B.81) determines the time needed to restore this equilibrium from the side of the R species and the third component determines the time needed for the same process, but from the side of the P species (recrossing the border). From (B.81), it follows that k_{+}^{eq} is always larger than the exact rate constant k_{+} (Fig. B.3). If all three components in (B.81) are comparable (as in the case of reactions of small molecules in the gas phase), the reaction rate constant is well described by the transition state theory, possibly with a certain transmission coefficient smaller than unity. The initial stage of the reaction is then practically absent. If, on the contrary, the second and third components prevail, the reaction is said to be *controlled* by processes of intramolecular dynamics and the transition state theory fails. In the latter case, the initial stage of the reaction can even appear to dominate.

B.4 Reaction Rate and the First-Passage Time Problem

One should note that (Widom, 1965): "the rate constants k_{+} and k_{-} are not the probabilities per unit time of an R molecule making the R $\rightarrow$ P transition and a P molecule making the P $\rightarrow$ R transition, and $k_{+}P_{R}$ and $k_{-}P_{P}$ are not the separate P $\rightarrow$ R and R $\rightarrow$ P fluxes." This holds only for *imagined* irreversible reactions

$$R \rightarrow P \quad \text{or} \quad P \rightarrow R ,$$

with an *absorbing* boundary between the R and P subsets of microstates, which can be realized by adding an imagined totally absorbing *limbo state* (Fig. B.4). The stochastic theory of such imagined or real irreversible reactions is identical to the first-passage time problem for the corresponding stochastic processes (Gardiner, 1983; Montroll and West, 1987; van Kampen, 2001). For irreversible reactions, it is the addition of the limbo state (Fig. B.4) that introduces the smallest finite value τ_1^{-1} into the spectrum of reciprocal relaxation times (the dynamics in the sets R or P alone is characterized by the next larger value τ_2^{-1}, see Fig. B.2).

Later on we shall consider only the irreversible reaction R$\rightarrow$P. The case of the irreversible reaction P$\rightarrow$R is analogous. By definition, transition probabilities per unit time from the limbo state $*$ to any microstate l in the transition state $R^{\ddagger}$ vanish:

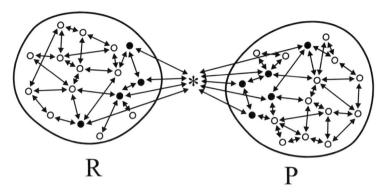

Fig. B.4. Any reversible reaction can be formally divided into two irreversible reactions after introducing the imagined *limbo state* (∗)

$$w_{l,*} = 0 . \tag{B.82}$$

Consequently, the occupation probability of the limbo state tends in time to unity:

$$\lim_{t \to \infty} p_*(t) = 1 . \tag{B.83}$$

In the presence of the limbo state, the quantity

$$P(t) \equiv \sum_{l \in R} p_l(t) = 1 - p_*(t) \tag{B.84}$$

has the meaning of the *survival probability* in R over time t (the mole fraction of molecules R that survived through time t). In various contexts, the time t in (B.84) is referred to as the *dwell time* in R, the *waiting time* for transition to P, the *first-exit time* from R, or the *first-passage time* to the limbo state. The quantity $1 - P(t)$ is the cumulative probability of the first-passage time being shorter than t, so that its derivative

$$-\dot{P}(t) = f(t) \tag{B.85}$$

has the meaning of the *first-passage time distribution*. Knowing the first-passage time density distribution, one can calculate the *mean first-passage time*:

$$\tau \equiv \int_0^\infty dt\, t\, f(t) = -\int_0^\infty t\, dt\, \frac{dP(t)}{dt} = \int_0^\infty dt\, P(t) , \tag{B.86}$$

provided of course that it is finite. In the last equality we used integration by parts.

In general the average survival probability P does not obey a kinetic equation, at least at the beginning, but one can always determine a *time-dependent rate parameter* $k(t)$ formally through the equation

$$\dot{P}(t) = -f(t) = -k(t)P(t) \,, \tag{B.87}$$

or equivalently,

$$k(t) = \frac{f(t)}{P(t)} \,. \tag{B.88}$$

If the reaction considered is the activated process, $k(t)$ in (B.88) reaches the long-lasting stationary value

$$k = \frac{f(t)^{st}}{P(t)^{st}} \,. \tag{B.89}$$

The flux-over-population formula (B.89) is usually simpler in applications than the time correlation function formula (B.76), which requires calculations of the full reactive flux (B.77). This method was used in the pioneering papers by Smoluchowski and Kramers. Substituting the solution of (B.87) with constant k,

$$P(t) = e^{-kt} \,, \tag{B.90}$$

into the last integral in (B.86), we get the relation

$$\tau = k^{-1} \,, \tag{B.91}$$

which means that the reaction rate constant for the irreversible reaction can also be calculated as the reciprocal of the mean first-passage time to the limbo state.

The formula (B.89) includes both the process of crossing the boundary, assuming the local equilibrium between initial and transition state, and the process of restoring this equilibrium from the R side, although of course it neglects the process of recrossing the boundary [see (B.81)]. One can, however, take into account the effects of the latter process by considering the backward irreversible reaction P → R. Because the forward and reverse transition state theory reaction rates are related by (B.74) [see Sect. 6.8 and (B.78)], the reciprocal reaction rates for both irreversible reactions are of the form

$$k_R^{-1} = (k_R^{eq})^{-1} + \tau_R \,, \qquad k_P^{-1} = K(k_R^{eq})^{-1} + \tau_P \,. \tag{B.92}$$

Knowing k_R^{eq}, one can express τ_R and τ_P in terms of k_R and k_P and, after substituting into (B.81) and (B.74), obtain the complete forward and reverse reaction rate constants:

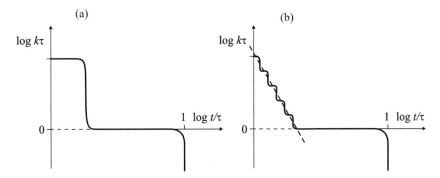

Fig. B.5. Log–log plot of the time-dependent rate parameter (the irreversible reactive flux) $k(t)$. (**a**) The case of a single time characterizing the intramolecular relaxation process (see Fig. B.3). (**b**) A whole scaling spectrum of intramolecular relaxation times

$$k_+^{-1} = k_R^{-1} + K^{-1}k_P^{-1} - (k_R^{eq})^{-1} \tag{B.93}$$

and

$$k_-^{-1} = k_P^{-1} + Kk_R^{-1} - K(k_R^{eq})^{-1} . \tag{B.94}$$

In the formulas (B.81) and (B.92), we assumed that the processes of intramolecular equilibration within R and P were characterized by the single relaxation time. The same was assumed when constructing Fig. B.3 which, introducing the natural unit of time τ, can be redrawn more correctly in a log–log plot (Fig. B.5a).

However, the processes of macromolecular relaxation are usually characterized by a whole spectrum of relaxation times (see Fig. B.2). Very often this spectrum scales, i.e., it has a self-similarity symmetry (Appendixes D.2 and D.3). In Fig. B.5b, such a scaling is sketched symbolically in the form of stairs which can be replaced by a sloping straight line. Thus, the time dependence of the rate parameter (B.88) in the initial stage of the reaction can be approximated by an algebraic function

$$k(t)\tau = (t/\tau)^{\alpha-1} , \tag{B.95}$$

where the exponent α varies between 0 and 1. The solution to (B.87) with a time-dependent k is

$$P(t) = e^{-\int_0^t dt\, k(t)} . \tag{B.96}$$

For $k(t)$ given by (B.95), the integral

$$\int_0^t dt\, k(t) = \tau^{-\alpha} \int_0^t dt\, t^{\alpha-1} = (t/\tau)^\alpha , \tag{B.97}$$

and hence the initial time course of the reaction, has the form of a *stretched exponential*:

$$P(t) = e^{-(t/\tau)^\alpha} \, . \tag{B.98}$$

Such an initial time course is often observed in the case of biochemical rations involving protein enzymes (Sect. 9.1).

B.5 One-Dimensional Diffusion in the Presence of a Sink

The process of continuous one-dimensional diffusion is described by the partial differential equation (see Appendix B.2):

$$\frac{\partial}{\partial t} p + \frac{\partial}{\partial x} j = 0 \, , \tag{B.99}$$

where t and x denote time and position (the value of the process), respectively, $p(x,t)$ is the probability density, and $j(x,t)$ the diffusion flux density.

As a complement to (B.99), we assume the general *reactive* boundary condition at $x = 0$:

$$j(0,t) = -\eta\, p(0,t) \, , \qquad p(x,t) = 0 \text{ for } x < 0 \, , \tag{B.100}$$

or

$$j(0,t) = \eta\, p(0,t) \, , \qquad p(x,t) = 0 \text{ for } x > 0 \, . \tag{B.101}$$

In the limit $\eta \to \infty$, the conditions (B.100) or (B.101) determine the *absorbing* boundary, viz.,

$$p(0,t) = 0 \, , \tag{B.102}$$

whereas in the limit $\eta \to 0$, the *reflecting* boundary is

$$j(0,t) = 0 \, . \tag{B.103}$$

For $\eta \neq 0$ there is always a negative jump in the flux density j at $x = 0$, whereupon (B.99) with the boundary condition (B.100) or (B.101) is to be replaced by the single equation

$$\frac{\partial}{\partial t} p + \frac{\partial}{\partial x} j = -\eta\delta(x)p \, , \tag{B.104}$$

with a delta-type sink. In fact, (B.104) is more general as the jump does not necessarily have to take place from or to the zero value of the diffusion flux density. Consequently, (B.104) describes both the processes of absorption and reflection, and the process of *transmission* from the region $(-\infty, 0)$ to $(0, \infty)$ or the reverse.

A general solution to (B.104) with an arbitrary initial probability density $p(x, 0)$ can be written in the form of the integral (B.45), where the kernel $p(x, t|x')$ is the solution to (B.104) with the delta-type initial probability distribution:

$$p(x, 0|x') = \delta(x - x') . \tag{B.105}$$

Our goal is to calculate the survival probability in the whole region $R = (-\infty, \infty)$ over time t, under the assumption that the process at $t = 0$ starts from the value x:

$$P(t|x) = \int_{-\infty}^{\infty} \mathrm{d}x' \, p(x', t|x) . \tag{B.106}$$

The probability $P(t|x)$ obeys the equation [see (B.85)]

$$\dot{P}(t|x) = -\eta \, p(0, t|x) \equiv -f(t|x) , \tag{B.107}$$

and therefore can also be calculated as a time integral

$$P(t|x) = 1 - \eta \int_0^t \mathrm{d}t' \, p(0, t'|x) . \tag{B.108}$$

Assuming that the solution $p^0(x, t|x')$ to the equation (B.99) for free diffusion, without any boundary condition or a sink, is known, one can find the solution $p(x, t|x')$ to the full equation (B.104) by treating the sink term formally as an external time-dependent perturbation. Following the theory of temporal Green functions [see, e.g., Byron and Fuller (1968)], the propagator $p(y, t|x)$, satisfies the self-consistent integral equation

$$p(y, t|x) = p^0(y, t|x) - \eta \int_0^t \mathrm{d}t' \, p^0(y, t-t'|0) \, p(0, t'|x) . \tag{B.109}$$

The particular propagator we need, i.e., $p(0, t|x)$, satisfies the integral equation

$$p(0, t|x) = p^0(0, t|x) - \eta \int_0^t \mathrm{d}t' \, p^0(0, t'|0) \, p(0, t - t'|x) . \tag{B.110}$$

The latter can be solved by introducing the Laplace transform

$$\tilde{p}(0, s|x) \equiv \int_0^{\infty} \mathrm{d}t \, e^{-st} p(0, t|x) . \tag{B.111}$$

In terms of Laplace transforms, (B.110) reads

$$\tilde{p}(0,s|x) = \tilde{p}^0(0,s|x) - \eta \tilde{p}^0(0,s|0)\tilde{p}(0,s|x) , \qquad (B.112)$$

whence

$$\tilde{p}(0,s|x) = \frac{\tilde{p}^0(0,s|x)}{1 + \eta \tilde{p}^0(0,s|0)} . \qquad (B.113)$$

The exact solution to (B.110) can be obtained in the case of *homogeneous* diffusion with the reactive boundary condition, described by the equation

$$\frac{\partial}{\partial t}p - D\frac{\partial^2}{\partial x^2}p = -\alpha\delta(x)p , \qquad (B.114)$$

where D denotes the diffusion constant and α, the transition probability to the sink per unit time. In this problem there are natural units of length,

$$\xi \equiv 4D/\alpha , \qquad (B.115)$$

and time,

$$\eta^{-1} \equiv 4D/\alpha^2 . \qquad (B.116)$$

Passing to the dimensionless position variable,

$$\xi^{-1}x \to x , \qquad (B.117)$$

equation (B.114) reads

$$\frac{\partial}{\partial t}p - \frac{\eta}{4}\frac{\partial^2}{\partial x^2}p = -\eta\delta(x)p . \qquad (B.118)$$

The free propagator, in the absence of a sink, is the Gaussian [see (B.48) and (B.49)]

$$p^0(0,t|x) = \frac{1}{\sqrt{\pi\eta t}}e^{-x^2/\eta t} , \qquad (B.119)$$

and both the direct and inverse Laplace transformations of (B.109) can be performed exactly using, e.g., the tables by Abramowitz and Stegun (1964). The result is

$$p(0,t|x) = \frac{1}{\sqrt{\pi\eta t}}e^{-x^2/\eta t} - \exp(\eta t + 2x)\,\mathrm{erfc}\left(\sqrt{\eta t} + \frac{x}{\sqrt{\eta t}}\right) , \qquad (B.120)$$

where the symbol erfc denotes the complementary error function:

$$\mathrm{erfc}\,z \equiv \frac{2}{\sqrt{\pi}}\int_z^\infty dy\, e^{-y^2} \approx \begin{cases} \dfrac{1}{\sqrt{\pi z^2}}e^{-z^2} & \text{for } z \gg 1 , \\[2mm] e^{-2z/\sqrt{\pi}} & \text{for } 0 < z \ll 1 . \end{cases} \qquad (B.121)$$

In this way, after integrating the propagator (B.120) following (B.108), we get the exact expression for the survival probability for the model considered:

$$P(t|x) = \exp(\eta t + 2x)\,\text{erfc}\left(\sqrt{\eta t} + \frac{x}{\sqrt{\eta t}}\right) + 1 - \text{erfc}\,\frac{x}{\sqrt{\eta t}} \,. \qquad \text{(B.122)}$$

For $x = 0$ we get from (B.120) the probability density of returning to the initial state,

$$p(0, t|0) = \frac{1}{\sqrt{\pi \eta t}} - e^{\eta t}\,\text{erfc}(\sqrt{\eta t}) \,, \qquad \text{(B.123)}$$

and from (B.122), the corresponding survival probability,

$$P(t|0) = e^{\eta t}\,\text{erfc}(\sqrt{\eta t})$$
$$\approx \begin{cases} \exp\left[-2(\eta t)^{1/2}/\sqrt{\pi}\right] & \text{for } t \ll \eta^{-1}\,, \\ (\eta t)^{-1/2}/\sqrt{\pi} & \text{for } t \gg \eta^{-1}\,. \end{cases} \qquad \text{(B.124)}$$

Because the region of diffusion is unbounded, the mean first-passage time (B.86) for the survival probability (B.122) or (B.124) is infinite. In other words, the survival probability has a *long-time tail*, i.e., the reaction modeled is not the activated process. However, the space of microstates of real molecules is bounded, which makes any gated reaction the activated process. Consequently, the reaction will eventually reach the stage of exponential decay with relaxation time equal to the reciprocal rate constant κ^{-1}. The crossing between the time course of reaction $P(t|x)$ described by (B.122) and the exponential decay can be described by the simple corrected formula for the survival probability (Kurzyński, 1997a)

$$\bar{P}(t|x) = \left[(1-a)P(t|x) + a\right]e^{-\kappa t} \,, \qquad \text{(B.125)}$$

where a denotes the level from which the exponential decay begins.

Following (B.107), we get the corresponding first-passage time distribution:

$$\bar{f}(t|x) = (1-a)\left\{f(t|x) + \kappa\left[P(t|x) + \frac{a}{1-a}\right]\right\}e^{-\kappa t} \,, \qquad \text{(B.126)}$$

and following (B.86), the corresponding mean first-passage time from x through the sink at 0 to the totally absorbing limbo state:

$$\bar{\tau}(x) = (1-a)\int_0^\infty dt\, P(t|x)e^{-\kappa t} + a\kappa^{-1} \,. \qquad \text{(B.127)}$$

The integral has the meaning of the Laplace transform of the survival probability $P(t|x)$. The finite value of κ secures the cutoff of its long-time tail. After integrating by parts and taking into account (B.107), we obtain

$$\bar{\tau}(x) = \kappa^{-1} - (1-a)\frac{\eta}{\kappa} \int_0^\infty dt\, e^{-\kappa t} p(0, t|x) . \tag{B.128}$$

For the Laplace transform of the transition probability density (B.120), equation (B.128) reads

$$\bar{\tau}(x) = \kappa^{-1} - (1-a)\kappa^{-1}\frac{e^{-2|x|\sqrt{\kappa/\eta}}}{\sqrt{\kappa/\eta}+1} . \tag{B.129}$$

This simple formula can be used in a more general context to describe the mean first-passage time dependence on some *effective* distance $|x|$ from the gate for homogeneous diffusion on lattices of arbitrary dimensions (Chełminiak and Kurzyński, 2004). Let us note that the mean first-passage time (B.129) is always shorter than the value determined by the reaction rate constant κ.

B.6 Diffusion in a Parabolic Potential

In the case of *inhomogeneous* one-dimensional diffusion in a certain potential $G(x)$, the diffusion flux density $j(x,t)$ consists of two components [Gardiner, 1983; van Kampen, 2001; see also (B.59)]:

$$j = -D\left[\beta\left(\frac{\partial G}{\partial x}\right) + \frac{\partial}{\partial x}\right]p , \tag{B.130}$$

where $\beta \equiv 1/k_B T$ denotes the inverse temperature.

For diffusion in the parabolic potential

$$G(x) = \frac{1}{2}K(x - x_0)^2 , \tag{B.131}$$

there are natural units of length,

$$\xi \equiv (\beta K/2)^{-1/2} , \tag{B.132}$$

and time,

$$\gamma^{-1} \equiv (\beta K D)^{-1} . \tag{B.133}$$

Passing to the dimensionless position variable,

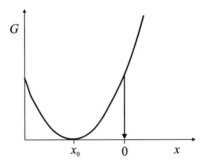

Fig. B.6. Diffusion in the parabolic potential in the presence of a point sink

$$\xi^{-1}x \to x , \tag{B.134}$$

equation (B.99) with j determined by (B.130) reads

$$\frac{\partial}{\partial t}p(x,t) = \frac{1}{2}\gamma\frac{\partial}{\partial x}\left[\frac{\partial}{\partial x} + \frac{\partial G(x)}{\partial x}\right]p(x,t)$$

$$= \frac{1}{2}\gamma\frac{\partial}{\partial x}\left\{e^{-G(x)}\frac{\partial}{\partial x}\left[e^{G(x)}p(x,t)\right]\right\} , \tag{B.135}$$

where our parabolic potential is given in $k_{\mathrm{B}}T$ units by (Fig. B.6)

$$G(x) = (x - x_0)^2 . \tag{B.136}$$

In the case discussed, (B.109) can no longer be solved analytically. However, if the potential $G(x)$ tends to infinity for $x \to -\infty$, the mean first-passage time $\tau(x)$ becomes finite without any additional assumptions. In the limit $\eta \to \infty$ of a totally absorbing boundary, the formula

$$\tau(x) = 2\gamma^{-1}\int_x^0 dy\, e^{G(y)}\int_{-\infty}^y dz\, e^{-G(z)} \tag{B.137}$$

follows from (B.135) for an arbitrary potential $G(x)$ (Gardiner, 1983; Montroll and West, 1987; Hänggi et al., 1990). In particular, for the parabolic potential (B.136), equation (B.137) reads

$$\tau(x) = 2\gamma^{-1}\int_x^0 dy\, e^{(y-x_0)^2}\int_{-\infty}^y dz\, e^{-(z-x_0)^2}$$

$$= 2\gamma^{-1}\int_{x-x_0}^{-x_0} dy\, e^{y^2}\int_{-\infty}^y dz\, e^{-z^2} . \tag{B.138}$$

The diffusion down the potential can be several orders of magnitude faster than the diffusion up it. We approximate the corresponding formulas using the asymptotic expansions of the integral that defines the error function (Abramowitz and Stegun, 1964):

$$\int_{-\infty}^{y} dz\, e^{-z^2} \equiv \frac{\sqrt{\pi}}{2}(1 + \operatorname{erf} y) \equiv \frac{\sqrt{\pi}}{2}(2 - \operatorname{erfc} y)$$

$$= \begin{cases} \sqrt{\pi} - \cdots & \text{for } y \to +\infty, \\ \dfrac{1}{2|y|}\, e^{-y^2} + \cdots & \text{for } y \to -\infty. \end{cases} \tag{B.139}$$

According to the expansion for $y \to \infty$, the time of diffusion up the parabolic potential from its minimum at $x = x_0$ to the gate at $x = 0$ can be expressed as

$$\tau_{\text{up}}(x_0 \to 0) = 2\gamma^{-1} \int_0^{|x_0|} dy\, e^{y^2} \int_{-\infty}^{y} dz\, e^{-z^2}$$

$$\approx 2\sqrt{\pi}\gamma^{-1} \int_0^{|x_0|} dy\, e^{y^2} \approx \sqrt{\pi}\gamma^{-1}|x_0|^{-1} e^{x_0^2}. \tag{B.140}$$

The first approximate equality results from the fact that the main contribution to the integral over y comes from large y values and the second approximate equality results from the asymptotic expansion of the Dawson integral (Abramowitz and Stegun, 1964):

$$\int_0^x dy\, e^{y^2} = \frac{1}{2x}e^{x^2} - \cdots \quad \text{for } x \to +\infty. \tag{B.141}$$

For x close to the gate and $x_0 < 0$, the diffusion still proceeds up the potential and, according to the expansion (B.139) for $y \to \infty$,

$$\tau_{\text{up}}(x \to 0) = 2\gamma^{-1} \int_{|x_0|}^{|x_0|+x} dy\, e^{y^2} \int_{-\infty}^{y} dz\, e^{-z^2}$$

$$\approx 2\sqrt{\pi}\gamma^{-1} \int_{|x_0|}^{|x_0|+x} dy\, e^{y^2} \approx 2\sqrt{\pi}\gamma^{-1}|x|e^{x_0^2}. \tag{B.142}$$

Conversely, for x close to the gate and $x_0 > 0$, the diffusion proceeds down the potential and, according to the expansion (B.139) for $y \to \infty$,

$$\tau_{\text{dn}}(x \to 0) = 2\gamma^{-1} \int_{-|x_0|+x}^{-|x_0|} dy\, e^{y^2} \int_{-\infty}^{y} dz\, e^{-z^2}$$

$$\approx \gamma^{-1} \int_{-|x_0|+x}^{-|x_0|} dy\, \frac{1}{|y|} = \gamma^{-1} \ln \frac{|x_0|}{|x_0| - x}$$

$$\approx \gamma^{-1} \frac{|x|}{|x_0|}. \tag{B.143}$$

The times (B.142) and (B.143) are short compared to the time (B.140), which is to a good approximation the longest time in the system. Its reciprocal equals the reaction rate constant k [see (B.91)]. Equation (B.140) reconstructs the original Kramers result for the reaction controlled by the overdamped motion in the parabolic potential along the reaction coordinate (Hänggi et al., 1990). Equations (B.142) and (B.143) present corrections to the Kramers theory for the initial stage of a reaction (Kurzyński, 1997b), or determine steady-state flux–force dependences (Sects. 9.3 and 9.5) in the protein-machine model of intramolecular enzyme dynamics (see Appendix D.4).

C Structure of Biomolecules

C.1 Elementary Building Blocks

Animate matter is almost exclusively built from six elements:

- hydrogen H (60.5%),
- oxygen O (25.7%),
- carbon C (10.7%),
- nitrogen N (2.4%),
- phosphorous P (0.17%),
- sulfur S (0.13%).

The percentages of atomic abundance in the soft tissues of the mature human body are given in brackets (Bergethon and Simons, 1990). The remaining 0.4% are ions that control the electrolytic equilibrium:

- calcium Ca^{2+} (0.23%),
- sodium Na^+ (0.07%),
- potassium K^+ (0.04%),
- magnesium Mg^{2+} (0.01%),
- chloride Cl^- (0.03%).

Two transition metal ions that are electron carriers, i.e.,

- iron $Fe^{3+} \longleftrightarrow Fe^{2+}$,
- copper $Cu^{2+} \longleftrightarrow Cu^+$,

and trace elements

- Mn, Zn, Co, Mo, Se, J, F, ... ,

are much less abundant.

At the lowest level of chemical organization, animate matter is also composed of a rather limited number of standard building blocks. One can divide them into seven classes (Pauling and Pauling, 1975):

- carboxylic acids characterized by general formula R−COOH, dissociated to anions R−COO$^-$ in a neutral water environment,

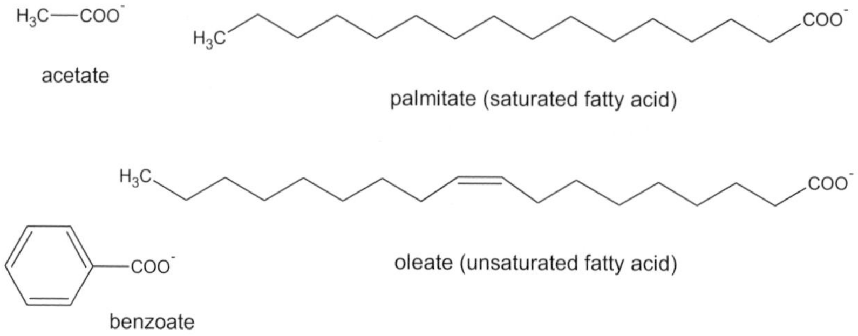

Fig. C.1. Four typical examples of carboxylic acids with purely hydrocarbon substituents

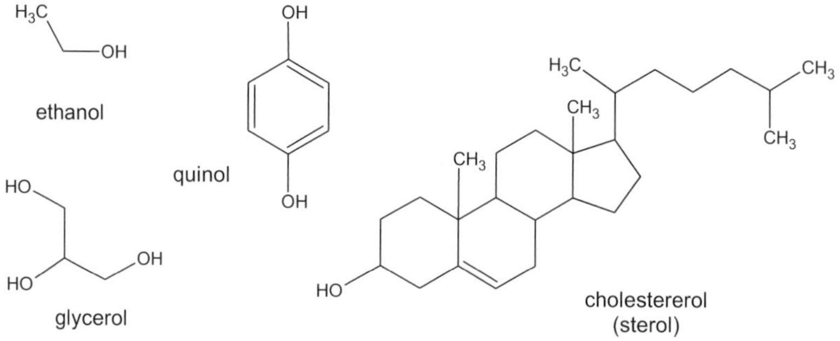

Fig. C.2. Four typical examples of alcohols with purely hydrocarbon substituents

- alcohols characterized by the general formula $R-OH$,
- monosaccharides characterized by the general formula $(CH_2O)_n$ with $n = 5$ (pentoses) or $n = 6$ (hexoses) and containing either an aldehyde group $-CHO$ (aldoses) or a ketone group $-CO-$ (ketoses),
- amines characterized by general formula NH_2-R, protonated to cations NH_3^+-R in a neutral water environment,
- nitrogen heterocycles,
- phosphates characterized by the general formula $R-O-PO_3^{2-}$, doubly dissociated in a neutral water environment,
- hydrosulfides characterized by the general formula $R-SH$.

The Rs are substituents that distinguish individual compounds. They can be shorter or longer, open or closed hydrocarbon chains. Several examples of carboxylic acids and alcohols are shown in Figs. C.1 and C.2, respectively.

A chemical compound is definitely identified only if the spatial structure of its system of covalent bonds (*constitution*) is described explicitly, because two or more different compounds can happen to have the same atomic composition (the phenomenon of *isomerism*). In organic chemistry, to simplify the notation of structural formulas characterizing the molecular constitution, we often omit the carbon C and hydrogen H symbols and leave only a lattice of covalent bonds between the carbon atoms. When interpreting schemes like those presented in Figs. C.1 and C.2, it is assumed that, at all vertices of lattices shown, the carbon atoms are completed by an appropriate number of hydrogen atoms to preserve the fourfold carbon valency.

Figure C.3 shows the three most important examples of monosaccharides: glucose, fructose and ribose. Under physiological conditions, the majority of monosaccharide molecules form a ring structure closed by an oxygen atom that originates from breaking the double carbon–

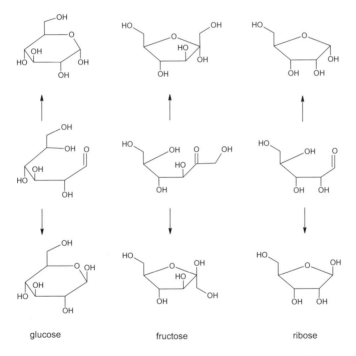

Fig. C.3. Three important examples of monosaccharides: glucose, fructose (hexoses) and ribose (pentose). The glucose and ribose rings are closed by an oxygen coming from an aldehyde group. The fructose ring is closed by an oxygen coming from a ketone group. α anomers are shown in the *upper row* and β anomers in the *lower row*

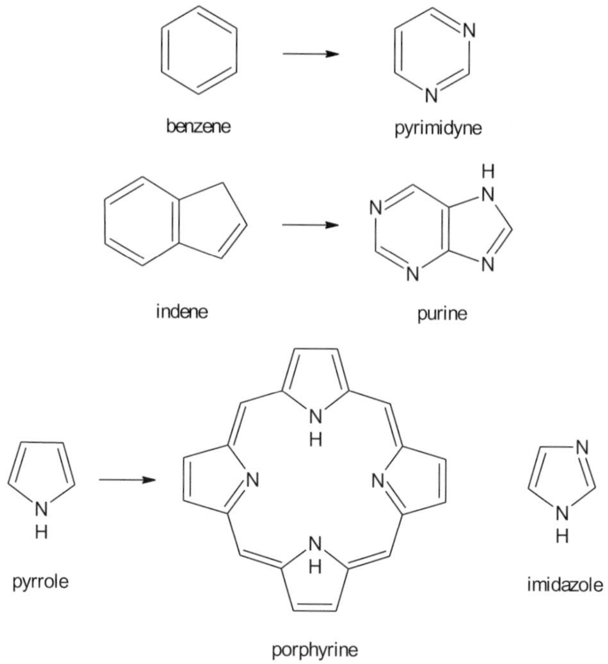

Fig. C.4. Several important examples of heterocycles. Pyrimidine is a direct modification of benzene, and purine is a direct modification of indene. No purely hydrocarbon homologues of pyrrole and imidazole exist. Porphyrin, the main component of heme and chlorophyll, is formed from four pyrrole rings

oxygen bond in an aldehyde or a ketone group. Just before closing the monosaccharide ring, a rotation can take place about the single carbon–oxygen bond, and this results in the formation of two ring isomers, α and β. Because of the low activation energy, they are often called *anomers*. At physiological temperatures, the β anomer is twice as abundantly occupied as the α anomer.

Nitrogenous heterocycles are modifications of aromatic hydrocarbons in which a singly hydrogenated carbon (CH) is replaced by N. Figure C.4 shows the most important examples.

The simplest examples of phosphates are the *orthophosphate* $H-O-PO_3^{2-}$, a doubly dissociated anion of orthophosphoric acid known in biochemistry as an *inorganic phosphate* (P_i), and the *pyrophosphate* $H-O-PO_2^+-O-PO_3^{2-}$, a triply dissociated anion of pyrophosphoric acid known in biochemistry as an *inorganic diphosphate*

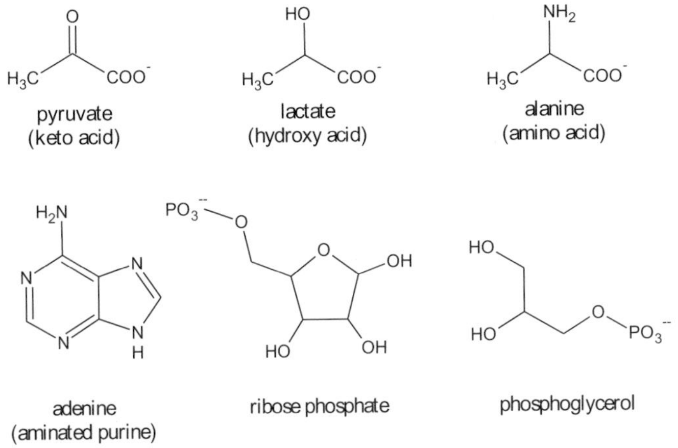

pyruvate
(keto acid)

lactate
(hydroxy acid)

alanine
(amino acid)

adenine
(aminated purine)

ribose phosphate

phosphoglycerol

Fig. C.5. Examples of organic compounds with two functional groups

(PP$_i$). Phosphates will be considered further in the context of phosphodiester bonds.

Hydrosulfides R$-$SH are structurally similar to hydroxides (alcohols) R$-$OH. However, they differ by forming not only thioester R$-$CO$-$S$-$R$'$, but also (after reduction) disulfide R$-$S$-$S$-$R$'$ bonds.

Elementary organic building blocks very often belong simultaneously to two or more classes listed above. This means that the substituent R, besides a hydrocarbon component, comprises also an additional functional group or serves as such a group. Thus, we can have simple carboxylic acids with ketonic $-$CO$-$, hydroxylic $-$OH or protonated aminic $-$NH$_3^+$ groups (*keto*, *hydroxy* and *amino* acids, respectively) or, e.g., aminated heterocycles or phosphorylated saccharides and alcohols (Fig. C.5).

C.2 Generalized Ester Bonds

The reaction between a carboxylic acid and an alcohol is of special importance in organic chemistry. The product is referred to as an *ester*, with a characteristic ester bond $-$CO$-$O$-$ (Fig. C.6a). Alcohol in the esterification reaction can be replaced by an amine to form an *amide* with an amide bond $-$CO$-$NH$-$ (Fig. C.6b). Carboxylic acid, on the other hand, replaced by phosphate forms a *phosphodiester bond* $-$O$-$PO$_2^-$$-O-$ (Fig. C.6c). *Phospholipids*, the main component of biological membranes, are double esters of fatty acids and glycerol and, simultaneously, phosphodiesters (Fig. C.7).

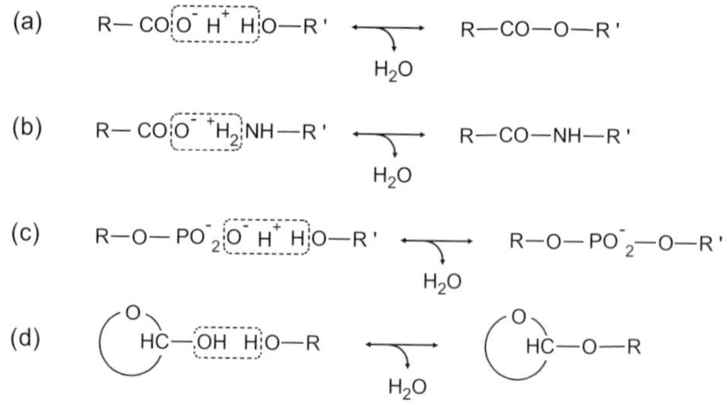

(a)   R— CO[O⁻ H⁺ H]O—R'   ⟶   R—CO—O—R'
                                H₂O

(b)   R— CO[O⁻ ⁺H₂]NH—R'   ⟶   R—CO—NH—R'
                                H₂O

(c)   R—O— PO⁻₂[O⁻ H⁺ H]O—R'   ⟶   R—O— PO⁻₂—O—R'
                                H₂O

(d)   HC—[OH H]O—R   ⟶   HC—O—R
                                H₂O

Fig. C.6. Formation of the ester bond (**a**) and its generalization to amide (**b**), phosphodiester (**c**) and glycosidic (**d**) bonds. The circular ring represents any monosaccharide ring closed by an oxygen atom neighboring a carbon atom taking part in the glycosidic bond. The carboxylic and phosphate groups are assumed to be dissociated

Monosaccharides behave both as alcohols and as carboxylic acids. From their open structure (Fig. C.3), it follows that the carboxylic properties are characteristic of $-OH$ groups bound to the carbon atom neighboring the oxygen atom that closes the saccharide ring. The generalized ester bond formed by such a group with another hydroxylic $-OH$ group with alcoholic properties is referred to as a *glycosidic bond* (Fig. C.6d). The alcohol hydroxylic group can be replaced by the $-NH_2-$ group of some nitrogenous heterocycles that are derivatives of pyrimidine and purine. Such a bond is called an N-glycosidic bond. Compounds of ribose with nitrogenous heterocycles (or nitrogenous bases as the $-NH_2-$ groups are good proton acceptors) are referred to as *nucleosides* and their phosphates as *nucleotides* (Fig. C.8).

The chemical equilibrium of all the four reactions presented in Fig. C.6 is strongly shifted to the left, toward hydrolysis rather than synthesis of the bond. The formation of the generalized ester bond needs an additional source of free energy. It therefore proceeds along

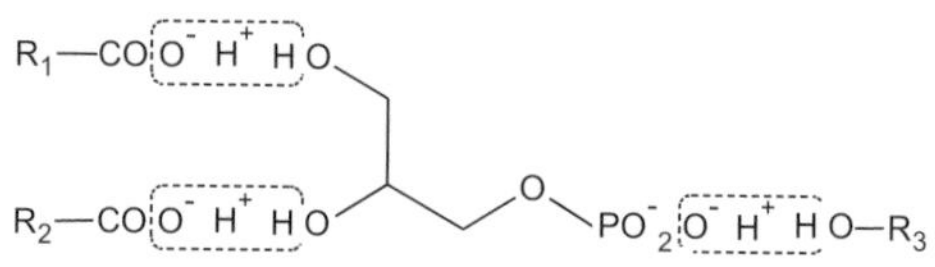

R_1—CO[O⁻ H⁺ H]O

R_2—CO[O⁻ H⁺ H]O   PO⁻₂[O⁻ H⁺ H]O—R_3

Fig. C.7. Phospholipids originate as a result of triple esterification of phosphoglycerol (see Fig. C.5)

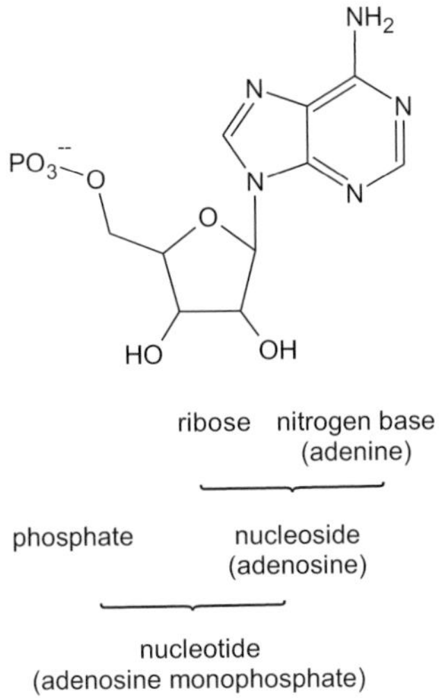

ribose   nitrogen base
(adenine)

phosphate   nucleoside
(adenosine)

nucleotide
(adenosine monophosphate)

Fig. C.8. The nucleotide, an elementary component unit of nucleic acids

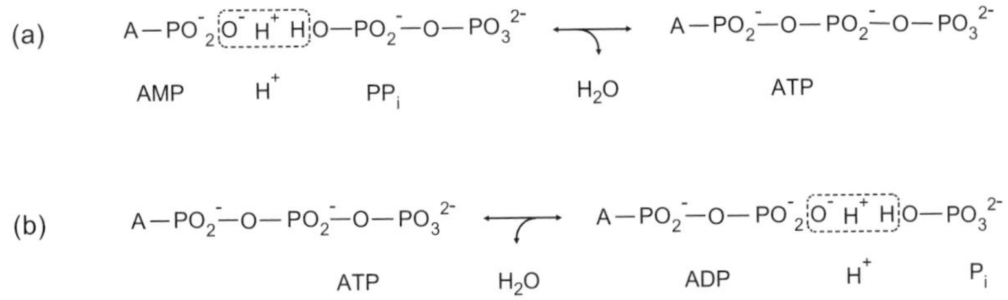

(a) $A-PO_2^- \boxed{O^- H^+ H}O-PO_2^--O-PO_3^{2-}$ ⟶ $A-PO_2^--O-PO_2^--O-PO_3^{2-}$

       AMP H^+ PP_i H_2O ATP

(b) $A-PO_2^--O-PO_2^--O-PO_3^{2-}$ ⟶ $A-PO_2^--O-PO_2 \boxed{O^- H^+ H}O-PO_3^{2-}$

              ATP H_2O ADP H^+ P_i

Fig. C.9. (a) Formation of ATP (adenosine triphosphate) from AMP (adenosine monophosphate) and PP_i (inorganic diphosphate). (b) Hydrolysis of ATP to ADP (adenosine diphosphate) and P_i (inorganic phosphate)

a more complex pathway than the one presented in Fig. C.6. Universal donors of biological free energy are nucleoside triphosphates, mainly ATP (adenosine triphosphate). Hence, the formation of the generalized ester bonds usually proceeds simultaneously with the hydrolysis of ATP to ADP (adenosine diphosphate) (see Fig. C.9b). The ATP itself originates from AMP (adenosine monophosphate) as a result of phosphodiester bond formation with inorganic diphosphate PP_i (Fig. C.9a).

All three kinds of biological macromolecule, i.e., polysaccharides, proteins and nucleic acids, are built from the elementary entities

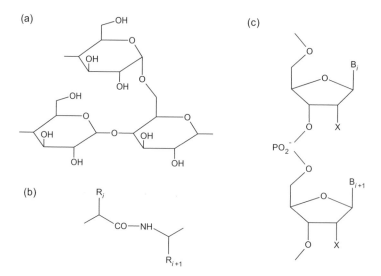

Fig. C.10. (**a**) Polysaccharides (the case of glycogen) are polymers of monosaccharides linked by glycosidic bonds. (**b**) Proteins are linear polymers of amino acids linked by amide bonds. R_i denote side chains of the successive amino acids. (**c**) Nucleic acids are linear polymers of nucleotides linked by phosphodiester bonds. B_i denote the successive nitrogenous bases. X = OH in the case of RNA (ribonucleic acids) and X = H in the case of DNA (deoxyribonucleic acids)

monosaccharides, amino acids and nucleotides, respectively, linked by generalized ester bonds.

Polysaccharides are polymers of monosaccharides linked by glycosidic bonds (Fig. C.10a). The polymers can be either linear (e.g., cellulose, chitin or amylose – a type of starch which is the main storage polysaccharide in plants) or branched (e.g., amylopectin, another type of starch, or glycogen, the main storage polysaccharide in animals). Small polysaccharides are called *oligosaccharides* and very small ones are called *disaccharides*, *trisaccharides*, etc.

Proteins are linear polymers of amino acids linked by amide bonds (Fig. C.10b). Small proteins are called *peptides*, and amide bonds are often called *peptide bonds*.

Nucleic acids are linear polymers of nucleotides linked by phosphodiester bonds (Fig. C.10c). Note that in *dinucleotites*, which often appear as cofactors of protein enzymes, two component nucleotides are linked by phosphotriester rather than phosphodiester bonds. Examples are NAD$^+$ (nicotinamide adenine dinucleotide) and FAD (flavin adenine dinucleotide), two molecules that carry electrons jointly with protons in the form of hydrogen atoms, considered in Chaps. 4 and 5.

A more systematic but nevertheless compact introduction to the foundations of organic chemistry can be found in Chap. 13 of the textbook by Pauling and Pauling (1975).

C.3 Directionality of Chemical Bonds

The spatial structures of biomolecules, so important for their function, are related to directional properties of chemical bonds. These properties are in turn determined by the spatial distribution of electronic states of the constituent atoms. The hydrogen atom contributes only one s-orbital to chemical bonding, while carbon, nitrogen and oxygen atoms contribute one s-orbital and three p-orbitals, and phosphorous and sulfur atoms can contribute another five d-orbitals (Atkins, 1998, Chap. 14).

Figure C.11a outlines the spatial distribution of the electron probability density in the s- and p-orbitals. One s-orbital and three p-orbitals can hybridize into four orbitals of tetrahedral symmetry (sp^3 hybridization), and one s-orbital and two p-orbitals can hybridize into three orbitals of trigonal symmetry (sp^2 hybridization) (see Fig. C.11b). Two orbitals of digonal symmetry coming from sp hybridization play no practical role in biochemistry.

The carbon atom has four electrons on the outer shell, organized into four sp^3 hybridized orbitals. It needs four additional electrons for those orbitals to be completely filled. As a consequence, it can bind four hydrogen atoms, each giving one electron and admixing its own s-orbital to the common bonding σ-orbital with axial symmetry (Fig. C.12).

The nitrogen atom has five electrons in four sp^3 hybridized orbitals. It needs only three additional electrons and thus binds three hydrogen atoms. The ammonia molecule formed, like the methane molecule, has tetrahedral angles between the bonds. However, one bond is now replaced by a lone electron pair (Fig. C.12). The oxygen atom has six electrons in four sp^3 hybridized orbitals. It binds two hydrogen atoms and has two lone electron pairs (Fig. C.12). As a consequence of sp^3 hybridization, the nitrogen and oxygen atoms have highly directed negative charge distributions, leading as we shall see to hydrogen bonding, essential for the spatial structures of all biomolecules.

By replacing one hydrogen atom H in methane, ammonia and water by the methyl group CH_3, one obtains ethane, methylamine and

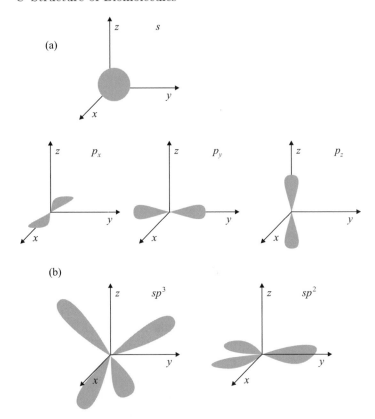

Fig. C.11. (a) Outline of the spatial distribution of electron probability density in an s-orbital and three p-orbitals. (b) Four orbitals of tetrahedral sp^3 hybridization and three orbitals of trigonal sp^2 hybridization

methanol, respectively (Fig. C.13). The CH_3 can rotate around the C−C, C−N, or C−O bonds. The lowest value of the potential energy of the molecule is found for the electron densities of one triple of atoms or lone electron pairs located between the electron densities of the other triple of atoms when looking along the rotation axis.

In the cases of ethane, methylamine, and methanol, a $120°$ $(2\pi/3)$ rotation does not lead to a structural change of the molecule. However, if the other triple has free electron pairs or atoms other than hydrogen, a rotation around the central covalent bond may lead to a new *conformational state* of the molecule (Pauling and Pauling, 1975, Chap. 7). Such a state cannot be reconstructed from the original state by either a translation or a rigid body rotation. Various conformational states become geometrically significant for long molecular chains, e.g., hy-

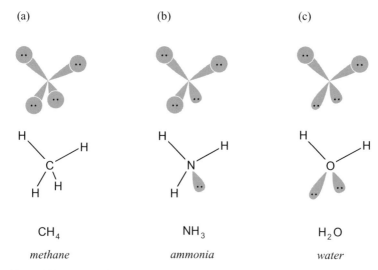

Fig. C.12. The spatial structures of methane, ammonia and water molecules are determined by tetrahedral sp^3 hybridization of the central atom orbitals. A 'cloud' with two dots denotes a lone electron pair

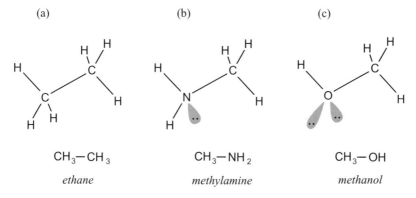

Fig. C.13. Spatial structures of ethane, methylamine, and methanol

drocarbons, which can exist both in maximally stretched linear forms and as folded clusters. A rotation around each covalent bond in the chain allows one conformational *trans* state and two *gauche* states (Fig. C.14a).

If we ignore the steric constraints (excluded volume effects) that rapidly emerge in longer chains and constitute a separate and significant problem, the differences between equivalent conformational states of a single bond amount to several kJ/mol, while the potential energy barrier height is on the order of 10 to 20 kJ/mol (Fig. C.14b). This

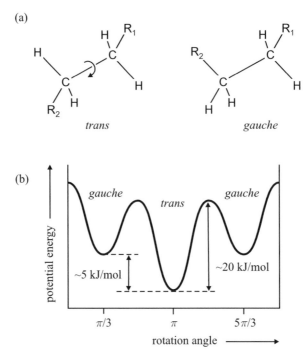

Fig. C.14. (**a**) Conformational states related to rotation around one covalent bond. R_1 and R_2 are atoms or molecular groups (in the case of longer chains) other than atomic H. (**b**) The potential energy related to the rotation around a single covalent bond. The three minima correspond to the three conformational states: one *trans* and two *gauche*. They are separated by energy barriers with heights of 10 to 20 kJ/mol

corresponds to four to eight times the value of the mean thermal energy $k_BT = 2.5$ kJ/mol at physiological temperatures. The probability of a random accumulation of such an amount of energy in one degree of freedom, defined by the Boltzmann factor, i.e., the exponential of its ratio to k_BT, $\exp(-\Delta/k_BT)$, equals 10^{-2} to 10^{-4}. The latter value multiplied by the average frequency of thermal vibrations, 10^{13} s^{-1}, gives 10^{11} to 10^9 random local conformational transitions per second at physiological temperatures. A conformational state of a molecule is not therefore very stable.

In the case of closed chains (unsaturated cyclic hydrocarbons or monosaccharides), 120° rotations around individual bonds are not possible without breaking them. Hence, conformational transitions involve much smaller rotations that are simultaneously applied to many bonds. The process is called ring puckering since an entirely flat conformation

envelope   half-chair   chair   twist   boat

Fig. C.15. Envelope and half-chair conformations of five-atom rings, and chair, twist and boat conformations of six-atom rings

becomes energetically unstable. In the case of a five-atom ring, we distinguish the conformations of an *envelope* and a *half-chair*. In the case of a six-atom ring we distinguish the conformations of a *chair*, a *twist* and a *boat* (Fig. C.15).

Not all transitions between different energetically stable geometrical structures can be achieved by simple rotations around covalent bonds involving small energy barrier crossings. A mutual exchange of hydrogens with hydroxyl groups of monosaccharide rings (see Fig. C.3) is not possible without covalent bond breaking and its subsequent restoration. Table C.1 lists some of the biologically important bond lengths and bond energies (Atkins, 1998, Chap. 14). It can be seen that the process of a single covalent bond breaking and restoration requires an energy of about 300 kJ/mol. This is one and a half orders of magnitude greater than the energies mentioned earlier. Various geometrical forms of chemical molecules with the same summary formulas but requiring bond breaking and bond restoration are called *isomers*. It is easy to imagine a large diversity of such isomers even for simple monosaccharides. Isomers differ in the location of hydrogen atoms and hydroxyl groups and in the position of the oxygen bridges. Hexoses can form both five- and six-atom rings (see the structures of glucose and fructose in Fig. C.3).

Half the monosaccharides are simple mirror images of their counterparts. Such isomers are called *enantiomers* and the phenomenon itself is called *chirality*. This refers to the symmetry-breaking of handedness (left/right invariance), from the analogy with human hands. In fact the term comes from the Greek word 'cheir' meaning 'hand'. Ignoring specific relations with double bonds, steroisomerism is usually linked to the existence of at least one carbon atom in the molecule whose four covalent bonds are all inequivalent in that they lead to different atoms or groups of atoms. Chiral molecules are optically active, twisting the light polarization plane to the left or right.

Enantiomers are divided into L- and D-types. Their definitions are based on the simplest monosaccharide, glyceraldehyde (Fig. C.16). L-

Table C.1. Lengths and energies of biologically important covalent bonds. Note: 1 kJ/mol = 0.24 kcal/mol = 1.0×10^{-2} eV/bond = 0.40 kT/bond at 300 K

	Bond length [nm]	Energy [kJ mol^{-1}]
C–C	0.154	350
C=C	0.134	610
C–N	0.147	300
C=N	0.126	610
C–O	0.143	360
C=O	0.114	740
C–S	0.182	260
S–H	0.135	340
C–H	0.114	410
N–H	0.101	390
O–H	0.096	460
H–H	0.074	440

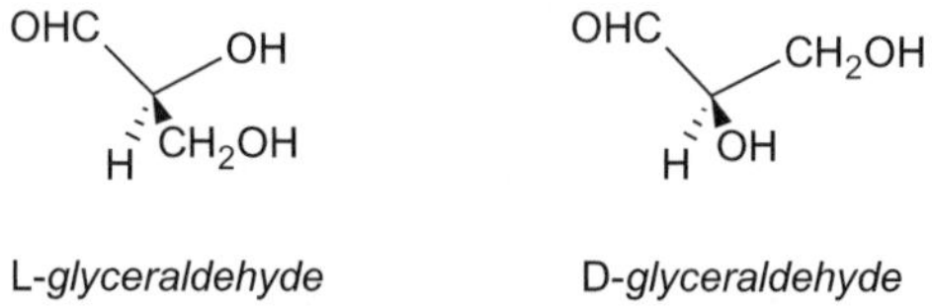

L-glyceraldehyde   D-glyceraldehyde

Fig. C.16. Two enantiomers of glyceraldehyde. *Thick continuous lines* represent bonds that point to the observer, whereas *thick dashed lines* represent bonds that point from the observer

type monosaccharides are derivatives of L-glyceraldehyde. Similarly, D-type monosaccharides are derivatives of D-glyceraldehyde. By definition D-glyceraldehyde twists the polarization plane to the right, but this is not generally the case for all D-type monosaccharides. Furthermore, all amino acids except for glycine are chiral, which is linked to the fact that the central carbon C^α has four inequivalent bonds. Therefore, we have both D-amino acids and L-amino acids. Chemists have adopted an unambivalent way of transferring the determination of D- and L-type enantiomers from the defining glyceraldehydes onto other compounds different from monosaccharides. This is a little too complicated for our purposes, so we will not dwell on it here. We only wish to note that all biologically active monosaccharides are D-types, while all biologically active amino acids are L-types.

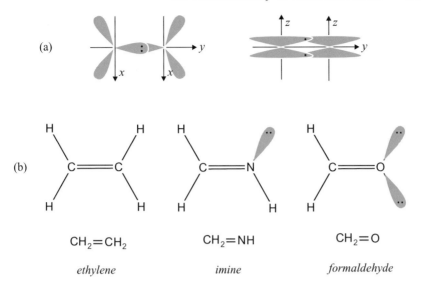

Fig. C.17. (a) A double bond consists of a σ-bond formed by orbitals in sp^2 hybridization and a π-bond formed from p-orbitals perpendicular to the previous ones. (b) Planar molecules of ethylene, imine, and formaldehyde

Let us now consider the bonds formed by electron orbitals in trigonal hybridization sp^2. Using these orbitals, two carbon atoms can form σ-bonds (with axial symmetry) among themselves and with the other four atoms that lie in the same plane, for example hydrogen. The remaining p-orbitals perpendicular to this plane, one for each carbon atom, then form a second, slightly weaker bond between these atoms. This is called a π-bond (see Fig. C.17a). A planar molecule of ethylene is formed in which two carbon atoms are bound together via a double bond. If one carbon atom is replaced by a nitrogen atom, with an extra electron, and simultaneously one hydrogen atom is replaced by a lone electron pair, we obtain an imine molecule. If, on the other hand, one carbon atom is replaced by an oxygen atom with two excess electrons and simultaneously two hydrogen atoms by two lone electron pairs, we obtain a formaldehyde molecule (Fig. C.17b). Table C.1 lists the most important double bond lengths and energies.

Double bonds can exist in longer chain molecules, for example, unsaturated hydrocarbons. For each double bond, two energetically stable spatial structures can occur, each obtained from the other by a $180°$ rotation. A rotation around a double bond requires a temporary breakage of the π-bond, and hence an energy of nearly 300 kJ/mol (see

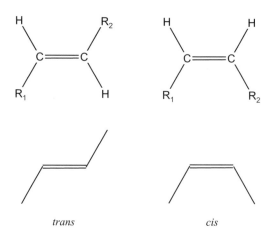

Fig. C.18. Configurational *trans* and *cis* isomers are related via rotation through one double bond. R_1 and R_2 denote atoms or molecular groups other than H

Table C.1). Hence, both structures are isomers rather than conformers. They are called *trans* and *cis configurational isomers* (Fig. C.18).

The π-bonds often occur in a delocalized form. In the famous historical case of the benzene molecule C_6H_6 (Fig. C.19a), one can imagine two symmetrical Kekule structures in which every other carbon–carbon bond is a double bond. Which of these two structures is actually adopted by the molecule? Are both represented statistically in the population of identical molecules with the same probability? Answers to these questions come from quantum mechanics. The actual state of each benzene molecule is a quantum-mechanical linear combination of the two states defined by the Kekule structures. The π-bond is not localized on every other carbon pair, but rather delocalized over the entire ring. An electric current circulates around the ring and can be induced by magnetic fields and other electric currents. Due to such planar interactions in benzene rings and also in other aromatic heterocycles (e.g., in purine and pyrimidine bases), their compounds have a tendency for parallel *stacking*. Effectively, each carbon pair is allocated one half of a π-bond, which justifies the commonly accepted notation (Fig. C.19a).

The π-bond between carbon and oxygen in a dissociated carboxyl group $-COO^-$ is also delocalized (Fig. C.19b). This group can be envisaged as a planar, negatively charged plate with a rotational degree of freedom around the axis that connects one carbon atom to the next atom. The phosphodiester bond (Fig. C.19c) exhibits similar behav-

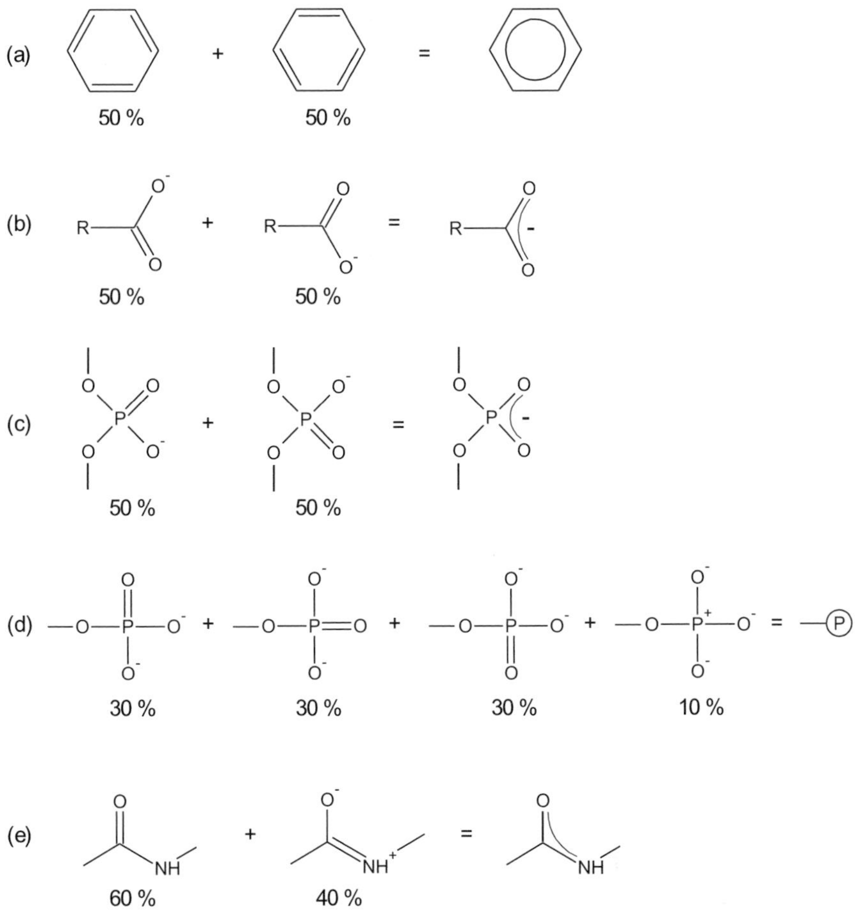

Fig. C.19. Delocalization of π-bonds in the case of benzene (**a**), a carboxylic anion (**b**), a phosphodiester bond (**c**), an orthophosphate anion (**d**), and an amide bond (**e**)

ior, except that the phosphorus atom is connected to the rest of the molecule via two single bonds, each of which provides a rotational axis.

The delocalization of the π-bond in the case of the orthophosphate group $-PO_3^{2-}$ is more complicated. A double bond can be established between carbon and each of the three oxygen atoms. Another possibility is a state involving transfer of an electron from the phosphorus atom onto a hitherto neutral oxygen atom. The actual state of the group is a linear combination of all four possibilities (Fig. C.19d). Biochemists represent the entire orthophosphate group using the symbol P drawn within a small circle.

The delocalization of the π-bond in the orthophosphate group is related to a high energy which is stored in a phosphodiester bond in ATP. As a result of ATP hydrolysis into ADP and an inorganic phosphate P_i, two phosphodiester bonds and one orthophosphate group are replaced by one phosphodiester bond and two orthophosphate groups (see Fig. C.9b). In the latter arrangement π-electrons are more delocalized. The negative delocalization energy in connection with additional negative energy of hydration causes the products of ATP hydrolysis to have lower energy than the products of hydrolysis of other generalized ester bonds.

An electron transfer also takes place in the case of amide (peptide) bonds (Fig. C.19e). Due to partial delocalization of the π-bond, the amide bond structure becomes planar. All four O, C, N and H atoms lie in one plane, which is an important element of the protein structure discussed later in this appendix.

C.4 Hydrogen Bond. Amphiphilic Molecules in Water Environments

The fact that nitrogen and oxygen atoms possess one and two lone electron pairs, respectively, has enormous consequences for the spatial organization of the four most important classes of biomolecules: lipids, polysaccharides (carbohydrates), proteins and nucleic acids. The negative lone electron pairs distributed on tetrahedrally oriented σ-orbitals attract positively polarized hydrogen atoms of other molecules forming *hydrogen bonds* with them (Atkins, 1998). The energy of hydrogen bonds is comparable to the potential barrier heights for rotations around single covalent bonds, i.e., it ranges between 10 and 20 kJ/mol. The processes of reorganization of the system of hydrogen bonds, their breakage and restoration possibly in new locations, take place at a rate that is comparable to conformational transitions and to all intents and purposes are indistinguishable from them.

Water is the simplest system in which the structure and dynamics of hydrogen bonds play fundamental roles (Eisenberg and Kautzmann, 1969). Each oxygen atom, in addition to covalent bonds within the same water molecule, can form two hydrogen bonds with other water molecules (Fig. C.20). The structure of crystalline ice with completely saturated hydrogen bonds is highly ordered but has lower density than

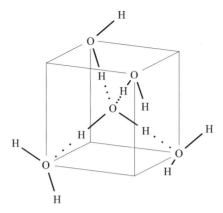

Fig. C.20. Locally saturated system of hydrogen bonds (*dotted lines*) in water

the disordered structure of liquid water with globally unsaturated hydrogen bonds. To determine which structure is more stable at a given temperature T, one uses the condition of free energy minimum, where the free energy is (see Chap. 3)

$$F = E - TS \, , \tag{C.1}$$

and the internal energy E favoring order competes with the entropy S favoring disorder. The higher the density of the system, the lower the internal energy E, and the more hydrogen bonds are formed the lower the entropy S, hence the greater the value of the term $-TS$.

The system of hydrogen bonds in water determines its main biologically significant properties: high specific heat capacity, high latent heat of melting, high electric susceptibility, and specific dynamic properties that will be discussed in Appendix D. This system also determines the water solubility properties of various molecules.

Molecules capable of forming hydrogen bonds with water, e.g., sugars or alcohols, increase the disorder in the system of hydrogen bonds and hence increase entropy leading to free energy reduction. The process of solvation is thermodynamically favorable in this case. Molecules that do not form hydrogen bonds but are electrically charged, or at least have high dipole moments, reduce the electrostatic energy of the system, and their solvation is also thermodynamically favorable in spite of introducing order into the hydrogen bond distribution. Molecules which do not form hydrogen bonds and are uncharged and non-polar, e.g., long hydrocarbon chains or aromatic rings, only order the water environment (Fig. C.21), but do not contribute to the energy

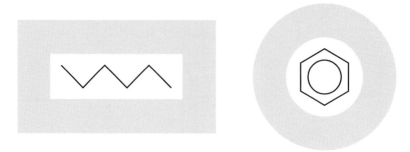

Fig. C.21. In water surroundings of non-polar hydrocarbon chains or aromatic rings, the hydrogen bond system is ordering and local 'icebergs' are formed. This reduces entropy and hence increases free energy

of the system and hence are not water soluble. We refer to them as *hydrophobic* (water fearing), in contrast to *hydrophilic* soluble particles (Tanford, 1980).

Interesting physical phenomena take place when *amphiphilic* molecules, containing both a hydrophilic and a hydrophobic moiety, are placed in an aqueous environment (Tanford, 1980; Evans and Wennerström, 1999; Hamley, 2000). Among biological systems, examples of such amphiphilic molecules are phospholipids (see Fig. C.7) and similar to them sphingolipids (Stryer et al., 2002, Chap. 12). Their polar head is hydrophilic and their two hydrocarbon tails are hydrophobic. *Surfactants* (superficially active molecules, e.g., soap) have a hydrophilic head and one hydrophobic tail. In a water environment, to minimize free energy, amphiphilic molecules spontaneously organize into spherical *micelles* (usually the case of one-tail molecules) or *bilayers* (usually the case of two-tail molecules). These structures allow the amphiphilic molecules to have their hydrophilic head groups facing outside and hydrophobic tails inside (Fig. C.22a). Bilayers can close up to form three-dimensional *vesicles* which, when sufficiently large, contain a hierarchy of internal vesicles and are referred to as *liposomes*. When the amount of solvent becomes too small, liposomes unfold to form *lamellae*, in which consecutive bilayers are placed parallel to each other (Fig. C.22b).

Micelles, vesicles and liposomes are *lyophilic* (they like solvents) *colloidal* (5 to 500 nm in diameter) particles, and when dispersed in water form a spatially inhomogeneous dispersive structure called a *sol*. A decrease in water content results in an unfolding process of liposomes into lamellae and a transition of the sol into a spatially homogeneous lamellar phase with successive bilayers of amphiphiles

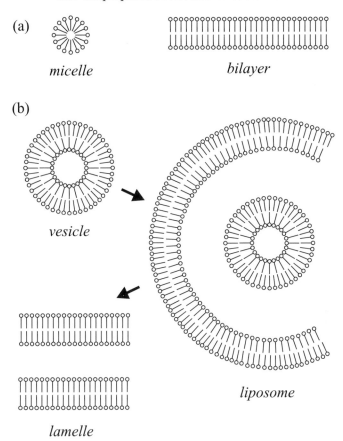

Fig. C.22. (a) Structures created by amphiphilic molecules in a water environment: micelles and bilayers. (b) From vesicles through liposomes to lamellae. *White circles* denote hydrophilic head groups, while *line segments* denote one or two hydrophobic tails

divided by monomolecular layers of water. In the lamellar structure, molecular orientation is ordered, but molecular spatial arrangement is not. Hence, the lamellar phase is a special example of a *liquid crystal* (Fig. C.23). A spatial ordering of the molecular heads is possible at lower temperatures when translational degrees of freedom become frozen. The liquid crystal transforms into a *solid crystal*. Lowering the temperature also causes translational degrees of freedom to freeze in the sol phase. The liposomes and vesicles become unfolded and water is allowed to penetrate their interior. The sol undergoes a phase transition into a *gel*. Figure C.23 presents only the main characteristics of the phase diagram of an amphiphilic molecule–water system.

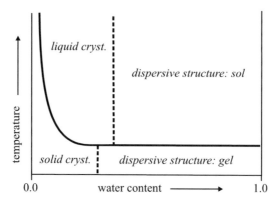

Fig. C.23. Simplified phase diagram of an amphiphilic molecule–water system as a function of temperature and water content

In fact, the diagram is much more complex and depends on particular properties of the amphiphilic molecules (Tanford, 1980; Evans and Wennerström, 1999; Hamley, 2000).

For biological systems, the sol phase is optimal since it contains vesicles or liposomes of appropriate sizes. The stability and mechanical properties of a bilayer at physiological temperatures are controlled by an appropriate chemical composition of phospholipids and sphingolipids, and also by cholesterol (see Fig. C.2), which stiffens the bilayer. Such a lipid bilayer integrated with built-in protein molecules forms a *biological membrane*. Proteins of the membrane can perform various functions such as those of immobilized enzymes, channels, pumps, receptors, signal generators, and constituents of a membrane skeleton. An important immunological role is played by carbohydrates of glycolipids and glycoproteins on the outer face of the cytoplasmic membrane.

C.5 Protein Structures

Proteins are linear polymers of amino acids (to be more exact, of L-enantiomers of α-amino acids, with a side chain from the central C^α atom) linked by amide (peptide) bonds $-CONH-$ (Fig. C.24a). In each polypeptide chain we distinguish the N (amine) and the C (carboxyl) termini. Successive amino acids are numbered starting from the N terminus. The formation of a peptide bond is an endoergic reaction and needs free energy, which is usually released in GTP (guanine triphosphate) hydrolysis. The sequence in which the individual amino

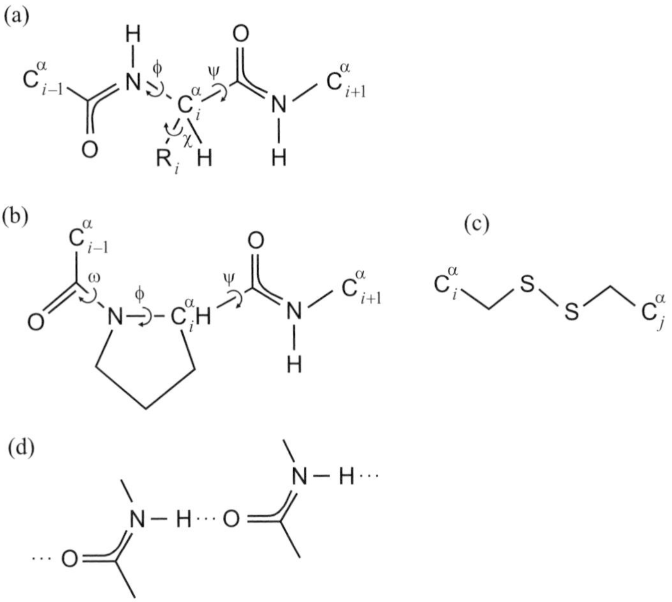

(a)

(b)

(c)

(d)

Fig. C.24. Chemical structure of proteins. (**a**) The main chain. Consecutive amino acids with central carbon atoms C_i^α and characteristic side chains R_i are connected by amide bonds $-\mathrm{CONH}-$ in the planar *trans* configuration. (**b**) The way proline is attached to the main chain. (**c**) Disulfide bridges made by cysteine side chains. (**d**) Hydrogen bonds between amide groups of the main chain. The notation of dihedral angles is shown

acids of definite side chains occur along the main chain of a protein is strictly fixed and genetically determined. We refer to it as the *primary structure* of the protein.

There are 20 'canonical' amino acids. Figure C.25 shows their side chains and the notation used. Three amino acids play special roles in the spatial structure of proteins. We refer to them as *structural* amino acids. The first is *glycine* (Gly or G); it has a side chain reduced to a single hydrogen atom. The small side chain produces no serious steric hindrance and enables almost free rotation of the main chain about the neighboring $C^\alpha-N$ and $C^\alpha-C$ bonds. The site at which glycine is situated behaves like a ball joint in the polypeptide chain. *Proline* (Pro or P) is an imino acid. It attaches the main chain through the two bonds (Fig. C.24b) and makes it locally rigid and looped in a defined manner. The third structural amino acid, *cysteine* (Cys or C), forms relatively strong covalent bonds (disulfide bridges) after oxidation and

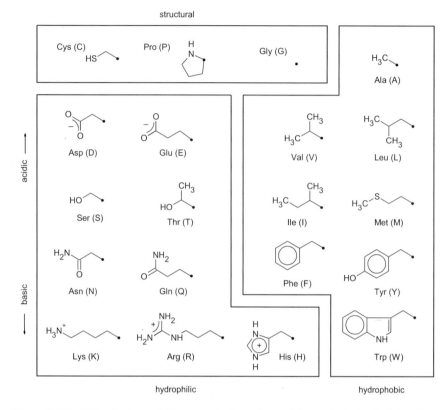

Fig. C.25. Side chains of 20 canonical amino acids and their division into three main groups. *Dots* denote central C^α atoms

pieces together distant sites of a single polypeptide chain and separate chains composing the protein macromolecule.

The concept of primary structure is identical to that of chemical structure (constitution). This completely determines the system of covalent bonds in the protein macromolecule and hence also includes information about the disulfide bridges which form spontaneously during protein synthesis on the ribosome. However, after completion of this process, most proteins are subjected to additional chemical modifications. The modification may involve cutting off fragments of the main chain, methylation of some charged side chains leading to charge neutralization, or phosphorylation of side chains ending with hydroxyl groups which endows an originally neutral chain with a negative charge. Enzymatic proteins often form permanent bonds with different prosthetic groups (coenzymes), while the external proteins of cell membranes undergo glycolization, as a result of which glycoproteins with up to 80% saccharide component are formed. The process of protein biosynthesis ends with a spontaneous formation of supramolecular structures, e.g., multienzyme complexes.

The spatial structure of a protein macromolecule of a given primary structure is determined by local conformations of the main and side chains as well as a system of noncovalent (secondary) bonds (Schulz and Schirmer, 1979; Creighton, 1992; Fersht, 1999). The commonly accepted notation of dihedral angles in the polypeptide chain is shown in Fig. C.24. In the absence of steric hindrances, rotation about the single bonds $C-C$, $C-N$, $C-O$ and $C-S$ allows three stable local conformations (one *trans* and two *gauche*, see Fig. C.14), unless the rotation involves planar carboxylic, amidic or aromatic rings (Fig. C.19) which would allow two stable conformations. Individual conformations can differ in energy by a few kJ/mol and are separated by barriers with heights from 10 to 20 kJ/mol originating from van der Waals and electrostatic multipolar interactions (Fig. C.14).

Only rotations around covalent bonds distant from C^α atoms (branching sites of the polypeptide chain) are usually unhindered, if one does not take into account the long-range excluded volume effects. This applies to angles χ^2, χ^2, χ^3, etc., describing internal conformations of longer side chains if not branched (Fig. C.24).

The peptide bond $CO-NH$, as discussed in Appendix C.3, is in part a double bond and may exist as two configurational *trans* and *cis* isomers (Fig. C.18). The potential energy barrier height is determined by the delocalization energy and approaches 80 kJ/mol. This is why peptide bonds occur as a single local configurational isomer, almost exclusively in *trans* form. The only exception is a peptide bond neighboring a proline residue (Fig. C.24b) in which the part with the polar structure is smaller, thereby lowering the *trans* to *cis* transformation barrier to 50 kJ/mol.

The presence of bulky side chains of amino acids always brings about steric hindrance for angles ϕ and ψ, making the corresponding local conformations mutually dependent. Figures C.25a and b show the region sterically allowed for angles ϕ and ψ (the Ramachandran map) separately, for glycine and arbitrary side chains different from glycine. Glycine behaves just like a ball joint: the sterically allowed region is connected and exceeds 50%. For the remaining side chains, the three distinct, much smaller, sterically allowed regions correspond to three distinct, cooperative conformations of the pair (ϕ, ψ). Strictly speaking, because rotation about the angle χ^1 is hindered for some

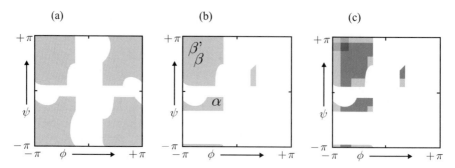

Fig. C.26. Sterically allowed regions for angles ϕ and ψ (Ramachandran map) for glycine (**a**) and an arbitrary side chain different from glycine (**b**). All atoms involved in peptide bonds are assumed to be rigid spheres of an appropriate van der Waals radius. Configurations of the α-helix as well as parallel and antiparallel β-pleated sheets (β and β', respectively) are indicated. (**c**) Number of sterically allowed conformations for the angle χ^1 depending on the value of ϕ and ψ. For residues with one side carbon atom third from C^α, *blackened*, *heavily shaded* and *lightly shaded* regions correspond to three, two and one conformation, respectively. For residues with two side carbon atoms third from C^α (threonine, valine and isoleucine), shadings correspond to three, two, and zero conformations, respectively. For alanine three rotational conformations are allowed everywhere

values of angles ϕ and ψ, the whole triple (ϕ, ψ, χ^1) should be treated as a single unit. The number of sterically allowed local conformations for the angle χ^1 is shown against the background of the Ramachandran map in Fig. C.25c.

The secondary bonds in proteins are mainly hydrogen bonds. Their energy, 10 to 20 kJ/mol, is comparable to potential barriers for transitions between the local conformations of protein chains. It is reasonable to distinguish hydrogen bonds within the polypeptide backbone from those formed by side chains.

The hydrogen bonds within the main chain link nitrogen and oxygen atoms of distinct amide groups (Fig. C.24d). A regular pattern in which hydrogen bonds are organized in the polypeptide backbone without reference to side chain types is traditionally known as a *secondary structure* of the protein. Two main secondary structures are distinguished. The α-helix is a helical arrangement of a single polypeptide chain with hydrogen bonds between each carbonyl group at position i and a peptide amine at position $i+4$ (Fig. C.27a). In β-pleated sheets, hydrogen bonds are also realized in completeness but they link differ-

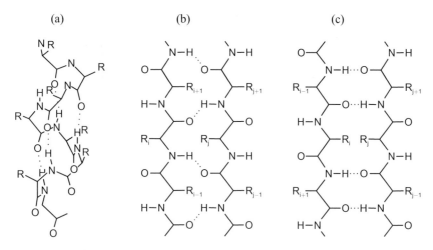

Fig. C.27. (a) α-helix structure. (b) and (c): Parallel and antiparallel β-pleated sheets, respectively

ent strands of a polypeptide chain placed in a parallel or antiparallel manner (Fig. C.27b and c).

Assuming all peptide bonds occur in *trans* conformation, the values of the dihedral backbone angles are $\phi_i = -57°$, $\psi_i = -47°$ for the α-helix, $\phi_i = -119°$, $\psi_i = 113°$ for the parallel β-pleated sheet, and $\phi_i = -139°$, $\psi_i = 135°$ for the antiparallel β-pleated sheet. These positions lie within two distinct sterically allowed regions of angles ϕ and ψ (Fig. C.26b) and coincide almost exactly with two main minima of the potential energy as a function of these angles, although it is energetically profitable for planar β-sheet structures to be slightly twisted. The system of hydrogen bonds in the α-helix and β-pleated sheet structures considerably stabilize local backbone conformations. This is why fragments of both secondary structures are so abundant in protein macromolecules.

However, there is also a need for the main chain to form reverse turns in addition to participating in regular α and β secondary structures. Some turns are standard, but most turns are nonstandard and contain no backbone hydrogen bonds. In these, the structural amino acids glycine and proline play this part. A complete distribution of all hydrogen bonds within the polypeptide backbone can be considered as composed of fragments of secondary structures linked by standard reverse turns and shorter or longer nonstandard sections of the main chain. An example of such an organization is shown in Fig. C.28d. Hydrogen bonds formed by side chains compete with backbone hydrogen

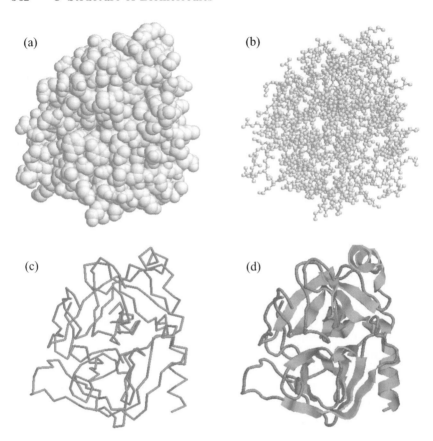

Fig. C.28. Various ways of presenting the spatial structure of the chymotrypsin molecule. (**a**) The atoms different from hydrogen are represented by spheres of the corresponding van der Waals radius. (**b**) Atoms are represented as small balls and bonds as sticks. (**c**) The route of the main chain (backbone). (**d**) The component secondary structure elements. The regions of α-helices are represented by helical ribbons and strands of β-structure are represented by broad arrows pointing to a C-terminal. All the drawings were made using the program Rasmol on the basis of Protein Data Bank (pdb) entry 1AB9 (Yennavar et al., 1994)

bonds and this fact, together with the position of structural amino acids, largely determines which sections of the main chain (chains) do not participate in secondary structures. All charged and polar amino acid side chains (see Fig. C.25) are involved in hydrogen bond formation. They bind mainly with surrounding water molecules, but also with carbonyl and amidic groups of the main chain.

Charged amino acids: aspartate (Asp, D), glutamate (Glu, E), lysine (Lys, K), arginine (Arg, R), as well as charged C and N termini of the main chains are also able to form ionic bonds (salt bridges), apart from the usual hydrogen bonds. Ionic bonds can be formed both with free counter-ions occurring in the surroundings and between the mentioned groups. All ionic bonds, when immersed in an environment of high dielectric constant (water, or regions of the protein interior with high polar group concentration) are only slightly stronger than the usual hydrogen bonds. Other secondary bonds between the residues set out along the main chain, the van der Waals bonds, are much weaker still (0.1 to 0.2 kJ/mol).

The side chains of all amino acids involved determine the native conformation of a protein. Some canonical amino acid side chains are hydrophobic and some are hydrophilic (Fig. C.25). All the hydrophobic side chains try to concentrate in the centers of protein molecules, away from the aqueous environment. Hydrophilic side chains are attracted to the outsides of molecules in order to lower the free energy by making contact with the environment.

The fundamental structural unit of a protein is a *domain* composing roughly one to two hundred amino acid residues (Janin and Wodak,1983; Creighton, 1992). It has a hydrophobic interior and a more or less hydrophilic surface. If the surface is completely hydrophilic, protein can occur as a water-soluble, single-domain molecule. Partly hydrophobic surfaces ensure domain contacts in multi-domain protein macromolecules. Figure C.28 shows an example of the water-soluble two-domain protein, chymotrypsin. Specific organization of surface hydrophobic amino acids also ensures the formation of protein supramolecular complexes or proteins joining the lipid membranes. On the other hand, specific organization of surface hydrophilic amino acids can result in non-covalent binding of proteins to nucleic acids.

C.6 Nucleic Acid Structures

Nucleic acids are linear polymers of nucleotides linked by phosphodiester bonds $-O-PO_2^- -O-$ (Fig. C.29). Two kinds of nucleic acid are distinguished: *ribonucleic acid* (RNA) and *deoxyribonucleic acid* (DNA). The sugar ribose that occurs in RNA is replaced by the sugar deoxyribose with one missing oxygen atom in DNA. Five carbon atoms in ribose and deoxyribose are numbered from 1' to 5'. Accordingly, in each nucleic acid chain, we distinguish the 5' (phosphate) and the 3' (hydroxyl) termini.

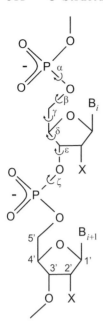

Fig. C.29. Chemical structure of nucleic acids. Dihedral angle notation and carbon atom numbering of the ribose are shown. X = OH for RNA and X = H for DNA

In DNA there are 4 'canonical' nitrogenous bases, guanine (G) and adenine (A), both derivatives of pyrimidine, and cytosine (C) and thymine (T), both derivatives of purine. In RNA, thymine is replaced by uracil (U). Crucial for molecular biology was the discovery by Watson and Crick in 1953 that specific pairs of nitrogenous bases can make unique hydrogen bonds (Fig. C.30).

The sequence in which the individual nucleotides of definite nitrogenous bases occur along a given nucleic acid chain is called its *primary structure*. In both DNA and RNA, successive nucleotides are numbered starting from the 5' terminus. DNA stores the genetic information. It occurs in the form of two complementary strands linked by the Watson–Crick hydrogen bonds. This form makes possible both DNA replication and transcription on RNA (Stryer et al., 2002, Chap. 5). In order for all the Watson–Crick hydrogen bonds to be saturated, the dihedral angles in complementary chains have to take values giving the whole DNA molecule a characteristic *double-helix form*. Among three possible types of double helix the B type is the most stable under physiological conditions (Fig. C.31).

RNA transmits information and can play a catalytic role. In contrast to DNA, it occurs in the form of single strands. Several types of RNA can be distinguished. Messenger RNA (mRNA) is transcribed from DNA and later translated into protein chains. Transfer RNA

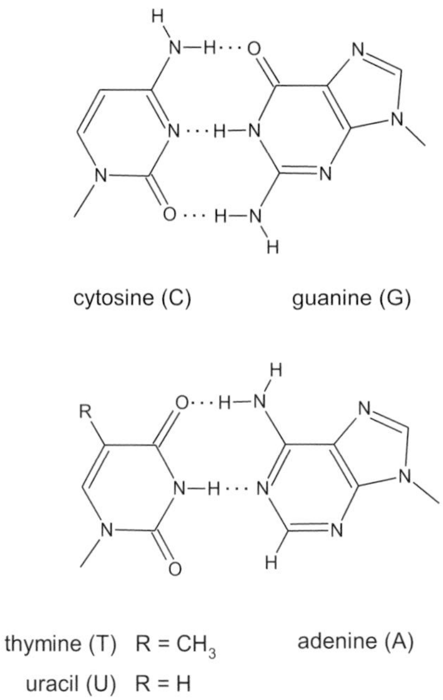

cytosine (C)   guanine (G)

thymine (T) R = CH$_3$ adenine (A)

uracil (U) R = H

Fig. C.30. Four canonical nitrogenous bases of nucleic acids and Watson–Crick pairing. Thymine (T) in DNA is replaced by uracil (U) in RNA

(tRNA) molecules carry particular amino acids to the growing protein chains. Ribosomal RNA (rRNA) is used in the building of ribosomes. Small nuclear RNA (smRNA) plays various roles in processing the three former types of RNA, e.g., it is used in the building of spliceosomes that excise introns and splice exons in divided genes transcribed onto mRNA, tRNA or rRNA. Micro RNA (miRNA) consists of processed parts of excised introns that have several regulatory functions (Dennis, 2002; Mattick, 2004).

Single-strand RNA can form a double helix only locally. Regions of the Watson–Crick pairing are indicated in a *secondary structure* of RNA. Figure C.32 shows the secondary structure of phenylalanine tRNA from yeasts, obtained for the first time by Holley in 1965, and Fig. C.33 shows the complete spatial structure of that tRNA.

Figure C.34 shows the spatial structure of rRNA of the large ribosomal subunit and Fig. C.35 shows the spatial structure of rRNA of the small ribosomal subunit.

(a)   (b)

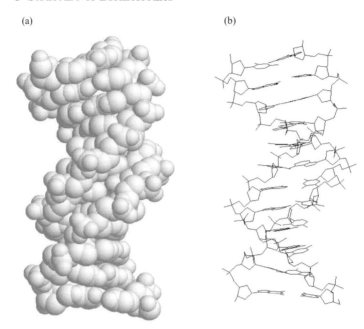

Fig. C.31. Spatial structure of the B-DNA double helix. (**a**) Atoms different from hydrogen are represented by spheres of the corresponding van der Waals radius. (**b**) Bonds are represented as wires. Both drawings were made using the program Rasmol on the basis of pdb entry 2BNA (Drew et al., 1982)

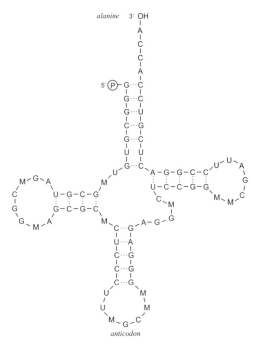

Fig. C.32. Secondary structure of phenylalanine tRNA. M denotes a post-transcriptionally modified nucleotide. Three-nucleotide anticodon and a place of amino acid binding are shown

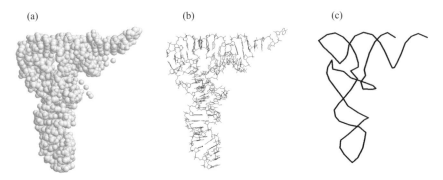

Fig. C.33. Spatial structure of phenylalanine tRNA. (**a**) Atoms different from hydrogen are represented by spheres of the corresponding van der Waals radius. (**b**) Bonds are represented as wires. (**c**) Route of the main chain. Drawings were made using the program Rasmol on the basis of pdb entry 6TNA (Sussman et al., 1978)

(a)   (b)

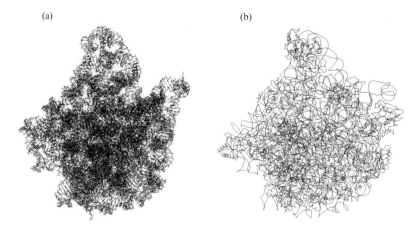

Fig. C.34. Spatial structure of large ribosomal subunit RNA (23S). (**a**) Bonds are represented as wires. (**b**) Route of the main chain. Drawings were made using the program Rasmol on the basis of pdb entry 1FFK (Ban et al., 2000)

(a)   (b)

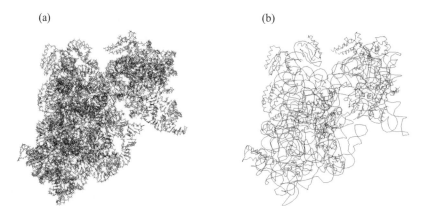

Fig. C.35. Spatial structure of small ribosomal subunit RNA (16S). (**a**) Bonds are represented as wires. (**b**) Route of the main chain. Drawings were made using the program Rasmol on the basis of pdb entry 1FKA (Schluenzen et al., 2000)

D Dynamics of Biomolecules

D.1 Vibrations Versus Conformational Transitions

Since atomic nuclei are at least two thousand times more massive than electrons, nuclear dynamics is very well separated from electronic dynamics and can be described with the help of the so-called *Born–Oppenheimer adiabatic potentials* for individual quantum mechanical electronic states of a molecule (Atkins, 1998, Chap. 14). We assume that the ground electronic state of biomolecules is always well separated in energy from the excited states. The ground state adiabatic potential energy is a function of all nuclear degrees of freedom or their linear combinations.

Besides three translational and three rotational degrees of freedom, a typical structural subunit of biological macromolecules has approximately 10^4 internal degrees of freedom. This number corresponds to a small protein or a protein domain consisting of some 200 amino acids, each of a dozen or so atoms, a polysaccharide crystallite composed of 150 monosaccharides, or a transfer RNA consisting of fewer than 100 nucleotides. The internal degrees of freedom are defined by values of covalent bond lengths and angles as well as dihedral angles of rotations about the single bonds (see Figs. C.24 and C.29). The ability to perform such rotations (limited only to some degree by steric hindrance), combined with the possibility of hydrogen bond break-up and reformation, makes the many-dimensional landscape of the potential energy of internal degrees of freedom extremely complex, with a huge number of local minima separated by higher or lower energy barriers of a non-covalent nature (Schulz and Schirmer, 1979; McCammon and Harvey, 1987; Brooks et al., 1988; Creighton, 1992).

As in the stereochemistry of low-molecular weight organic compounds (see Appendix C.3), regions of the configurational space surrounding the local minima can be referred to as *conformational states* (*substates* in particular contexts) or, more simply, *conformations* of biomolecules. Because the distribution of barrier heights spreads practically from zero (McCammon and Harvey, 1987; Brooks et al., 1988; Frauenfelder et al., 1991; 1999), the stereochemical notion of a conformational state is, however, not as well defined for biomolecules as for small molecules. To make the stereochemical definition of a biomolecular conformational state more precise and at the same time to make it consistent with the thermodynamic definition referring this concept to a certain free energy level (Hill, 1989), we shall consider a well in the potential energy to be a conformational state only if it is surrounded by barriers high enough to ensure internal equilibration of microstates preceding each transition to another conformational state (Kurzyński, 1998). As a lower bound for the interconformational barrier heights, one can assume a few units of $k_B T$, say 10 to 20 kJ/mol, which is a typical energy barrier height for a local rotation about a single covalent bond in the absence of any steric constraints (see Appendix C.3) and, simultaneously, a typical energy of a hydrogen bond (see Appendix C.4).

We have devoted so much space to semantic considerations because in the current biochemical literature the term 'biomolecule conformation' is rather poorly defined and often used with quite different meanings. We would like to point out a slight difference between our concept of a conformational state (substate) and that of a conformational substate due to Frauenfelder et al. (1991, 1999), who use this term with reference to *any* local minimum of the configurational potential energy.

Each global conformational state of a biomolecule can be represented by a sequence $(s_1, s_2, \ldots, s_M)$ of local conformational states of the component monomers, each described by a generalized 'spin' variable s_i, which takes two or more discrete values $s_i = 0, 1, 2, \ldots, m_i - 1$, where the m_i are not necessarily equal to each other. For $s_i = 0, 1$ (the local states labeled by a single bit), the global states represent vertexes of an M-dimensional cube (Fig. D.1) and for larger values of m_i, they represent sites of a more complex M-dimensional lattice. For typical biological monomers, m_i are of the order of 10 (e.g., for proteins, two conformational states of the peptide bond times five conformational states of the side chain, see Fig. C.26). For $M=200$ we thus obtain the astronomical number of 10^{200} global conformational states.

However, only a minute fraction of these states are occupied under physiological conditions. Due to steric constraints and environmental entropic effects, free energies of many global conformational states are exceedingly high, making these states practically negligible. Low conformational mobility is characteristic for polysaccharides, which are branched polymers (Fig. C.10a), so that the steric constraints are espe-

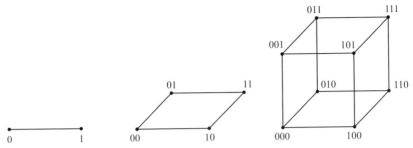

Fig. D.1. For two local states labeled by a single bit $s_i = 0, 1$, the global states of a system composed of two units represent vertexes of a square. Those of a system composed of three units represent vertexes of a 3-dimensional cube, and those of a general system composed of M units represent 2^M vertexes of an M-dimensional cube

cially large. In contrast, in the absence of interaction with proteins, the double-strand DNA is highly mobile and behaves like a statistical coil. Intermediate conformational mobility is characteristic for proteins and ribonucleic acids, both occurring in globular or multi-globular form (Figs. C.28, and C.33–C.35). The dynamics of ribonucleic acids is similar to that of proteins (McCammon and Harvey, 1987), but the latter is much more intensively studied at present. This, combined with the fact that DNA in vivo is always bound to proteins which determine its dynamics, is the reason why we confine our considerations to proteins below.

The more specific concept of conformational state that we have adapted is a good starting point for an approximate description of protein internal dynamics. A distinction is thus to be made between *vibrations* within particular conformational states and *conformational transitions* (Kurzyński, 1998). The former are approximated by damped harmonic oscillations subjected, according to the fluctuation–dissipation theorem, to weaker or stronger stochastic perturbations. The latter are purely stochastic activated processes described by a set of master equations like (B.37).

The vibrational dynamics is characterized by a spectrum of *periods* (or frequencies) of vibrational normal modes. Their number equals the number of degrees of freedom ($\sim 10^4$). The conformational transition dynamics is characterized by a spectrum of *relaxation times* (or their reciprocals), equal in number to the huge number of conformational states.

The spectrum of vibrational periods ranges from 10^{-14} s [weakly damped localized N–H or C–H stretching modes, easily observed spec-

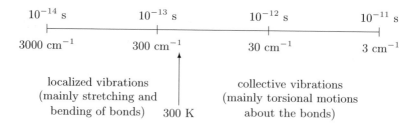

Fig. D.2. Frequency spectrum of protein vibrations and its conventional division into two ranges. Particular vibrational periods are related to corresponding wave numbers of infrared radiation

troscopically (Krimm and Bandekar, 1986)] to 10^{-11} s [overdamped collective modes involving the whole domains, directly studied only by numerical methods (Parak, 2003a)]. It can be conventionally divided into two ranges (Fig. D.2). A reasonable dividing point is the period 2×10^{-13} s. Equilibrium vibrations with this period have energy corresponding to temperature 300 K. Modes in the high-frequency range mainly involve the stretching and bending of bonds, whereas those in the low-frequency range mainly involve the collective torsional motions in dihedral angles about the bonds. Because of the effect of quantization, only low-frequency vibrations are, in principle, thermally excited, whence only they contribute to the thermal properties of proteins (the value of entropy and specific heat).

Free energies, and hence probabilities of equilibrium occupation, and transition probabilities between protein conformational states both depend strongly on temperature and environmental conditions (the influence of the latter is mainly of entropic nature, Schulz and Schirmer, 1979; Creighton, 1992). In physiological conditions, the spectrum of relaxation times of conformational transitions begins at 10^{-11} s (local side chain rotations or hydrogen bond rearrangements related to overcoming the energy barrier of order of 10 kJ/mol) and its upper limit is discussed further on.

The assumed picture of protein dynamics is somewhat oversimplified due to the anharmonicity of actual vibrations in the period range 10^{-12}–10^{-11} s (McCammon and Harvey, 1987; Brooks et al., 1988). This anharmonicity, already directly observed in the first molecular dynamics simulation (McCammon et al., 1977), also comprises local minima divided by barriers lower than 10 kJ/mol. In the simplest approach, the anharmonicity of vibrations is taken into account by assuming a finite correlation time of stochastic forces and, accordingly, a time-dependent friction (Schlitter, 1988). However, under special conditions it can result in non-ergodic, long-lived excitations such as *solitons* in proteins (Davydov, 1982; Yomosa, 1985; Yakushevich, 1993).

Until now, the actual existence of solitons in proteins has remained an open question (Careri et al., 1988; Yakushevich, 1993). Solitons are propagating, highly energetic non-thermal coherent motions of *local* groups of atoms. The problem as to whether they exist or not is only a question of their lifetime in real (not exactly periodic and influenced by thermal fluctuations) chains of bonds in proteins (Cruzeiro-Hansson and Takeno, 1997). The situation is quite different with coherent low-frequency *collective* vibrations of proteins, considered in a number of theoretical papers and reported to have been observed experimentally (Fröhlich and Kremer, 1983; Del Guidice et al., 1988). In our opinion they cannot occur, simply because the notion of coherence makes sense only for an *ensemble* of identical or almost identical modes of vibrations, and such an ensemble of low-frequency collective normal modes does not exist in highly inhomogeneous biomolecular structures.

D.2 Conformational Transitions Within the Protein Native State

A characteristic feature of biologically active proteins in physiological conditions is their well-defined spatially folded structure. Protein folding is a process of discontinuous thermodynamic phase-transition, independent for each domain (Privalov, 1989; Onuchic at al., 1997; Mirny and Shakhnovich, 2001), i.e., all the particular conformational states of each domain in physiological conditions are in principle to be divided unambiguously between the *native* (folded) and *unfolded* states. The mean waiting time of folding lies in the range 10^{-1}–10^2 s, and the mean waiting time of spontaneous unfolding in physiological conditions should be longer by the same factor as the equilibrium population of the native state is higher than the equilibrium population of the unfolded state. Careful estimation (Creighton, 1992) yields a value in the range 10^3–10^{12} s, and this value is to be considered as the upper limit of the relaxation time spectrum of protein conformational transitions.

Up until the end of the 1970s, the native state of protein was commonly considered to be a single conformational state, identified with the protein *tertiary structure*, and only a few scientists with Blumenfeld (1974), Careri (Careri et al., 1979) and Williams (1979) in the forefront, opposed this view. Most biochemists have disregarded the fact that the penetration of water or oxygen molecules into the protein interior, observed in historical experiments by Linderstrom-Lang (Linderstrom-Lang and Schellman, 1959) and Perutz (Perutz and Mathews, 1966), requires much more complex motions than simple vibrations. The first distinct experimental evidence that the native state of protein is in fact a dynamical mixture of a multitude of conformational substates came from Frauenfelder and coworkers' studies of the low-temperature dispersive kinetics of ligand rebinding to myoglobin after laser flash photolysis (Austin et al., 1975).

A veritable avalanche of observations of conformational transition dynamics within the protein native state was sparked off in the 1980s. Techniques used include Mössbauer spectroscopy (Parak, 2003a), quasi-elastic neutron scattering (Doster et al., 1989; Bicout and Zaccai, 2001), fluorescence depolarization and quenching (Milar, 1996), saturation transfer (Berger and Thomas, 1994; Adhikari et al., 1997) and hyperfine splitting (Ostap et al., 1995; Columbus and Hubbell, 2002) in electron paramagnetic resonance of spin labels and, last but not least, liquid-solution (Wüttrich, 1986; Dayie at al., 1996; Wider and Wüttrich, 1999) and solid-state (de Groot, 2000) high-resolution nuclear magnetic resonance.

Studies of the dispersive rebinding kinetics of small ligands to heme proteins after photolysis were continued under a broad and varied range of conditions (Nienhaus et al., 1997; Parak, 2003a). To distinguish the dynamics of ligand binding from that of conformational relaxation, multiple flash photolysis experiments appeared effective ('kinetic hole burning', Ormos et al., 1990; 1998).

Various techniques enabling observations of single biomolecules in time have turned out to be an even more powerful tool (see Sect. 9.1). These include recording the ionic current which flows through individual protein channels (the patch-clamp technique, Sackman and Naher, 1995; Liebovitch et al., 1987), single fluorophore detection using confocal fluorescence microscopy (Eigen and Rigler, 1994) or total internal reflection fluorescence microscopy (Funatsu at al., 1995), fluorescence resonance energy transfer (Haran et al., 1992; Weiss, 1999; Xu and Root, 2000, Margittai et al., 2003), and single-molecule fluorescence polarization (Warshaw et al., 1998; Corrie et al., 1999).

Conformational transitions within the native state of proteins were observed directly in molecular dynamics simulations. Since the first 8.5 ps simulation by McCammon et al. (1977), the computational power and quality of algorithms has improved so much that at present simulations of over 10 ns are more and more frequently reported (Hannson et al., 2002). With the help of special procedures like conformational flooding (Grubmüller, 1995), the simulation time can even be extended to 1 s for molecules that are not too large (Doniach and Eastman, 1999). Using the technique called *jumping-among-minima*, Kitao et al. (1998) showed that the separation of the protein intramolecular dynamics into vibrations and conformational transitions is quite good. A key step in the methodology of molecular dynamics simulations was introduced by Garcia (1992) and Amadei et al. (1993), namely the concept of the *molecule optimal dynamical coordinates* or the *essential modes* of motion. These modes, now simply called *collective modes* (Kitao and Gō, 1999), diagonalize not the force matrix (as the *normal modes* do), but the covariance matrix of atomic displacements.

The slowest processes are studied with the help of hydrogen-to-deuterium exchange (Englander et al., 1996; Hernandez et al., 2000). The time scale of conformational transition dynamics observed with the help of particular techniques is given in Fig. D.3.

As far as the X-ray diffraction technique is concerned, the classical evidence for conformational mobility comes from the often observed diffusion of some regions of electron density (Creighton, 1992) and from a comparison of different crystalline structures, when available (see, e.g., Clothia and Lesk, 1985). More subtle evidence is provided by an analysis of the Debye–Waller factors (Parak, 2003a) and a refinement of structures with an assumed heterogeneity (Rejto and Freer, 1996). One should also mention diffuse scattering observations (Thüne and Badger, 1995). The X-ray diffraction method is not included in Fig. D.3 because it does not provide information on time resolution. This situation is gradually changing, however, as a result of the availability of strong white beam sources of synchrotron radiation (Burgeois at al., 2003).

Single-molecule X-ray diffraction (Hajdu, 2000) remains with no time resolution. Force-induced conformational transitions in single molecules are studied, also without time resolution, by scanning (Bustamante et al., 1997; Heynmann et al., 1997) and force-clamp (Fisher et al., 2000) atomic spectroscopy, and also using optical tweezers (Finner et al., 1994; Kitamura et al., 1999; Mehta et al., 1998).

Both structural and dynamical studies indicate that conformational transitions within the protein native state do not take place in the entire body of the globule, but are limited to liquid-like regions surrounding solid-like fragments of secondary structure (α-helices or β-pleated sheets, Fig. D.4). These fragments survive the transition to the un-

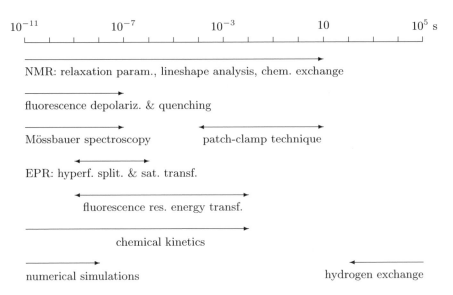

Fig. D.3. Time scale of conformational transitions within the protein native state observed with the help of various experimental techniques. The time period 10^{-11} s at one end of the spectrum characterizes localized conformational transitions on the protein surface. The time period 10^5 s at the other end is a (rather underestimated) value of the waiting time for spontaneous unfolding of the protein in physiological conditions. Note that the typical reciprocal turnover number of enzymatic reactions, 10^{-3} s, lies exactly in the middle of the scale

folded state in physiological conditions ('molten globule', Creighton, 1990; Onuchic et al., 1995, Mirny and Shakhnovich, 2002). However, the experimentally provided picture of conformational dynamics is still far from complete. We know that conformational transitions occur over the whole time scale from 10^{-11} s to 10^5 s or more, but we do not in fact know the size of the population of conformational states composing the native state.

In many experiments, only a few, often two, conformational states are apparent. Obviously, this is a result of the observational methodology applied. From this point of view, numerical simulation results appear important, pointing to the existence of a whole quasi-continuum of conformational substates within the protein native state (Kitao et al., 1998; Kitao and Gō, 1999). Although this has only yet been proved for time periods shorter than a few tens of nanoseconds, there is no reason to doubt that conformational states visited on the longer time-

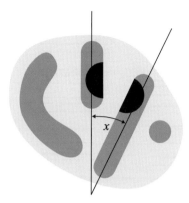

Fig. D.4. Schematic cross-section of a domain, the fundamental structural unit of protein. Solid-like fragments of secondary structure (α-helices or β-sheets) are *darkly shaded* and surrounding liquid-like regions are *weakly shaded*. *Black areas* indicate the catalytic center, usually localized in two neighboring solid-like elements. In models of protein-machine type, the dynamics of conformational transitions is treated as a quasi-continuous diffusive motion of solid-like elements relative to each other. Alternatively, in models of protein-glass type, this dynamics is treated as a diffusion of structural defects through the liquid-like medium. The picture can be reinterpreted on a higher structural level: solid-like elements then represent whole domains moving in a multidomain enzymatic complex (see Fig. 9.6)

scale also form some kind of quasi-continuum. The number of different conformational states observed seems to increase proportionally to the square root of the simulation time (Kitao and Gō, 1999; Hannson et al., 2002). Clear evidence of a quasi-continuous distribution of interdomain distances, hence conformational states of the whole protein, comes from studies of fluorescence energy transfer between donor and acceptor centers located on different domains (Haran et al., 1992; Weiss, 1999; Xu and Root, 2000; Margittai et al., 2003). The non-exponential time course of processes discussed in Sect. 9.1 also indicates the existence of a quasi-continuum of conformational substates.

Because the experiments at hand cannot elucidate the nature of conformational transition dynamics within the protein native state in detail, the problem of modeling this dynamics is to some extent left open to speculation. In two classes of models provided hitherto in the literature, the speculative element seems to be kept within reasonable limits. We refer to them symbolically as *protein-glass* and *protein-machine* (Kurzyński, 1998). Both approaches have support from molecular dynamics simulations (Kitao et al., 1998; Kitao and

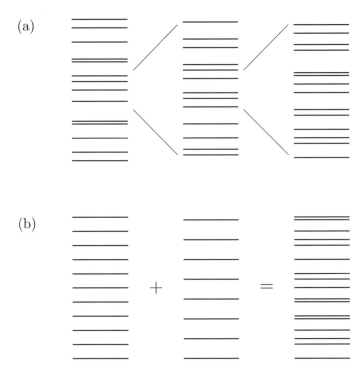

Fig. D.5. Schematic spectra of reciprocal relaxation times of conformational transition dynamics within the native state of a protein. (**a**) Protein-glass model: spectrum looks approximately alike on several successive time scales. (**b**) Protein-machine model: spectrum consists of a few more or less equidistant subspectra

Gō, 1999). In essence, the question concerns the form of the reciprocal relaxation time spectrum above the gap (see Fig. B.2). The simplest way to tackle problems without a well-defined time scale separation is to assume that the dynamics of a system looks alike on every time scale, i.e., the spectrum of relaxation times has a self-similarity symmetry. This assumption is the core of any protein-glass model (Fig. D.5a). An alternative is provided by the protein-machine class of models in which the variety of conformations composing the native state is supposed to be labeled by only a few 'mechanical' variables. The reciprocal relaxation time spectrum is then a sum of several more or less equidistant subspectra (Fig. D.5b).

D.3 Protein-Glass Model

There are two fundamental experimental facts indicating the glass-like dynamics of conformational transitions in native proteins. The first is the stretched-exponential time course of the initial stages of protein-involving reactions, during which partial thermodynamic equilibrium is established. Some examples of such reactions are presented in Sect. 9.1 and their explanation in terms of a hierarchy of relaxation times is given in Appendix B.4.

The second fact indicative of glassy dynamics is a characteristic temperature anomaly in the Lamb–Mössbauer or the Debye–Waller factors, observed at about 200 K in Mössbauer spectra, inelastic neutron scattering, temperature-dependent X-ray scattering (Parak, 2003a), as well as in specific heat spectroscopy and deuterium NMR relaxation (Lee and Wand, 2001). This anomaly is interpreted as a transition to glassy phase (Parak, 2003a; 2003b). There are some prerequisites for this transition to be a slaved process driven by the solvent (Fenimore et al., 2002; Tournier and Smith, 2003). The glass transition is a kinetic phenomenon which depends on an experimental time scale rather than a thermodynamic phase transition (Jäckle, 1986; Götze and Sjögren, 1992). This means that, below the glass-transition temperature, the protein is frozen in a conformational substate of a specific level of the hierarchy, not necessarily having the lowest free energy.

Time scaling, considered to be a generic property of glassy materials (Palmer, 1982; Götze and Sjögren, 1988), can originate either from a hierarchy of barrier heights in the potential energy landscape or from a hierarchy of bottlenecks in the network joining conformations between which direct transitions take place (the 'fractal time' and the 'fractal space', respectively; Blumen and Schnörer, 1990). All the experimental observations mentioned above were interpreted in terms of the hierarchy of barrier heights, but most of them can also be interpreted in terms of the hierarchy of bottlenecks (Kurzyński, 1998).

A hierarchy (tiers) of interconformational barrier heights (Fig. D.6) was originally proposed by Frauenfelder and coworkers in order to give a unified explanation of the results of various experiments concerning the process of ligand binding to myoglobin (see the reviews by Frauenfelder et al., 1991, 1999, and by Nienhaus et al., 1997). Such an organization of barrier heights was directly confirmed by numerical simulations (Troyer and Cohen, 1995; Garcia et al., 1997, Nymeyer et al., 1998; Wales et al., 2000).

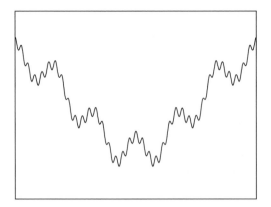

Fig. D.6. An example of a one-dimensional potential with a hierarchy of barrier heights. The curve represents a superposition of three sinusoids of appropriately scaled periods and amplitudes. Addition of more and more subtle components leads in the limit to the Weierstrass function, everywhere continuous but nowhere differentiable

The potential energy landscape with a hierarchy of barrier heights (Fig. D.6) forms what is known in mathematics as an *ultrametric* space (Rammal et al., 1986). *Spin glasses* are a reasonable mathematical realization of ultrametric structures in the context of application to proteins (Stein, 1985; Byngelson and Wolynes, 1987; Garel and Orland 1988; Shakhnovich and Gutin, 1989; Stein, 1992; Bryngelson et al., 1995). In models of spin-glass type, each *global* conformational state of a protein molecule is represented by a sequence $(s_1, s_2, \ldots, s_N)$ of *local* conformational states, each described by a spin s_i which can assume two or more discrete values (see Fig. D.1). The number N can be equal, e.g., to the number of amino acid residues. The free energy of a particular global conformational state is assumed to be determined by the energies of (possibly all) individual pairs of spins. An important feature is the randomness of pair interactions 'frustrating' the spins because of the impossibility of finding a single well-defined state of lowest energy.

Most spin-glass models display a discontinuous phase transition to the spin-glass phase (Binder and Young, 1986). This is to be interpreted as the protein-folding transition (Stein, 1992; Bryngelson et al., 1995). In standard spin-glass models, the matrix of interactions between spins is assumed to be completely random (Binder and Young, 1986). This assumption is not in fact justified in the case of proteins because the well-defined primary structure of the amino acid sequence determines, as we believe, the folding pattern of the native state and reduces degeneracy of the ground state. Following the Hopfield (1982) theory of associative memory, Wolynes and coworkers proposed a procedure for incorporating into the interaction matrix information about how the primary structure is related to the tertiary structure in a set of known proteins included in the database, thus obtaining the 'associative memory Hamiltonian' (Friedrichs et al., 1991) or the 'perfect funnel model' (Onuchic and Wolynes, 2004).

The dynamics of spin glasses is usually assumed to involve changes in the state of a single spin at a time with a transition probability of Glauber or Metropolis type (Binder and Young, 1986), satisfying the detailed balance condition. Spin transitions take place within both the disordered and the spin-glass phase. The latter consists of a number of nearly degenerate low-energy conformational states divided by a hierarchical system of energy barriers. Unfortunately, only the spin dynamics within the disordered phase has been considered as yet, for simulation of the time course of the protein-folding process.

An alternative to the hierarchy of barrier heights in the potential energy landscape is the hierarchy of bottlenecks in the network joining neighboring conformational states. Mathematical realizations of hierarchical networks are *fractal lattices* (Mandelbrot, 1982). Figures D.7 and D.8 show two examples of such lattices: the planar Sierpiński gasket and the planar percolation cluster. *Fractals* are defined as objects with a fractional value of the fractal dimension, but the hierarchical dynamical properties of lattices are related to the spectral rather than the fractal dimension. It is worth distinguishing the two concepts clearly here.

The notion of *fractal* (Hausdorff–Besicovitch) dimension $\bar{d}$ of a given lattice is simple (Mandelbrot, 1982). It is the exponent in the power law determining how the number of sites n changes with the scale (size) s:

$$n = s^{\bar{d}} . \tag{D.1}$$

Consequently,

$$\bar{d} = \frac{\log n}{\log s} . \tag{D.2}$$

For instance, for the planar Sierpiński gasket shown in Fig. D.7, a two-fold change in the scale entails a three-fold increase in the number of sites, i.e., $\bar{d} = \log 3/\log 2 \approx 1.585$.

The idea of *spectral* or *fracton* dimension is more complex (Nakayama et al., 1994). It resorts to the functional dependence of the density of vibrational normal modes vs. the frequency when a given lattice is considered to consist of massive points with elastic coupling between

Fig. D.7. Sierpiński gasket. Three small equilateral triangles are combined in a larger triangle, three larger triangles in an even larger one and so on. Here a finite order Sierpiński gasket is shown with imposed periodic boundary conditions (identification of outgoing bonds from one external vertex with those incoming at two other external vertexes)

nearest-neighbors. Quite generally, the Hamiltonian dynamics of a system of coupled harmonic oscillators is described by the equation

$$\dot{a} = -i\Omega a \;, \tag{D.3}$$

where a is the vector of complex numbers with real and imaginary parts corresponding to positions and momenta, respectively, of particular harmonic oscillators:

$$a_l = \frac{1}{\sqrt{2}}(q_l + ip_l) \;, \tag{D.4}$$

and Ω is the frequency matrix. In the coordinates of the normal modes of vibrations, the frequency matrix becomes diagonal and the set of equations (D.3) decouples into a set of independent equations

$$\dot{a}_k = -i\omega_k a_k \;. \tag{D.5}$$

If the density of vibrational modes in the spectrum of frequencies ω behaves regularly, according to a certain power law

$$\rho(\omega) \propto \omega^{\tilde{d}-1} \;, \tag{D.6}$$

the number $\tilde{d}$ is referred to as the spectral dimension of the lattice. The relation (D.6) can be considered as a generalization of the Debye

(a)

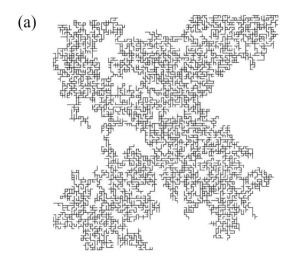

(b)

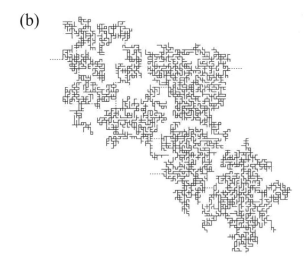

Fig. D.8. Percolation cluster. (**a**) Bonds on a square lattice are realized stochastically with probability $1/2$ and then clusters which are not connected to the largest one are removed. (**b**) Hierarchical structure of of the cluster from the picture above. The subcluster of highest order with three external bonds is singled out. Bonds joining subclusters of the two next lowest orders in the hierarchy are distinguished. The finite number of conformational substates in real proteins requires the hierarchy to be bounded both from below and from above (periodic or reflecting boundary conditions imposed on external bonds)

relation for acoustical phonons in crystal lattices of integer Euclidean dimension $\tilde{d}$ (Kittel, 1996). The normal modes of vibration in lattices of fractional dimension $\tilde{d}$ are referred to as *fractons* (Nakayama et al., 1994).

It is important for the present problem that the set of master equations (B.37) describing a random walk on a given lattice can be rewritten in a form analogous to (D.3):

$$\dot{p} = -\Gamma(p - p^{\text{eq}}) , \qquad (D.7)$$

where Γ is the matrix of transition probabilities $w_{ll'}$. The corresponding set of decoupled equations for the relaxational normal modes reads

$$\dot{p}_k = -\gamma_k(p_k - p_k^{\text{eq}}) . \qquad (D.8)$$

There are in fact twice as many equations (D.5) as equations (D.8) (the a_k are complex variables, whereas the p_k are real). Hence, following the relation (D.6), the density of relaxational modes in the spectrum of reciprocal relaxation times γ should behave as

$$\rho(\gamma) \propto \gamma^{\tilde{d}/2-1} . \qquad (D.9)$$

This is an alternative definition for the spectral dimension.

Time scaling takes place only if the density of relaxational normal modes increases with decreasing reciprocal relaxation time γ (Fig. D.5a). Consequently, the hierarchy of bottlenecks is characteristic only for lattices with spectral dimension smaller than 2, i.e.,

$$\tilde{d} < 2 . \qquad (D.10)$$

For a Sierpiński gasket embedded in d-dimensional Euclidean space $\tilde{d} = \log(d+1)/\log(d+3)$, and hence for the planar Sierpiński gasket ($d = 2$, Fig. D.7), the spectral dimension $\tilde{d} = \log 3/\log 5 \approx 1.365$. The spectral dimension of any percolation cluster (in particular that embedded in 2-dimensional Euclidean space, Fig. D.8) is very close to the value $\tilde{d} = 4/3$ (the Alexander–Orbach conjecture).

The spectral dimension influences two physically very important quantities. The first is the *probability of return to the original point*, which in the case of free diffusion (without any boundary conditions) behaves asymptotically in time as

$$p_{0|0}(t) \propto t^{-\tilde{d}/2} . \qquad (D.11)$$

This equation is a generalization of the well-known result for free diffusion in Euclidean spaces (van Kampen, 2001). The second quantity is the *mean number of distinct sites visited by a random walker*, which in the case of free diffusion behaves asymptotically in time as

$$S(t) \propto \begin{cases} t^{\tilde{d}/2} & \text{if} \quad \tilde{d} < 2 , \\ t & \text{if} \quad \tilde{d} > 2 . \end{cases} \tag{D.12}$$

The preexponential initial stage of a large class of gated reactions with intramolecular conformational transition dynamics of random-walk type on fractal lattices and the initial substate reduced to the gate itself is well described by the simple expression (Kurzyński et al., 1998)

$$P_{\text{ini}}(t) = \exp(\eta t)^{2\alpha}\text{erfc}(\eta t)^{\alpha} , \tag{D.13}$$

where the symbol erfc denotes the complementary error function, η^{-1} is a characteristic unit of time, and α is an exponent with value less than unity. In the limit of short times, (D.13) represents the stretched-exponential law and in the limit of long times, the algebraic power law:

$$P_{\text{ini}}(t) \approx \begin{cases} \exp\left[-2(\eta t)^{\alpha}/\sqrt{\pi}\right] & \text{for} \quad t \ll \eta^{-1} , \\ (\eta t)^{-\alpha}/\sqrt{\pi} & \text{for} \quad t \gg \eta^{-1} . \end{cases} \tag{D.14}$$

The crossover from nonexponential decay (D.13) to exponential decay with chemical relaxation time κ^{-1}, equal to the reciprocal of the irreversible reaction rate constant, is related to the finite number of conformational substates and can be described with the help of the formula

$$P(t) = \left[(1 - a)P_{\text{ini}}(t) + a\right]e^{-\kappa t} , \tag{D.15}$$

where a denotes the level (concentration) from which exponential decay begins.

Equation (D.13) with exponent $\alpha = 1/2$ is the exact solution of the continuous one-dimensional problem [see (B.124)], and for $\alpha > 1/2$ it has been carefully verified by computer simulations of random walks on fractal lattices (Kurzyński et al., 1998). The exponent α was found to be related to the spectral dimension of the lattice $\tilde{d}$:

$$\alpha = 1 - \tilde{d}/2 . \tag{D.16}$$

The approximation of the results of simulations with the help of the combined analytical formulas (D.13) and (D.15) is very good (Fig. D.9).

Depending on the value of the time constant η^{-1}, the initial stage of the reaction described by (D.13) can proceed according to either

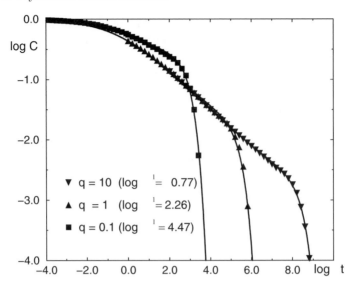

Fig. D.9. Fit of the results obtained in computer simulations of a random walk on the planar Sierpiński gasket with periodic boundary conditions (Fig. D.7) to the combined analytical formulas (D.13) and (D.15). The fitted curves are plotted as *continuous lines* and simulation data are represented by *points*. Not involved in the fitting procedure, the fixed value of $\alpha = 1 - \log 3/\log 5 = 0.317$ [see (D.16)] was assumed. Different values of the ratio q of the probability of leaving the lattice to the probability of transition between the neighboring sites determine different values of the time unit η^{-1}. After Kurzyński et al. (1998)

the stretched-exponential law or the algebraic power law, or both. All three types of behavior were observed in CO rebinding experiments after laser flash photolysis (Fig. 9.1) as well as in the patch-clamp experiments (Fig. 9.3). In single fluorophore detection experiments, only the initial stretched-exponential time course was observed (Fig. 9.4). Apart from the α exponent, (D.13) and (D.15) comprise two dimensionless parameters, the level (concentration) a related somehow to the conformational relaxation time γ^{-1}, determined by the fractal (not spectral) dimension of the lattice and responsible for a possible plateau preceding the exponential decay, and the ratio $b \equiv \kappa/\eta$. Both parameters depend on temperature according to the Arrhenius relation. As a consequence, one should have no problem describing the time course of any experimentally observed reaction in such terms, including its variation with temperature, especially when taking into account certain time variations of α and b allowed by the model with somewhat extended gate (Kurzyński et al., 1998). However, the success of the fit should be treated very cautiously, comparable fits can be obtained by the alternative one-dimensional model with fluctuating barriers (Agmon and Sastry, 1996).

Furthermore, (B.129) determining the mean first-passage time for one-dimensional diffusion in the presence of boundary conditions can be generalized to the case of diffusion on fractal lattices (Chełminiak and Kurzyński, 2004). This result is hoped to be important for modeling the steady-state fluxes in enzymatic reactions controlled by the intramolecular dynamics of protein-glass type (see Sects. 9.3 and 9.5).

D.4 Protein-Machine Model

The concept (and the name) *protein-machine* was proposed by Chernavsky, Khurgin and Shnol in 1967 (Chernavsky et al., 1987), but a similar picture of protein dynamics, rather speculative at that time, had also been considered independently by McClare (1971), Blumenfeld (1974), Williams (1993) and Gavish (1986) (for a review, see Kováč, 1987). All these authors use a similar notion of the *machine* as "a device which uses energized motion to bring about transformation" (Williams, 1993), "a structure which displays high mobility in certain directions and rigidity in others" (Gavish, 1986), or "a device with mechanically constrained parts predetermined to give some effects by restricting motion along one or several degrees of freedom" (Kováč, 1987). According to Kováč after Blumenfeld (1974): "Any machine exhibits a particular degree of freedom which, when excited by an input of energy, exchanges its energy with other degrees of freedom very slowly; in other words, its rate of relaxation is low." The latter statement coincides, however, with a definition of any macroscopic system (Callen, 1985). For the protein-machine concept, it is important that the distinguished variable is intensive rather than extensive and that the system is not macroscopic but mesoscopic, with stochastic rather than deterministic dynamics.

A quasi-continuum of conformational transitions of the character of a relative motion of the secondary structure elements has been observed directly in numerical simulations (Rojewska and Elber, 1990). More recent simulations also indicate a quasi-continuous, almost harmonic motion of whole domains relative to each other (de Groot et al., 1998; Hayward and Berensen, 1998; see review by Kitao and Gō, 1999). In the simplest case, the mechanical variables can be identified with angles or distances describing the mutual orientation of rigid fragments of secondary structure or larger structural elements (see

Fig. D.4). The mechanical coordinate may also be identified with a 'reaction coordinate', if this can be determined. There is serious evidence that the mechanical coordinates are related to certain collective variables in molecular dynamics simulations (Kitao and Gō, 1999).

The first, although implicit, practical use of the protein-machine picture of dynamics was the application of Kramers' theory of reaction rates in the position diffusion limit (Hänggi et al., 1990) to interpret two particular protein reactions studied by Gavish and Werber (1979) and Frauenfelder and coworkers (Beece et al., 1980) in solvents of various viscosities. Explicitly, the protein-machine model was formalized in terms of diffusion in an effective potential along the mechanical coordinate by Shaitan and Rubin (1982). This kind of dynamics with the simplest, parabolic potential was applied to describe a single irreversible protein reaction by Agmon and Hopfield (1983). Since those times, many authors have used similar language to describe various dynamical processes in proteins. The author of this book applied the model to describe a reversible three-step enzymatic reaction in both the transient and the steady-state stages (Kurzyński, 1997b). Diffusion in a parabolic potential is described in more detail in Appendix B.6.

Assumptions which have to be made in order to approximate the dynamics of conformational transitions by a process of limited diffusion along a mechanical coordinate are presented schematically in Fig. D.10. By definition, any conformational state of the protein is characterized by a definite value of the free energy; thermodynamic averaging is taken over equilibrated vibrational degrees of freedom. One can always label a set of all 'free energy levels' of the protein with the help of one or several 'conformational' coordinates (Fig. D.10a). What distinguishes the *mechanical* coordinates from the others is that direct transitions take place only between the nearest neighbor conformational states along them (Fig. D.10b). Variation of, e.g., the angle between two structural elements of a protein molecule goes through a series of well-defined successive conformational transitions involving break-up and reformation of hydrogen bonds in the liquid-like interior of the protein, as well as in the surrounding water (see Fig. D.4). Smoothing of the free energy (Fig. D.10c) and transition to the continuous limit (Fig. D.10d) result in replacement of a chain of conformational jumps by diffusion in an effective, approximately parabolic conformational potential.

The mathematical equation of diffusion in a parabolic potential also describes overdamped low-frequency collective vibrations of domains. Moreover, numerical analysis indicates that these vibrations also have

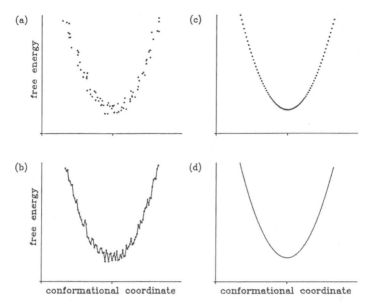

Fig. D.10. Formulation of the protein-machine model. (**a**) Conformational states (free energy levels) are labeled with the help of a conformational (mechanical) coordinate. (**b**) Transitions are assumed to be possible only between neighboring conformational states. (**c**) Smoothing of the conformational potential. (**d**) Transition to the limit of continuous diffusion. After Kurzyński (1998)

the character of mutual motions of relatively rigid fragments of secondary structure (Nishikawa and Gō, 1987). In fact, the crossover between sequences of conformational transitions along the mechanical coordinates and overdamped low-frequency collective vibrations is more or less a matter of convention, since one can introduce *effective* normal modes corresponding to the principal axes of the envelope of a ragged (with many local minima) potential (Kidera and Gō, 1990). On the other hand, the dynamics of large proteins, not studied in molecular dynamics simulations, can be described in terms of a diffusive, almost harmonic hinge-bending motion of whole multidomain subunits relative to one another, as suggested by fluorescence resonance excitation transfer experiments (Weiss, 1999; Xu and Root, 2000).

We do not claim that the description of protein dynamics in terms of diffusion in a parabolic potential represents a comprehensive version of the one-dimensional protein-machine model. Firstly, the effective potential can differ considerably from the parabolic one. This is not a problem in itself, as various approximation techniques well known from quantum mechanics can then be applied in order to solve the diffusion equation (which is equivalent to the Schrödinger equation with imaginary time, Kurzyński, 1994). Secondly, the smoothing procedure [transition from (b) to (c) in Fig. D.10] can appear too rough an approximation. In that case the model known as a random walk in a random potential (Zwanzig, 1988) is more appropriate. And thirdly, the restriction to transitions between nearest neighbors in the conformational coordinate (Fig. D.10b) may turn out to be insufficient. This is technically the most difficult problem. When the insufficiency arises from too small a number of the considered degrees of freedom, one can try to replace the neglected degrees of freedom by memory effects (Schlitter, 1988). There are many ways to generalize the described version of the protein-machine model and some of them certainly merit detailed study.

D.5 Calculation of Mean First-Passage Time

In time-course kinetics (Appendix B.4) and in steady-state kinetics (Sects. 9.3 and 9.5), values of the mean first-passage times between certain conformational substates of a protein enzyme are important. Here we present a general method for calculating such times for an arbitrary network of conformational substates (Kurzyński and Chełminiak, 2003).

Let us consider a set of states $\mathbf{M}$ of arbitrary nature with stochastic dynamics determined by a set of master equations (B.37) with transition probabilities per unit time $w_{ll'}$ satisfying the detailed balance conditions. The set $\mathbf{M}$ can be considered as a *graph* (*diagram, lattice*): the substates of the system are represented by the *vertexes* (*points, sites*) and the direct transitions, determined by the non-zero w, by the *edges* (*lines, nearest neighbors*) [see Wilson (1996)]. By definition (see Appendix B.4), to find the mean first-passage time from some initial to some final state of the diagram $\mathbf{M}$, one has to put a statistical ensemble of the systems into the initial state and observe its stochastic evolution. Each system reaches the final state after a certain time. The average of these times is the mean first-passage time from the initial to final state. But one can also observe some equivalent infinite process for a single system, assuming that each time a given system reaches the final state, the same system appears anew at the initial state. After a long enough time this will be the stationary flux in a diagram that determines the sought mean first-passage time.

More precisely, following Hill (1989), the mean first-passage time $\tau_{\mathbf{M}}(l_0 \to l)$ from the state l_0 to l in the diagram $\mathbf{M}$, with absorption at the final state l, can be found as the reciprocal of a steady-state one-way cycle flux or a sum of such fluxes J in a modified diagram $\mathbf{M}_{l_0 \to l}$ in which the absorption transition or transitions are redirected to the starting state l_0 with a simultaneous elimination of the absorption state from the original diagram $\mathbf{M}$. An illustrative example of such a modification is shown in Figs. D.11a and b. In formal terms,

$$\tau_{\mathbf{M}}(l_0 \to l) = J^{-1} = \left(\sum_{l'} w_{ll'} p_{l'} \right)^{-1} , \qquad (D.17)$$

where the probabilities $p_{l'}$ are solutions of the set of master equations (B.37) for the modified diagram $\mathbf{M}_{l_0 \to l}$ under the steady-state boundary conditions, with some detailed balance conditions broken. A useful method for calculating such steady-state probabilities and fluxes is offered by the technique of summing up the directional diagrams described in an algorithmic way in the above-mentioned book by Hill (1989).

Hill's algorithm for finding the steady-state probability p_l (or the equilibrium probability in the case when the detailed balance condition is satisfied for each transition probability $w_{ll'}$) for an arbitrary diagram $\mathbf{S}$ comprises the following steps:

1. Construction of the complete set of *partial diagrams* for $\mathbf{S}$, each of which contains the maximum possible number of lines that can be included in the diagram without forming any cycle (closed path).
2. Construction of *directional diagrams* for each state l and each partial diagram, if possible, in which all connected paths are directed toward and end at the state l. The directional diagram is uniquely attributed a number equal to the product of transition probabilities corresponding to all directed lines involved.
3. Calculation of the sum of all directional diagrams for each state l in $\mathbf{S}$, denoted by $D_l(\mathbf{S})$, and then the sum of all directional diagrams in $\mathbf{S}$, viz.,

$$D(\mathbf{S}) \equiv \sum_{l \in \mathbf{S}} D_l(\mathbf{S}) . \qquad (D.18)$$

The steady-state (or equilibrium) occupation probability of the state l is determined by the ratio

$$p_l = \frac{D_l(\mathbf{S})}{D(\mathbf{S})} . \qquad (D.19)$$

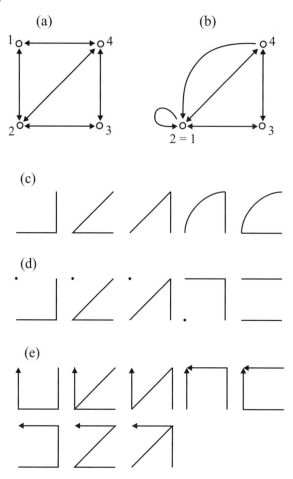

Fig. D.11. (a) Example of the diagram $\mathbf{M}$ determining certain stochastic dynamics. (b) The modification $\mathbf{M}_{2\to1}$ of the adjoining diagram, used in the calculation of the mean first-passage time $\tau_{\mathbf{M}}(2 \to 1)$. (c) A complete set of partial diagrams for the modified diagram $\mathbf{M}_{2\to1}$. (d) A complete set of partial diagrams for three possible dissections of the original diagram $\mathbf{M}$ into a subdiagram containing site 1 and a subdiagram containing site 2. If isolated sites counted as unity are disregarded, it can be seen that the set of diagrams in (d) is identical to the set in (c). (e) Multiplying diagrams in the set (d) by appropriate transition probabilities to the final site 1, one gets a set of diagrams which is identical to the complete set of all directional diagrams for site 1 in the original diagram $\mathbf{M}$

Applying this algorithm to the the the probabilities $p_{l'}$ in (D.17), we get

$$\tau_{\mathbf{M}}(l_0 \to l) = \frac{\sum_{l'} D_{l'}(\mathbf{M}_{l_0 \to l})}{\sum_{l'} w_{ll'} D_{l'}(\mathbf{M}_{l_0 \to l})} \; . \tag{D.20}$$

To proceed further, it is essential to note that the complete set of partial diagrams for the modified diagram $\mathbf{M}_{l_0 \to l}$ is identical to the complete set of partial diagrams for all possible dissections of $\mathbf{M}$ into sums of the form $\mathbf{M}_{l_0} \cup \mathbf{M}_l$, the subdiagram $\mathbf{M}_{l_0}$ containing site l_0 and the subdiagram $\mathbf{M}_l$ containing site l (see the example in Figs. D.11c and d). As a consequence, the numerator of (D.20) can be rewritten as the quantity

$$D_{l,l_0}(\mathbf{M}) \equiv \sum_{\mathbf{M}_{l_0} \cup \mathbf{M}_l} D_l(\mathbf{M}_l) D(\mathbf{M}_{l_0}) \; , \tag{D.21}$$

with the summation running over all possible dissections $\mathbf{M}_{l_0} \cup \mathbf{M}_l$ of $\mathbf{M}$, and the denominator of (D.20) simply equals the sum of all directional diagrams of the final state l in $\mathbf{M}$ (see Fig. D.11e). Hence,

$$\tau_{\mathbf{M}}(l_0 \to l) = \frac{D_{l,l_0}(\mathbf{M})}{D_l(\mathbf{M})} \; . \tag{D.22}$$

In the example considered in Fig. D.11, the numerator consists of 12 different terms in the form of products of two possible transition probabilities allowed in the five diagrams presented in Fig. D.11d (the points count as unity), whereas the denominator consists of 8 terms in the form of products of three transition probabilities, presented directly in Fig. D.11e.

From the formula (D.22), three general theorems of increasing complexity follow, useful in the theory of reactions controlled and gated by the intramolecular stochastic dynamics. Simple proofs are given in the paper by Kurzyński and Chełminiak (2003). The notation is explained in Figs. D.12a to c, where diagrams of intramolecular dynamics between an arbitrary number of conformational substates within a given chemical molecular state are represented by shaded boxes. Figures D.12a and b present irreversible gated reactions – the product state is replaced by the completely absorbing *limbo state* (see Appendix B.4) denoted by the asterisk. It should be stressed that the zero transition probability in the backward direction does not mean here that the detailed balance is broken, but that the equilibrium occupation probability of the initial chemical state P^{eq} is zero.

(a)   (b)

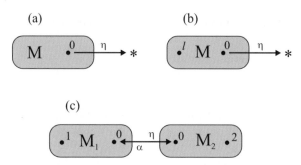

(c)

Fig. D.12. Illustrations for Theorem 1 (**a**), Theorem 2 (**b**), and Theorem 3 (**c**). *Shaded boxes* represent diagrams of an arbitrary number of sites and the direct transitions between them. The *asterisk* denotes the completely absorbing limbo state

Theorem 1. *The mean first-passage time between the transition state 0 of a reaction and the limbo state $*$ equals the reciprocal of the transition state theory rate constant:*

$$\tau(0 \to *) = (k_+^{eq})^{-1} , \qquad (D.23)$$

where

$$k_+^{eq} = \frac{p_0^{eq}}{P^{eq}} \eta , \qquad (D.24)$$

the ratio of p_0^{eq} to P^{eq} being the partial equilibrium probability of the transition state occupation (to be interpreted in terms of a conditional probability, otherwise it equals zero over zero), and η is the transition probability per unit time through the gate from the transition state 0 to the limbo state (Fig. D.12a).

Theorem 2. *For an arbitrary state l in $\mathbf{M}$ (see Fig. D.11b),*

$$\tau(l \to *) = \tau_{\mathbf{M}}(l \to 0) + (k_+^{eq})^{-1} . \qquad (D.25)$$

Theorem 3. *Let the two diagrams $\mathbf{M}_1$ and $\mathbf{M}_2$ (representing the reagent and the product chemical state of a molecule) be connected by a reversible transition between the gates denoted by 0, with probabilities per unit time η and α in the forward and backward directions, respectively (Fig. D.11c). Then, for an arbitrary state 1 in $\mathbf{M}_1$ and an arbitrary state 2 in $\mathbf{M}_2$,*

$$\tau(1 \to 2) = (k_+^{eq})^{-1} + \tau_{\mathbf{M}_1}(1 \to 0) + \tau_{\mathbf{M}_2}(0 \to 2) + K^{-1}\tau_{\mathbf{M}_2}(0 \leftrightarrow 2) , \qquad (D.26)$$

where $\tau_{\mathbf{M}_i}$ $(i = 1, 2)$ denote the mean first-passage times confined to the corresponding subdiagrams. The quantity

$$K = \frac{k_+^{\mathrm{eq}}}{k_-^{\mathrm{eq}}} \tag{D.27}$$

has the meaning of a reaction equilibrium constant,

$$k_+^{\mathrm{eq}} = \eta \frac{D_0(\mathbf{M}_1)}{D(\mathbf{M}_1)} , \qquad k_-^{\mathrm{eq}} = \alpha \frac{D_0(\mathbf{M}_2)}{D(\mathbf{M}_2)} , \tag{D.28}$$

corresponding respectively to the forward and backward transition state theory rate constants, respectively.

For a typical (highly occupied at equilibrium) state 2, the mean first-passage time $\tau_{\mathbf{M}_2}(0 \to 2)$ is much shorter than $\tau_{\mathbf{M}_2}(2 \to 0)$ and can be neglected. Then (D.26) reconstructs the general expression (B.81) for the complete forward rate constant:

$$k_+^{-1} = (k_+^{\mathrm{eq}})^{-1} + \tau_{\mathbf{M}_1}(1 \to 0) + K^{-1}\tau_{\mathbf{M}_2}(2 \to 0) , \tag{D.29}$$

and similarly for the backward rate constant.

The main application of the formula (D.22) is to express steady-state fluxes that occur in various enzymatic reactions controlled and gated by the intramolecular stochastic dynamics of proteins in terms of the mean first-passage times between certain distinguished conformational substates (Kurzyński and Chełminiak, 2003). Important results obtained in this way for the two special models are presented in Sects. 9.3 and 9.5.

D.6 Nonadiabatic Processes of Charge and Energy Transfer

In Appendix B and up to now in this appendix, we have studied reaction rates only in classical terms. However, if a considered reaction $R \longleftrightarrow P$ involves an intramolecular transfer of light particles, an electron between a donor D and an acceptor A,

$$\mathrm{D\,A} \longleftrightarrow \mathrm{D^+A^-} ,$$

or a proton between an acid A–H and a base B,

$$\mathrm{A–H}\cdots\mathrm{B} \longleftrightarrow \mathrm{A}\cdots\mathrm{H–B^+} ,$$

where dots denote a hydrogen bond, quantum effects cannot be neglected. Quantum dynamics must also be taken into account when the energy transfer occurs between different quantum states, either electronic,

$$D^*A \longleftrightarrow D A^* \, ,$$

or vibrational ones, e.g., in the complex of two amide groups[1] OCN–H linked by a hydrogen bond as shown in Fig. C.24d:

$$OCN^*\text{–}H \cdots OCN \longleftrightarrow OCN\text{–}H \cdots OCN^* \, ,$$

where the asterisk distinguishes excited states.

The ground adiabatic electronic state of biomolecules is well separated in energy from excited states. Only the latter may not be so well separated one from another. As a consequence, the electron or the energy transfer between two different electronic quantum states R and P must be preceded by an excitation process. Such an excitation can occur either at the expense of a light quantum absorption, i.e.,

$$R_0 + \gamma \longrightarrow R \longleftrightarrow P \longrightarrow \text{subsequent states} \, ,$$

or as a result of a bimolecular chemical reaction, i.e.,

$$S + R_0 \longrightarrow R \longleftrightarrow P \longrightarrow \text{subsequent states} \, ,$$

e.g., of redox type (Marcus and Sutin, 1985; see Sect. 6.5),

$$D_{\text{red}} + A_{\text{oxy}} \longleftrightarrow D A \longleftrightarrow D^+A^- \longleftrightarrow D^+_{\text{oxy}} + A^-_{\text{red}} \, .$$

Excitation of the amide group vibrational quantum state is also supposed to be preceded by some bimolecular chemical reaction (Davydov, 1982). A transition to the subsequent state can involve product detachment, another excitation or electron transfer, as well as another photon emission (fluorescence) or absorption (e.g., resulting in stimulated emission or transient absorption in pump–probe experiments, Mukamel, 1995).

---

[1]The amide group OCN–H has an exceptional property. Its high-frequency vibrational mode with excitation energy $1\,700$ cm^{-1} = 20 kJ/mol, thus not describable classically, is collective rather then localized, contrary to what was suggested in Fig. D.2. This collective aspect takes its origin in a π-bond delocalization (see Fig. C.19e), which results in a high dipolar electric moment able to interact over a long distance.

The proton transfer reaction occurs within a single electronic adiabatic potential. As a matter of fact, it is a transition between two different conformational substates that only additionally involves quantum tunneling effects (May and Kühn, 1984). It can be a stage of a bimolecular protolysis reaction (Eigen, 1964; see Sect. 6.4):

$$AH + B \longleftrightarrow A\text{–}H \cdots B \longleftrightarrow A^- \cdots H\text{–}B^+ \longleftrightarrow A^- + HB^+ \; ,$$

or it can be preceded by a photoexcitation to an excited electronic state (photoinduced tautomerization; Manz and Wöste, 1995).

Further on, we restrict our considerations to two adiabatic quantum states $|R\rangle$ and $|P\rangle$, corresponding to chemical states R and P, respectively, and to a single nuclear dynamic variable $\mathcal{Q}$ representing either the distance between the molecular centers at which the quantum states $|R\rangle$ and $|P\rangle$ are localized or a more collective variable such as a local polarization of the surrounding protein matrix. As usual, a value of the dynamic variable $\mathcal{Q}$ will be denoted by the corresponding small italic letter q. We shall refer to the states $|R\rangle$ and $|P\rangle$ as *excited states*, although it may not be the case for the proton transfer.

In the four-dimensional space of operators acting in the two-dimensional space spanned by states $|R\rangle$ and $|P\rangle$, it is convenient to distinguish two Pauli spin operators:

$$|P\rangle\langle P| - |R\rangle\langle R| = \sigma_z \; , \qquad |P\rangle\langle R| + |R\rangle\langle P| = \sigma_x \; . \qquad (D.30)$$

The condition of the system being in either excited state is written as

$$|P\rangle\langle P| + |R\rangle\langle R| = 1 \; . \qquad (D.31)$$

We assume that both adiabatic potentials corresponding to R and P have the same parabolic shape determining harmonic *vibrations*. We also assume that their minima are translated by Q_0 and their values by ϵ (Fig. D.13a). Accordingly, the excitation-vibrational Hamiltonian takes the form

$$\mathcal{H}_{\text{exc}} + \mathcal{H}_{\text{vib}} + \mathcal{H}_{\text{exc-vib}} = \frac{\epsilon}{2}\sigma_z + \frac{\omega}{2}\left[\mathcal{P}^2 + \left(\mathcal{Q} + \frac{1}{2}\sigma_z Q_0\right)^2\right] + \text{const.}$$

$$= \frac{\epsilon}{2}\sigma_z + \frac{\omega}{2}(\mathcal{P}^2 + \mathcal{Q}^2) + \frac{\omega}{2}Q_0\mathcal{Q}\sigma_z \; . \qquad (D.32)$$

The position $\mathcal{Q}$ and the conjugate momentum $\mathcal{P}$ are chosen to be dimensionless quantities, whence the frequency ω has the meaning of vibrational energy in units $\hbar = 1$. A translation Q_0 is related to a *reorganization energy* (see Fig. D.13a) (Marcus and Sutin, 1985):

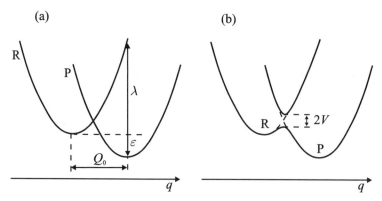

Fig. D.13. (a) Potentials of two localized excited states in the nonadiabatic transfer problem. (b) Potentials of two delocalized excited states in the adiabatic transfer problem. The lower state is well separated in energy from the higher one

$$\lambda = \frac{1}{2}\omega Q_0^2 .$$
(D.33)

Localized states $|R\rangle$ and $|P\rangle$ are not the exact excited quantum states. An excitation transfer is possible due to a delocalization component of the Hamiltonian, which we assume to be independent of the nuclear variable Q (the *Condon approximation*, May and Kühn, 2004):

$$\mathcal{H}_{\mathrm{tr}} = V\sigma_x ,$$
(D.34)

where V is a phenomenological parameter resulting from various possible mechanisms of long-range excitation transfer in proteins (Bixon and Jortner, 1999; May and Kühn, 2004).

We assume only a *vibrational relaxation* in the system, described by equations similar to (B.60) and (B.61):

$$\dot{Q} = \omega \mathcal{P} , \quad \dot{\mathcal{P}} = -\omega\left(Q + \frac{1}{2}\sigma_z Q_0\right) - 2\kappa\mathcal{P} + 2\sqrt{\frac{\kappa}{\omega}\frac{k_{\mathrm{B}}T}{\omega}}\mathcal{Y} ,$$
(D.35)

where $\mathcal{Y}$ is normalized Brownian white noise and the dimensionless momentum $\mathcal{P}$ is related to velocity $\mathcal{V}$ by the equation $\mathcal{V} = \omega\mathcal{P}$. Equations (D.35) can be treated either as the classical Poisson–Langevin equations or as the quantum Heisenberg–Langevin equations (Zwanzig, 2001).

The treatment of the problem depends on the ratio of rates of vibrational relaxation κ and electron transfer $2V$ (Bixon and Jortner, 1999; May and Kühn, 2004). If $2V \ll \kappa$, $\mathcal{H}_{\mathrm{el}} + \mathcal{H}_{\mathrm{vib}} + \mathcal{H}_{\mathrm{el-vib}}$ is to be treated as an unperturbed Hamiltonian and $\mathcal{H}_{tr}$ as a perturbation, and we have a *nonadiabatic theory* involving two adiabatic potentials (Fig. D.13a). If $\kappa \ll 2V$, it is $\mathcal{H}_{el} + \mathcal{H}_{tr} + \mathcal{H}_{vib}$ that should be treated as an unperturbed Hamiltonian and $\mathcal{H}_{el-vib}$ is a perturbation. We then have a typical *adiabatic theory*, because only the ground adiabatic potential is sufficient to be considered (Fig. D.13b). The problem is identical to that considered by Kramers (see Appendix B.3).

As in Kramers' adiabatic approach (Hänggi, 1990), we shall consider the nonadiabatic theory of reaction rates in two opposite limits of *high* and *low vibrational damping*, i.e., $\kappa \gg \omega$ and $\kappa \ll \omega$, respectively.

For high vibrational damping, the two equations (D.35) are replaced by a single equation of overdamped motion [see (B.62)]:

$$\dot{\mathcal{Q}} = -\frac{\omega^2}{2\kappa}\left(\mathcal{Q} + \frac{1}{2}\sigma_z Q_0\right) + \sqrt{\frac{k_B T}{\kappa}}\, y\,. \tag{D.36}$$

As a consequence, the positional representation of vibrational states is convenient and we choose the adiabatic states in the form of products $|R\rangle|Rq\rangle$ and $|P\rangle|Pq\rangle$, where $|Rq\rangle$ and $|Rq\rangle$ are vibrational states with well-determined positions.

In the classical approach to nuclear dynamics, (D.36) is equivalent to a system of two coupled Smoluchowski equations ($\sigma_z = -1$ or $+1$ for R and P, respectively; Wang et al., 1992):

$$\frac{\partial}{\partial t}p_R(q,t) = \frac{\omega^2}{2\kappa}\frac{\partial}{\partial q}\left[\left(q - \frac{1}{2}Q_0\right) + \frac{k_B T}{\omega}\frac{\partial}{\partial q}\right]p_R(q,t)$$
$$-w(q)\Big[p_R(q,t) - p_P(q,t)\Big]\,,$$

$$\frac{\partial}{\partial t}p_P(q,t) = \frac{\omega^2}{2\kappa}\frac{\partial}{\partial q}\left[\left(q + \frac{1}{2}Q_0\right) + \frac{k_B T}{\omega}\frac{\partial}{\partial q}\right]p_P(q,t)$$
$$+w(q)\Big[p_R(q,t) - p_P(q,t)\Big]\,, \tag{D.37}$$

where $p_R(q,t)$ and $p_P(q,t)$ are the probability densities of being in the states $|Rq\rangle$ and $|Pq\rangle$, respectively, at time t. The transition probability density $w(q)$ per unit time between the states $|Rq\rangle$ and $|Pq\rangle$ is defined by the first order perturbation theory (Fermi golden rule):

$$w(q) = 2\pi\, |\langle Rq|\langle R|\mathcal{H}_{tr}|P\rangle|Pq\rangle|^2\, \delta\left[\epsilon + \frac{\omega}{2}(q + Q_0/2)^2 - \frac{\omega}{2}(q - Q_0/2)^2\right]$$

$$= \frac{2\pi V^2}{\omega Q_0}\delta(q - Q_c)\,, \tag{D.38}$$

where Q_c is the position where the two potentials cross each other (Fig. D.13a). In the derivation of (D.38), we have taken into account the fact that $\langle Rq|Pq'\rangle = 1$ only for $q = q'$ and replaced the density of states with respect to energy by the density of states with respect to position.

When the positional relaxation is fast compared with the transfer rate, the system first reaches the equilibrium probability distribution in R:

$$p_R^{eq}(q) = \left(\frac{2\pi k_B T}{\omega} \right)^{-1/2} \exp \left[-\frac{\omega(q - Q_0/2)^2}{2k_B T} \right] . \qquad (D.39)$$

The rate constant for the next forward transition is given by

$$k_+^{eq} = \int_{-\infty}^{\infty} dq \, p_R^{eq}(q) w(q) , \qquad (D.40)$$

and substituting (D.38) and (D.39), we reconstruct the historical Marcus result (Marcus and Sutin, 1985):

$$k_+^{eq} = 2\pi V^2 (4\pi \lambda k_B T)^{-1/2} \exp \left[-\frac{(\epsilon + \lambda)^2}{4\lambda k_B T} \right] , \qquad (D.41)$$

where $(\epsilon + \lambda)^2/4\lambda$ is the value of the potential at the crossing point Q_c. The full transition rate constant is given by (B.81). The relaxation times for diffusion in the parabolic potential are calculated in Appendix B.6. The reverse transition rate constants are related to the forward ones by (B.74) and (D.27).

The Marcus equilibrium rate constant (D.41) for nonadiabatic transitions differs from the Eyring equilibrium rate constant (B.78) for adiabatic transitions. The preexponential factor does not increase with temperature like the mean frequency of transitions in the Eyring theory but, instead, decreases with temperature. This points to the contribution of tunneling-type quantum effects. And secondly, the expression divided by $k_B T$ in the exponent cannot be simply interpreted as the activation free energy since, depending on the relation between the energy drop $-\epsilon$ and the reorganization energy λ, three various cases can occur in the nonadiabatic transition, as shown in Fig. D.14.

For low vibrational damping, $\kappa \ll \omega$, it is convenient to choose the energy representation by introducing complex variables

$$a = \frac{1}{\sqrt{2}} (Q + iP) , \qquad a^* = \frac{1}{\sqrt{2}} (Q - iP) . \qquad (D.42)$$

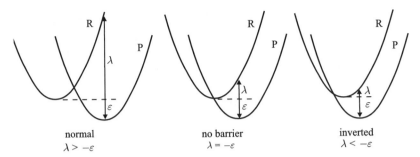

normal
$\lambda > -\varepsilon$

no barrier
$\lambda = -\varepsilon$

inverted
$\lambda < -\varepsilon$

Fig. D.14. Three cases in the Marcus theory of nonadiabatic transitions

Then the vibrational Hamiltonian reads

$$\mathcal{H}_{\mathrm{vib}} = \frac{\omega}{2}(a^*a + aa^*) \, . \tag{D.43}$$

In the quantum mechanical approach, a^* and a represent bosonic creation and annihilation operators. The energy eigenstates are states
with a well-determined number of bosons, $n = 0, 1, 2, \ldots$:

$$\mathcal{H}_{\mathrm{vib}}|n\rangle = \omega\left(n + \frac{1}{2}\right)|n\rangle \, . \tag{D.44}$$

In the energy representation, the adiabatic states are products of the
form $|R\rangle|Rm\rangle$ and $|P\rangle|Pn\rangle$, where $|Rm\rangle$ and $|Pn\rangle$ are the energy eigen-
states of the oscillators with minima at $-Q_0/2$ and $Q_0/2$, respectively,
and m varies in the same way as n.

Despite introducing the discrete variables m and n, we shall treat
the vibrational subsystem classically (in the so-called *semiclassical approximation*). In this case, the density matrix equation of motion in
the energy representation (Suzuki et al., 1995) simplifies to the usual
set of master equations:

$$\dot{p}_{Rm}(t) = w_{m,m-1}p_{Rm-1}(t) + w_{m,m+1}p_{Rm+1}(t) \tag{D.45}$$
$$-(w_{m+1,m} + w_{m-1,m})p_{Rm}(t)$$
$$+ \sum_n \left[v_{m,n}p_{Pn}(t) - v_{n,m}p_{Rm}(t) \right] ,$$

$$\dot{p}_{Pn}(t) = w_{n,n-1}p_{Pn-1}(t) + w_{n,n+1}p_{Pn+1}(t) \tag{D.46}$$
$$-(w_{n+1,n} + w_{n-1,n})p_{Pn}(t)$$
$$+ \sum_m \left[v_{n,m}p_{Rm}(t) - v_{m,n}p_{Pn}(t) \right] .$$

The w are transition probabilities per unit time determining vibrational relaxation:

$$w_{m-1,m} = m\kappa , \qquad w_{m,m-1} = m\kappa e^{-\beta\omega} , \qquad \text{(D.47)}$$

and the v are transition probabilities between different excited states determined in first order perturbation theory by the Fermi golden rule (Bixon and Jortner, 1999; May and Kuhn, 2004):

$$v_{n,m} \equiv 2\pi V^2 |S_{nm}|^2 \delta[\epsilon + (n-m)\omega] , \qquad \text{(D.48)}$$

where the *overlap integral* S_{nm} is given by

$$S_{nm} = \langle Pn | Rm \rangle = \langle n | e^{-iQ_0 P} | m \rangle = \langle n | e^{\sqrt{\gamma}(a^*-a)} | m \rangle , \qquad \text{(D.49)}$$

with γ a *coupling parameter* given by

$$\gamma \equiv \frac{\lambda}{\omega} . \qquad \text{(D.50)}$$

The overlap integral has the property

$$S_{nm} = (-1)^{n-m} S_{mn} . \qquad \text{(D.51)}$$

Some manipulations with the bosonic creation and annihilation operators result in a formula for the squared modulus of the overlap integral:

$$|S_{nm}|^2 = \frac{e^{\gamma} \gamma^{n-m}}{n!m!} \left[\frac{d^n}{d\gamma^n} (e^{-\gamma} \gamma^m) \right]^2 . \qquad \text{(D.52)}$$

Formulas like this are usually the starting point for calculating the equilibrium rate constant (Bixon and Jortner, 1999). However, we would like here to consider also a dynamical contribution to the rate constant originating in the vibrational relaxation, so we shall still deal with the complete (D.45) and (D.46).

For high temperatures, $\beta\omega \ll 1$, one can pass to the continuous energy limit by introducing probability densities:

$$p_R(e,t) = \omega^{-1} p_{Rm}(t) , \qquad e \equiv \omega m , \qquad \text{(D.53)}$$

and

$$p_P(e - \epsilon, t) = \omega^{-1} p_{Rn}(t) , \qquad e - \epsilon \equiv \omega n . \qquad \text{(D.54)}$$

Energy conservation during the transition has been taken into account in the above, as indicated by (D.48). In the continuous limit, (D.45) and (D.46) take the form

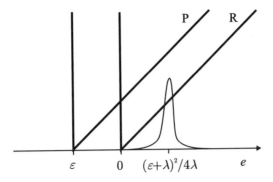

Fig. D.15. Potentials of the two localized excited states in the energy space. The transition probability density distribution between the two states is shown, with maximum at the energy value where the corresponding potentials in the positional space cross one another (see Fig. D.13a)

$$\frac{\partial}{\partial t} p_R(e,t) = \kappa \omega \frac{\partial}{\partial e} D(e) \left[\beta \frac{\partial U(e)}{\partial e} + \frac{\partial}{\partial e} \right] p_R(q,t)$$

$$- w(e) \left[p_R(e,t) - p_P(e,t) \right],$$

$$\frac{\partial}{\partial t} p_P(e,t) = \kappa \omega \frac{\partial}{\partial e} D(e-\epsilon) \left[\beta \frac{\partial U(e-\epsilon)}{\partial e} + \frac{\partial}{\partial e} \right] p_P(q,t)$$

$$- w(e) \left[p_P(e,t) - p_R(e,t) \right]. \tag{D.55}$$

For $e > 0$, the potential and the diffusion coefficient now depend on e:

$$U(e) = e, \qquad D(e) = e, \tag{D.56}$$

respectively, and for $e < 0$, V is infinite and D vanishes (Fig. D.15). The transition probability density per unit time can be approximated by the Gaussian

$$w(e) = 2V^2 (4\lambda\beta)^{-1/2} \gamma e^{-\gamma^2 [e-(\epsilon+\lambda)^2/4\lambda]}. \tag{D.57}$$

In the strong coupling limit, $\gamma \gg 1$, the distribution (D.57) narrows to the Dirac delta function:

$$w(e) = 2\pi V^2 (4\pi\lambda\beta)^{-1/2} \delta \left[e - (\epsilon+\lambda)^2/4\lambda \right]. \tag{D.58}$$

If the energy (vibrational) equilibration is fast compared with the transfer rate, the system reaches the equilibrium probability distribution

$$p_R^{\text{eq}}(e) = \beta e^{-\beta e}, \tag{D.59}$$

for $e > 0$, which determines the equilibrium rate constant

$$k_+^{\text{eq}} = \int_0^\infty \mathrm{d}e\, p_{\text{R}}^{\text{eq}}(e) w(e) \,. \tag{D.60}$$

The integral can be calculated in the strong coupling limit given by (D.58), and the Marcus formula (D.41) is again reconstructed. Owing to the energy dependence of the diffusion coefficient, the mean first-passage time to the energy gate, defining a nonequilibrium contribution to the rate constant, is determined by a slight modification of the formula (B.137) (Zwanzig, 1988):

$$\tau(0 \to e) = \frac{1}{\kappa \omega} \int_0^e \mathrm{d}x\, \mathrm{e}^{\beta U(x)} D(x)^{-1} \int_0^x \mathrm{d}y\, \mathrm{e}^{-\beta U(y)} \,. \tag{D.61}$$

Substituting in the functions (D.56), we get an integral

$$\tau(0 \to e) = \frac{1}{\kappa \omega \beta} \int_0^{\beta e} \mathrm{d}(\beta x) \frac{\mathrm{e}^{\beta x} - 1}{\beta x} = \frac{1}{\kappa \omega \beta} \Big[\mathrm{Ei}(\beta e) - \ln(\beta e) - \mathrm{Eu} \Big] \,,$$
$$\tag{D.62}$$

where Ei is the integral exponential function and Eu the Euler constant. For high energy barriers $\beta e \gg 1$, we can use the asymptotic expansion of the integral exponential (Abramowitz and Stegun, 1964) to obtain

$$\tau(0 \to e) \approx (\kappa \omega \beta^2 e)^{-1} \mathrm{e}^{\beta e} \,. \tag{D.63}$$

Treating the vibrational subsystem in a full quantum manner, i.e., replacing the set of master equations (D.45) and (D.46) by a set of equations of motion for the density matrix, one can get *quantum beats*, often observed for low vibrational damping (Suzuki et al., 1995). Quite generally, the rate constant for an arbitrary quantum model of charge or energy transfer, both in the high and low vibrational damping limits, is determined by a quantum version of the flux-correlation formula (B.76) derived by Yamamoto (1960). However, calculations with this formula are not straightforward as a rule (Fain, 1980).

Let us now generalize the model considered to the case of quantum states localized on a linear chain of many sites. The main goal is a possible application to a system of amide groups linked by hydrogen bonds in the α-helices (see Figs. C.24d and C.27a) (Davydov, 1982). In the present case, it is more convenient to express the dynamics of excitation in term of creation and annihilation operators of certain quasiparticles:

$$\begin{aligned}
|\text{P}\rangle\langle\text{P}| - |\text{P}\rangle\langle\text{P}| &= c_{\text{P}}^* c_{\text{P}} - c_{\text{R}}^* c_{\text{R}} \,, \\
|\text{P}\rangle\langle\text{R}| + |\text{R}\rangle\langle\text{P}| &= c_{\text{P}}^* c_{\text{R}} + c_{\text{R}}^* c_{\text{P}} \,,
\end{aligned} \tag{D.64}$$

and similarly for other sites. For two-state dynamics on a site, it is indifferent whether the operators c^* and c are bosonic or fermionic. For N identical sites, the excitation Hamiltonian with a transfer term reads

$$\mathcal{H}_{\text{exc}} + \mathcal{H}_{\text{tr}} = \Delta \sum_{l=1}^{N} c_l^* c_l - V \sum_{l=1}^{N} (c_l^* c_{l-1} + c_l^* c_{l+1}) . \tag{D.65}$$

The condition (D.31) is replaced by the condition that there be a single excitation (quasiparticle) present:

$$\sum_{l=1}^{N} c_l^* c_l = 1 . \tag{D.66}$$

The vibrational Hamiltonian is

$$\mathcal{H}_{\text{vib}} = \frac{\omega}{2} \sum_{l=1}^{N} \left[\mathcal{P}_l^2 + (\mathcal{Q}_l - \mathcal{Q}_{l-1})^2 \right] , \tag{D.67}$$

where $\mathcal{Q}_l$ is a displacement from the equilibrium position of site l and $\mathcal{P}_l$ is the conjugate momentum. The interaction Hamiltonian between quasiparticles and vibrations is

$$\mathcal{H}_{\text{exc}-\text{vib}} = \chi \sum_{l=1}^{N} (\mathcal{Q}_{l+1} - \mathcal{Q}_{l-1}) c_l^* c_l . \tag{D.68}$$

We introduce Fourier transforms,

$$c_l = N^{-1/2} \sum_{k} c_k e^{ikl\xi} \tag{D.69}$$

and

$$\mathcal{Q}_l = N^{-1/2} \sum_{k} \mathcal{Q}_k e^{ikl\xi} , \tag{D.70}$$

where k assumes N values in the first Brillouin zone from $-\pi/\xi$ to π/ξ, ξ being the equilibrium distance between the sites. The excitation and vibrational Hamiltonians (D.65) and (D.67) then take the form

$$\mathcal{H}_{\text{exc}} + \mathcal{H}_{\text{tr}} = \sum_{k} \epsilon_k c_k^* c_k \tag{D.71}$$

and

$$\mathcal{H}_{\text{vib}} = \sum_{k} \omega_k a_k^* a_k , \tag{D.72}$$

respectively, where we introduced the creation and annihilation operators of bosons related to the Fourier transforms of positions $\mathcal{Q}_k$ and momenta $\mathcal{P}_k$. Collective quasiparticles created by the operators c_k^* are called *excitons*, whereas those created by the operators a_k^* are *phonons*. The energies are

$$\epsilon_k = \Delta - 2V \cos k\xi \tag{D.73}$$

and

$$\omega_k = 2\omega \left| \sin \frac{k\xi}{2} \right| . \tag{D.74}$$

In terms of the phonon operators, the interaction Hamiltonian (D.68) reads

$$\mathcal{H}_{\text{exc-vib}} = N^{-1/2} \chi \sum_{lk} e^{ikl\xi} c_l^* c_l (b_k + b_{-k}^*) . \tag{D.75}$$

For low values of the wave number k, the exciton and phonon energies can be approximated by

$$\epsilon_k \approx \Delta - 2V + v(k)|k| \tag{D.76}$$

and

$$\omega_k \approx v_0 |k| , \tag{D.77}$$

where

$$v(k) = V\xi^2 |k| \tag{D.78}$$

and

$$v_0 = \omega\xi \tag{D.79}$$

are the exciton and phonon (longitudinal sound wave) velocities, respectively. If the exciton velocity is higher than the sound velocity,

$$v(k) > v_0 , \tag{D.80}$$

the exciton can lose energy by creation of phonons which then dissipate due to the process described by (D.35). However, if the condition (D.80) is not satisfied, the interaction (D.75) results in a local deformation rather then phonon creation, moving in a correlated way with the exciton. This deformation–exciton pair is called a *Davydov soliton*.

Under the semiclassical approximation, i.e., a quantum treatment of the excitons and a classical treatment of the deformation, the soliton state is sought in the form of a linear combination:

$$|\text{sol}\rangle = \sum_{l=1}^{N} \phi_l(\mathcal{Q}, \mathcal{P}, t) c_l^* |0\rangle , \tag{D.81}$$

where $|0\rangle$ is the exciton vacuum state and $\mathcal{Q}$ and $\mathcal{P}$ are sequences of positions and momenta of all deformations at the sites $l = 1, 2, \ldots, N$. In the continuous limit, assuming

$$x = l\xi \,, \tag{D.82}$$

and inserting the explicit time dependence $\mathcal{Q}_l(t)$ and $\mathcal{P}_l(t)$, the sequences $\mathcal{Q}$ and $\mathcal{P}$ are replaced by continuous fields $\mathcal{Q}(x,t)$ and $\mathcal{P}(x,t)$. Similarly, the coefficients $\phi_l(\mathcal{Q}, \mathcal{P}, t)$ in (D.81) are replaced by a continuous field $\phi(x,t)$. It has the meaning of a probability amplitude for finding the exciton at a given position and time.

The classical Hamilton equations for $\mathcal{Q}$s and $\mathcal{P}$s and the Schrödinger equation for the exciton jointly minimize the functional

$$\langle \text{sol} | \mathcal{H}_{\text{exc}} + \mathcal{H}_{\text{tr}} + \mathcal{H}_{\text{vib}} + \mathcal{H}_{\text{exc-vib}} | \text{sol} \rangle \,, \tag{D.83}$$

and result in a nonlinear Schrödinger equation for the field $\phi(x,t)$ (Davydov, 1982):

$$\left[\mathrm{i}\frac{\partial}{\partial t} - \Delta + 2V - W + V\frac{\partial^2}{\partial q^2} + 4\mu|\phi(q,t)|^2 \right] \phi(q,t) = 0 \,, \tag{D.84}$$

where W is the chain deformation energy,

$$W \equiv \frac{1}{2} \int \mathrm{d}x \left[\omega^{-1} \left(\frac{\partial \mathcal{Q}}{\partial t} \right)^2 + \omega \left(\frac{\partial \mathcal{Q}}{\partial x} \right)^2 \right] \,, \tag{D.85}$$

and μ is a parameter of nonlinearity,

$$\mu \equiv \frac{\chi^2}{\omega} \left(1 - \frac{v^2}{v_0^2} \right)^{-1} \,, \tag{D.86}$$

meaningful only if the inequality opposite to (D.80) holds.

The squared modulus of the exact solution to (D.84), viz.,

$$|\phi(q,t)|^2 = 2\mu \operatorname{sech}^2 \left[\mu(x - vt) \right] \,, \tag{D.87}$$

represents a bump-like soliton with width $\pi\mu^{-1}$, traveling with velocity $v(k)$ equal to the exciton velocity (Fig. D.16). The chain deformation is correlated with the probability amplitude of the exciton localization (D.87), i.e.,

$$\frac{\partial \mathcal{Q}(x,t)}{\partial x} = \frac{2\mu}{\chi}|\phi(q,t)|^2 \,. \tag{D.88}$$

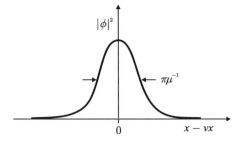

Fig. D.16. Davydov soliton. $|\phi|^2$ represents the probability amplitude of the exciton localization and, simultaneously, the spatial derivative of the chain deformation

The deformation energy (D.85) represents the binding energy of the exciton to the chain deformation and equals

$$W = \chi^4/3\omega^2 V \ . \tag{D.89}$$

It is this large energy that does not allow the exciton to emit phonons and makes the soliton exceptionally stable.

However, vibrational relaxation (D.35), not considered as yet, results in fast decoherence and decay of the exciton–deformation correlations on a timescale κ^{-1}, shorter than a split picosecond. Numerical simulations indicate that, despite the lack of correlations, the exciton itself remains localized and its motion acquires a diffusive character (Cruzeiro-Hannson and Takeno, 1997). This is not to be wondered at, since the excitation transfer between two centers studied at the beginning of this section did indeed have a stochastic character. Such a localized exciton could transfer energy on a timescale shorter than the lifetime of the excited amide vibrational mode, but we do not expect the distance traveled in a diffusive manner during that time to be especially long.

References

1. Aaronson, D.A., Horvath, C.M. (2002): A road map for those who don't know JAK-STAT, Science **296**, 1653–1655
2. Abramowitz, M., Stegun, I.A., Eds (1964): *Handbook of Mathematical Functions* (Natl. Bur. Stand., Dover)
3. Adhikari, B., Hideg, K., Fajer, P.G. (1997): Independent mobility of catalytic and regulatory domains of myosin heads, Proc. Natl. Acad. Sci. USA **94**, 9643–9647
4. Agmon, N., Doster, W., Post, F. (1994): The transition from inhomogeneous to homogeneous kinetics in CO binding to myoglobin, Biophys. J. **66**, 1612–1622
5. Agmon, N., Hopfield, J.J. (1983): CO binding to heme proteins: A model for barrier height distribution and slow conformational changes, J. Phys. Chem. **79**, 2042–2053
6. Agmon, N., Sastry, G.M. (1996): A temperature-dependent effective potential explains CO binding to myoglobin, Chem. Phys. **212**, 207–219
7. Agre, P., et al. (2002): Aquaporin water channels: From atomic structure to clinical medicine, J. Physiol. **542**, 3–16
8. Amitani, I., Sakamoto, T., Ando, T. (2001): Link between the enzymatic kinetics and mechanical behaviour in an actomyosin motor, Biophys. J. **80**, 379–397
9. Astumian, R.D. (1997): Thermodynamics and kinetics of a Brownian motor, Science **276**, 917–922
10. Atkins, P.W. (1998): *Physical Chemistry*, 6th edn. (Oxford University Press, Oxford)
11. Austin, R.H., Beeson, K.W., Eisenstein, L., Frauenfelder, H., Gunsalus, I.C. (1975): Dynamics of ligand binding to myoglobin, Biochemistry **14**, 5355–5373
12. Barber, J. (2002): Photosystem II: A multisubunit membrane protein that oxidizes water, Curr. Opin. Struct. Biol. **12**, 523–530
13. Barber, J., Kühlbrandt, W. (1999): Photosystem II, Curr. Opin. Struct. Biol. **9**, 469–475
14. Baumann, B.A.J., et al. (2004): Myosin regulatory domain orientation in skeletal muscle fibres, Biophys. J. **86**, 3030–3041

15. Baker, J.E., Brust-Mascher, I., Ramachandran, S., LaConte, L.E.W., Thomas, D.D. (1998): A large and distinct rotation of the myosin light chain domain occurs upon muscle contraction, Proc. Natl. Acad. Sci. USA **95**, 2944–2949

16. Baker, J.E., LaConte, L.E.W., Brust-Mascher, I., Thomas, D.D. (1999): Mechanochemical coupling in spin-labeled, active, isometric muscle, Biophys. J. **77**, 2657–2664

17. Baker, J.E., Thomas, D.D. (2000): Thermodynamics and kinetics of a molecular motor ensemble, Biophys. J. **79**, 1731–1736

18. Ban, N., Nissen, P., Hansen, J., Moore, P.B., Steitz, T.A. (2000): The complete atomic structure of the large ribosomal subunit at 2.4-Å resolution, Science **289**, 905–920

19. Beece, D., Eisenstein, L., Frauenfelder, H., Good, D., Marden, M.C., Reinisch, L., Reynolds, A.H., Sorensen, L.B., Yue, K.T. (1980): Solvent viscosity and protein dynamics, Biochemistry **19**, 5147–5157

20. Berger, Ch.J., Thomas, D.D. (1994): Rotational dynamics of actin-bound intermediates of the myosin adenosine triphosphate cycle in myofibrils, Biophys. J. **67**, 250–261

21. Bergethon, P.R., Simons, E.R. (1990): *Biophysical Chemistry. Molecules to Membranes* (Springer, Berlin)

22. Berry, M.V. (1978): Regular and irregular motion, AIP Conference Proceedings **46**, 16–120. Reprinted in: *Hamiltonian Dynamical Systems*, ed. by R.S. MacKay and J.D. Meiss (Hilger, Bristol) pp. 27–53

23. Bibricher, Ch.K., Gardiner, W.C. (1997): Molecular evolution of RNA in vitro, Biophys. Chem. **66**, 179–192

24. Bicout, D.J., Zaccai, D. (2001): Protein flexibility from the dynamical transition: A force constant analysis, Biophys. J. **80**, 1115–1123

25. Binder, K., Young, A.P. (1986): Spin glasses: Experimental facts, theoretical concepts, and open questions, Rev. Mod. Phys. **58**, 801–976

26. Bixon, M., Jortner, J. (1999): Electron transfer – From isolated molecules to biomolecules, Adv. Chem. Phys. **106**, 35–201

27. Blume-Jensen, P., Hunter, T. (2001): Oncogenic kinase signalling, Nature **411**, 355–365

28. Blumen, A., Schnörer, H. (1990): Fractals and related hierarchical models in polymer science, Angew. Chem. Int. Ed. **29**, 113–125

29. Blumenfeld, L.A. (1974): *Problems of Biological Physics* (Nauka, Moscow; English translation, 1981, Springer, Berlin)

30. Bolterauer, H., Limbach, H.J., Tuszyński, J.A. (1999): Models of assembly and disassembly of individual microtubules: Stochastic and average equations, J. Biol. Phys. **25**, 1–22

31. Böckmann, R.A., Grubmüller, H. (2002): Nanosecond molecular dynamics simulation of primary mechanical energy transfer steps in F_1-ATP synthase, Nature Struct. Biol. **9**, 198–202

32. Bränden, M., et al. (2001): On the role of the K-proton transfer pathway in cytochrome c oxidase, Proc. Natl. Acad. Sci. USA **98**, 5013–5018

33. Brejc, K., et al. (2001): Crystal structure of an ACl-binding protein reveals the binding domain of nicotine receptor, Nature **411**, 269–276

34. Brenner, B. (1991): Rapid dissociation and reassociation of actomyosin cross-bridges during force generation: A newly observed facet of cross-bridge action in muscle, Proc. Natl. Acad. Sci. USA **88**, 10490–10469

35. Brillouin, L. (1964): *Scientific Uncertainty and Information Theory* (Academic Press, New York)

36. Briggs, D.E., Crowther, P.R. (1990): *Paleobiology: A Synthesis* (Blackwell, Oxford)

37. Brooks III, C.L., Karplus, M., Pettitt, B.M. (1988): Proteins: A theoretical perspective of dynamics, structure and thermodynamics, Adv. Chem. Phys. **71**, 1–250

38. Bruns, C.M., Karplus, P.A. (1995): Refined crystal structure of spinach ferredoxin reductase at 1.7 Å, J. Mol. Biol. **247**, 125–145

39. Bryngelson, J.D., Onuchic, J.N., Socci, N.D., Wolynes, P.G. (1995): Funnels, pathways, and the energy landscape of protein folding: A synthesis, Proteins **21**, 167–195

40. Bryngelson, J.D., Wolynes, P.G. (1987): Spin glasses and the statistical mechanics of protein folding, Proc. Natl. Acad. Sci. USA **84**, 7524–7528

41. Burgeois, D., et al. (2003): Complex landscape of protein structural dynamics unveiled by nanosecond Laue crystallography, Proc. Natl. Acad. Sci. USA **100**, 8704–8709

42. Burns, L.D., McCormick, J.B., Boroni-Bird, C.E. (2002): Vehicle of charge, Sci. Amer. **287**(4), 64–73

43. Bustamante, C., Keller, D., Oster, G. (2001): The physics of molecular motors, Acc. Chem. Res. **34**, 412–420

44. Bustamante, C., Rivetti, C., Keller, D.J. (1997): Scanning force microscopy under aqueous solutions, Curr. Opin. Struct. Biol. **7**, 709–716

45. Byron, F.W., Fuller, R.W. (1968): *Mathematics of Classical and Quantum Physics*, Vol. 2, Chap. 7 (Addison-Wesley, Reading, Ma)

46. Cairns-Smith, A.G. (1990): *Seven Clues to the Origin of Life* (Cambridge University Press, Cambridge)

47. Callen, H.B. (1985): *Thermodynamics and Introduction to Thermostatistics*, 2nd edn. (Wiley, New York)

48. Cann, R.L., Stoneking, M., Wilson, A.C. (1987): Mitochondrial DNA and human evolution, Nature **325**, 31–36

49. Candau, R., Iorga, B., Travers, F., Barman, T., Lionne, C. (2003): At physiological temperatures the ATPase rates of shortening soleus and psoas myofibrils are similar, Biophys. J. **85**, 3132–3141

50. Cantor, C.R., Schimmel, P.R. (1980): *Biophysical Chemistry* (Freeman, San Francisco)

51. Capaldi, R.A., Aggeler, R. (2002): Mechanism of the F_1F_o-type ATP synthase, a biological rotary motor, TIBS **27**, 154–160

52. Careri, G., Fasella, P. and Gratton, E. (1979): Enzyme dynamics: The statistical physics approach, Ann. Rev. Biophys. Bioeng. **8**, 69–97

53. Careri, G., Gratton, E., and Shyamsunder, E. (1988): Fine structure of the amide I band in acetanilide, Phys. Rev. A **37**, 4048–4051

54. Cech, C.R. (1986): RNA as an enzyme, Sci. Amer. **255**, 76–84

55. Chandler, D. (1987): *Introduction to Modern Statistical Mechanics* (Oxford University Press, New York).

56. Chełminiak, P., Kurzyński, M. (2004): Mean first-passage time for diffusion on fractal lattices with imposed boundary conditions, Physica A **342**, 507–515

57. Chernavsky, D.S., Khurgin, Y.I., Shnol, S.E. (1987): Biophysics (Moscow) **32**, 775–781

58. Clothia, C., Lesk, A.M. (1985): Helix movement in proteins, TIBS **10**, 116–120

59. Columbus, L., Hubbel, W. (2002): A new spin on protein dynamics, TIBS **27**, 288–295

60. Cooke, R., White, H., Pate, E. (1994): A model of the release of myosin heads from actin in rapidly contracting muscle fibres, Biophys. J. **66**, 778–788

61. Cordova, N.J., Ermentrout, B., Oster, G.F. (1992): Dynamics of single-motor molecules: The thermal ratchet model, Proc. Natl. Acad. Sci. USA **89**, 339–343

62. Corrie, J.E.T., et al. (1999): Dynamic measurement of myosin light-chain-domain tilt and twist in muscle contraction, Nature **400**, 425–430

63. Cowen, R. (1990): *History of Life* (Blacklevel, Oxford)

64. Creighton, T.E. (1990): Protein folding, Biochem. J. **270**, 1–16

65. Creighton, T.E. (1992): *Proteins: Structures and Molecular Properties*, 2nd edn. (Freeman, New York)

66. Crick, F. (1981): *Life Itself. Its Origin and Nature* (Simon and Schuster, New York)

67. Crofts, A., Berry, E.A. (1998): Structure and function of the cytochrome bc_1 complex of mitochondria and photosynthetic bacteria, Curr. Opin. Struct. Biol. **8**, 501–509

68. Cruzeiro-Hansson, L., Takeno, S. (1997): Davydov model: The quantum, mixed quantum-classical, and full classical systems, Phys. Rev. E **56**, 894–906

69. Darnell, J., Lodish, H., Berk, S.L., Zipursky, D., Baltimore, D. (1999): *Molecular Cell Biology*, 4th edn. (Freeman, New York)

70. Darrouzet, E., Moser, C.C., Dutton, P.L., Daldal, F. (2001): Large scale domain movement in cytochrome bc_1: A new device for electron transfer in proteins, TIBS **26**, 445–451

71. Davydov, A.S. (1982): *Biology and Quantum Mechanics* (Pergamon, Oxford)
72. Dawkins, R. (1986): *The Blind Watchmaker* (Norton, New York)
73. Dawkins, R. (1989): *The Selfish Gene*, 2nd edn. (Oxford University Press, New York)
74. Dayie, K.W., Wagner, G., Lefèvre, J.F. (1996): Theory and practice of nuclear spin relaxation in proteins, Ann. Rev. Phys. Chem. **47**, 243–282
75. de Groot, H.J.M. (2000): Solid-state NMR spectroscopy applied to membrane proteins, Curr. Opin. Struct. Biol. **10**, 593–600
76. de Groot, B.L., Grubmüller, H. (2001): Water permeation across biological membranes: mechanism and dynamics of aquaporin-1 and GlpF, Science **294**, 2353–2357
77. de Groot, B.L., Grubmüller, H. (2005): The dynamics and energetics of water permeation and proton exclusion in aquaporins, Curr. Opin. Struct. Biol. **15**, 176–183
78. de Groot, B.L., Hayward, S., van Aalten, D.M.F., Amadei, A., Berendsen, H.J.C. (1998): Domain motions in bacteriophage T4 lysozyme: A comparison between molecular dynamics and crystallographic data, Proteins **31**, 116–127
79. Del Giudice, E., Doglia, S., Milani, M. (1988): Spontaneous symmetry breaking and electromagnetic interactions in biological systems, Physica Scripta **38**, 5005–5007
80. Dennis, C. (2002): Gene regulation: The brave new world of RNA, Nature **418**, 122–124
81. Dodson, G, Wlodawer, A. (1998): Catalytic triads and their relatives, TIBS **23**, 347–352
82. Dominiquez, R., Freyzon, Y., Trybus, K.M., Cohen, C. (1998): Crystal structure of a vertebrate smooth muscle myosin motor domain and its complex with the essential light chain: Visualization of the pre-powerstroke state, Cell **94**, 559–571
83. Doniach, S., Eastman, P. (1999): Protein dynamics simulations from nanoseconds to microseconds, Curr. Opin. Struct. Biol. **9**, 157–163
84. Doolittle, W.F. (1999): Phylogenetic classification and the universal tree, Science **284**, 2124–2128
85. Doster, W., Cusack, S., Petry, W. (1989): Dynamical transition of myoglobin revealed by inelastic neutron scattering, Nature **339**, 754–756
86. Doyle, D.A., et al. (1998): The structure of the potassium channel: Molecular basis of K^+ conduction and selectivity, Science **280**, 69–77
87. Duke, T.A.J. (1999): Molecular model of muscle contraction, Proc. Natl. Acad. Sci. USA **96**, 2770–2775
88. Duke, T., Leibler, S. (1996): Motor protein dynamics: A stochastic model with minimal mechanochemical coupling, Biophys. J. **71**, 1235–1247

89. Edman, L., Rigler, R. (2000): Memory landscapes of single-enzyme molecules, Proc. Natl. Acad. Sci. USA **97**, 8266–8271

90. Edman, L., Földes-Papp, Z., Wennmalm, S., Rigler, R. (1999): The fluctuating enzyme: a single molecule approach, Chem. Phys. **247**, 11–22

91. Eigen, M. (1964): Proton transfer, acid–base catalysis, and enzymatic hydrolysis, Angew. Chem. Internat. Edit. **3**, 1–18

92. Eigen, M. (1993): Virial quasispecies, Sci. Amer. **269**(1), 42–51

93. Eigen, M., Rigler, R. (1994): Sorting single molecules: application to diagnostics and evolutionary biotechnology, Proc. Natl. Acad. Sci. USA **91**, 5740–5747

94. Eigen, M., McCaskill, J., Schuster, P. (1989): The molecular quasi-pecies, Adv. Chem. Phys. **75**, 149–263

95. Eisenberg, D., Kautzmann, W. (1969): *The Structure and Properties of Water* (Oxford University Press, Oxford)

96. Elston, T., Wang, H., Oster, G. (1998): Energy transduction in ATP synthase, Nature **391**, 510–513

97. Englander, S.W., Sosnick, T.R., Englander, J.J., Mayne, L. (1996): Mechanisms and uses of hydrogen exchange, Curr. Opin. Struct. Bio. **6**, 18–23

98. *Enzyme Nomenclature* (1992): (Academic, San Diego). World Wide Web version including supplements: www.chem.qmul.ac.uk/iubmb/enzyme/ prepared by G.P. Moss

99. Ermler, U., Fritzsch, G., Buchanan, S.K., Michel, H. (1994): Structure of the photosynthetic reaction centre from *Rhodobacter sphaeroides* at 2.65-Å resolution: Cofactors and protein–cofactor interactions, Structure **2**, 925–936

100. Evan, G.I., Vousden, K.H. (2001): Proliferation, cell cycle and apoptosis in cancer, Nature **411**, 342–349

101. Evans, D.F., Wennerström, H. (1999): *The Colloidal Domain: Where Physics, Chemistry, Biology, and Technology Meet*, 2nd edn. (Wiley, New York)

102. Fain, B. (1980): *Theory of Rate Processes in Condensed Media*, Lecture Notes in Chemistry, Vol. 20 (Springer, Berlin)

103. Fenimore, P.W., Frauenfelder, H., McMahon, B.H., Parak, F.G. (2002): Slaving: Solvent fluctuations dominate protein dynamics and functions, Proc. Natl. Acad. Sci. USA **99**, 16047–16051

104. Ferreira, K., Iverson, T.M., Maghlaoui, K., Barber, J., Iwata, S. (2004): Architecture of the photosynthetic oxygen-evolving center, Science **308**, 1832–1838

105. Fersht, A. (1999): *Structure and Function in Protein Science* (Freeman, New York)

106. Feynman, R.P., Leighton, R.B., Sands, M. (1966): *The Feynman Lectures on Physics*, Vol. I, Chap. 46 (Addison-Wesley, Reading, Ma)

107. Finer, J.T., Simmons, R.M., Spudich, J.A. (1994): Single myosin molecule dynamics: Piconewton forces and nanometre steps, Nature **368**, 113–119

108. Fisher, M.E., Kolomeisky, A.B. (1999): The force exerted by a molecular motor, Proc. Natl. Acad. Sci. USA **96**, 6597–6602

109. Fisher, T.E., Marszalek, P., Fernandez, J.M. (2000): Stretching single molecules into novel conformations using the atomic force microscope, Nature Struct. Biol. **7**, 719–724

110. Fischmann, T.O., et al. (1991): Crystallographic refinement of the three-dimensional structure of the FabD1.3-lysozyme complex at 2.5-Å resolution, J. Biol. Chem. **266**, 12915–12920

111. Flomenbom, O., et al. (2005): Stretched exponential decay and correlations in the catalytic activity of fluctuating single lipase molecules, Proc. Natl. Acad. Sci. USA **102**, 2368–2372

112. Flomenbom, O., Klafter, J., Szabo, A. (2005): What can one learn from two-state single-molecule trajectories?, Biophys. J. **88**, 3780–3783

113. Frauenfelder, H., Sligar, S.G., Wolynes, P.G. (1991): The energy landscapes and motions of proteins, Science **254**, 1598–1602

114. Frauenfelder, H., Wolynes, P.G., Austin., R.H. (1999): Biological physics, Rev. Mod. Phys. **71**, S419–S430

115. Friedrichs, M.S., Goldstein, R.A., Wolynes, P.G. (1991): Generalized protein tertiary structure recognition using associative memory Hamiltonians, J. Mol. Biol. **222**, 1013–1023

116. Fröhlich, H., Kremer, F., Eds. (1988): *Coherent Excitations in Biological Systems* (Springer, Berlin)

117. Fujiyoshi, Y., Mitsuoka, K., de Groot, B.L., Philippsen, A., Grubmüller, H., Agre, P., Engel, A. (2002): Structure and function of water channels, Curr. Opin. Struct. Biol. **12**, 509–515

118. Fuller, G.M., Shields, D. (1998): *Molecular Basis of Medical Cell Biology* (Appleton & Lange)

119. Funantsu, T., Harada, Y., Tokunaga, M., Saito, K., Yanagida, T. (1995): Imaging of single fluorescent molecules and individual ATP turnovers by single myosin molecules in aqueous solution, Nature **380**, 451–453

120. Gabdoulline, R.R., Wade, R.C. (2002): Biomolecular diffusional association, Curr. Opin. Struct. Biol. **12**, 204–213

121. Garcia, A.E. (1992): Large-amplitude nonlinear motions in proteins, Phys. Rev. Lett. **68**, 2696–2699

122. Garcia, A.E., Blumenfeld, R., Hummer, G., Krumhansl, J.A. (1997): Multi-basin dynamics of a protein in a crystal environment, Physica D **107**, 225–239

123. Garcia-Viloca, M., Gao, J., Karplus, M., Truhlar, D.G. (2004): How enzymes work: Analysis by modern rate theory and computer simulations, Science **303**, 186–195

124. Gardiner, C.W. (1983): *Handbook of Stochastic Methods* (Springer, Berlin)

125. Garel, T., Orland, H. (1988): Mean-field model for protein folding, Europhys. Lett. **6**, 307–310

126. Gauld, S.B., Dal Porto, J.M., Cambier, J.C. (2002): B cell antigen receptor signalling: Roles in cell development and disease, Science **296**, 1641–1642

127. Gavish, B. (1986): Molecular dynamics and the transient strain model of enzyme catalysis. In: *The Fluctuating Enzyme*, ed. by G.R. Welch (Wiley, New York) pp. 267–339

128. Gavish, B., Werber, M.M. (1979): Viscosity-dependent structural fluctuations in enzyme catalysis, Biochemistry **18**, 1269–1275

129. Geeves, M.A., Holmes, K.C. (1999): Structural mechanism of muscle contraction, Annu. Rev. Biochem. **68**, 687–728

130. Gennis, R.B. (1989): *Biomembranes: Molecular Structure and Function* (Springer, Berlin)

131. Gesteland, R.E., Cech, T.R., and Atkins, J.F., Eds. (1999): *The RNA World* (Cold Spring Harbor Laboratory, New York)

132. Gibbons, C., Montgomery, M.G., Leslie, A.G., Walker, J.E. (2000): The structure of the central stalk in bovine F1-ATPase at 2.4-Å resolution, Nature Struct. Biol. **7**, 1055–1061

133. Goldbeter, A. (1996): *Biochemical Oscillations and Cellular Rhythms: The Molecular Bases of Periodic and Chaotic Behaviour* (Cambridge University Press, Cambridge)

134. Goldstein, L.S., Philip, A.V. (1999): The road less travelled: Emerging principles of kinesin motor utilization, Annu. Rev. Cell Dev. Biol. **15**, 141–181

135. Götze, W., Sjögren, L. (1988): Scaling properties in supercooled liquids near the glass transition, J. Phys. C **21**, 3407–3421

136. Götze, W., Sjögren, L. (1992): Relaxation processes in supercooled liquids, Rep. Prog. Phys. **55**, 241–376

137. Grabe, M., Wang, H., Oster, G. (2000): The mechanochemistry of V-ATPase proton pumps, Biophys. J. **78**, 2798–2818

138. Greider, C.W., Blackburn, E.H. (1996): Telomeres, telomerase and cancer, Sci. Amer. **274**, 92–97

139. Grigorieff, N. (1999): Structure of the respiratory NADH:ubiquinone oxidoreductase, Curr. Opin. Struct. Biol. **9**, 476–483

140. Gross, E.L. (1996): Plastocyanin structure, location, diffusion and electron transfer mechanisms. In: *Oxygenic Photosynthesis: The Light Reactions*, ed. by D.R. Ort and C.F. Yokun, Advances in Photophysics, Vol. 4 (Kluwer, Dordrecht)

141. Grubmüller, H. (1995): Predicting slow structural transitions in macromolecular systems: Conformational flooding, Phys. Rev. E **52**, 2893–2906

142. Hagan, S., Hameroff, S.R., Tuszyński, J. (2002): Quantum computation in brain microtubules. Decoherence and biological feasibility, Phys. Rev. E **65**, 61901, 1–10

143. Hahn, W.C., Weinberg, R.A. (2002): Modelling the molecular circuitry of cancer, Nature Rev. Cancer **2**, 331–341

144. Hajdu, J. (2000): Single-molecule X-ray diffraction, Curr. Opin. Struct. Biol. **10**, 569–573

145. Haken, H. (1990): *Synergetics. An Introduction. Nonequilibrium Phase Transitions and Self-Organization in Physics, Chemistry and Biology*, 3rd edn. (Springer, Berlin)

146. Hamley, I.W. (2000): *Introduction to Soft Matter: Polymers, Colloids, Amphiphiles and Liquid Crystals* (Wiley, Chichester)

147. Hanahan, D., Weinberg, R.A. (2002): The hallmarks of cancer, Cell **100**, 57–70

148. Hansson, T., Oostenbrink, Ch., van Gunsteren, W.F. (2002): Molecular dynamics simulations, Curr. Opin. Struct. Biol. **12**, 190–196

149. Haran, G., Haas, E., Szpikowska, B.K., Mas, M.T. (1992): Domain motions in phosphoglycerate kinase: Determination of interdomain distance distribution by site-specific labeling and time-resolved fluorescence energy transfer, Proc. Natl. Acad. Sci. USA **89**, 11764–11768

150. Harris, D.A. (1995): *Bioenergetics at a Glance* (Blackwell Science, Cambridge)

151. Hawking, S.W. (1996): *A Brief History of Time* (Bantam Books, New York)

152. Hayward, S., Berendsen, H.J.C. (1998): Systematic analysis of domain motions in proteins from conformational change: New results on citrate synthase and T4 lysozyme, Proteins **30**, 144–154

153. Hänggi, P., Talkner, P., Borkovec, M. (1990): Reaction-rate theory: Fifty years after Kramers, Rev. Mod. Phys. **62**, 251–341

154. He, Z.H., Bottinelli, R., Pellegrino, M.A., Ferenczi, M.A., Reggiani, C. (2000): ATP consumption and efficiency of human single muscle fibers with different myosin isoform composition, Biophys. J. **79**, 945–961

155. Heald, R., Walczak, C.E. (1999): Microtubule-based motor function in mitosis, Curr. Opin. Struct. Biol. **9**, 268–274

156. Heathcote, P., Fyfe, K.P., Jones, M.R. (2002): Reaction centres: The structure and evolution of biological solar power, TIBS **27**, 79–87

157. Hernández, G., Jenney, F.E., Adams, M.W.W., LeMaster, D.M. (2000): Milisecond time scale conformational flexibility in a hyperthermophile protein at ambient temperature, Proc. Natl. Acad. Sci. USA **97**, 3166–3170

158. Heymann, J.B., Müller, D.J., Mitsuoka, K., Engel, A. (1997): Electron and atomic force microscopy of membrane proteins, Curr. Opin. Struct. Biol. **7**, 543–549

159. Higuchi, H., Goldman, Y.E. (1995): Sliding distance per ATP molecule hydrolyzed by myosin heads during isotonic shortening of skinned muscle fibres, Biophys. J. **69**, 1491–1507

160. Hill, T.L. (1989): *Free Energy Transduction and Biochemical Cycle Kinetics* (Springer, New York)

161. Hille, B. (2001): *Ionic Channels of Excitable Membranes*, 3rd edn. (Sinauer, Sunderland)

162. Hillson, N., Onuchic, J.N., Garcia, A.E. (1999): Pressure-induced protein-folding/unfolding kinetics, Proc. Natl. Acad. Sci. USA **96**, 14848–14853

163. Hodsdon, M.E., Cistola, D.P. (1997): Ligand binding alters the backbone mobility of intestinal fatty acid-binding protein as monitored by ^{15}N NMR relaxation and ^{1}H exchange, Biochemistry **36**, 2278–2290

164. Hoff, A.J., Deisenhofer, J. (1997): Photophysics of photosynthesis, Phys. Repts. **287**, 1–247

165. Holmes, K.C., Angert, I., Kull, F.J., Jahn, W., Schroder, R.R. (2003): Electron cryomicroscopy shows how strong binding of myosin to actin releases nucleotide, Nature **425**, 423–427

166. Hoofnagle, A.N., Resing, K.A., Goldsmith, E.J., Ahn, N.G. (2001): Changes in protein conformational mobility upon activation of extracellular regulated protein kinase-2 as detected by hydrogen exchange, Proc. Natl. Acad. Sci. USA **98**, 956–961

167. Hopfield, J.J. (1982): Neural networks and physical systems with emergent collective computational abilities, Proc. Natl. Acad. Sci. USA **79**, 2554–2558

168. Houdusse, A., Sweeney, H.L. (2001): Myosin motors: Missing structures and hidden springs, Curr. Res. Struct. Biol. **11**, 182–194

169. Houdusse, A., Szent-Györgyi, A.G., Cohen, C. (2000): Three conformational states of scallop myosin S1, Proc. Natl. Acad. Sci. USA **97**, 11238–11243

170. Howard, J. (2001): *Mechanics of Motor Proteins and the Cytoskeleton* (Sinauer, Sunderland)

171. Huang, K. (1987): *Statistical Mechanics*, 2nd edn. (Wiley, New York)

172. Hu, X., Damjanović, A., Ritz, T., Schulten, K. (1998): Architecture and mechanism of the light-harvesting apparatus of purple bacteria, Proc. Natl. Acad. Sci. USA **95**, 5935–5941

173. Hummer, G., Garcia, A.E., Garde, S. (2000): Conformational diffusion and helix formation kinetics, Phys. Rev. Lett. **85**, 2637–2640

174. Huxley, A.F. (1957): Progr. Biophys. Molec. Biol. **7**, 255–318

175. Huxley, H.E. (1969): Science **164**, 1356–1366

176. Huxley, A.F., Simmons, R.M. (1971): Proposed mechanism of force generation in striated muscle, Nature **233**, 533–538

177. Ishijima, A., Yanagida, T. (2001): Single molecule nanobioscience, TIBS **26**, 438–444

178. Ishijima, A., Kojima, H., Funatsu, T., Tokunaga, M., Higuchi, K., Tanaka, H., Yanagida, T. (1998): Simultaneous observation of individual ATPase and mechanical events by a single myosin molecule during interaction with actin, Cell **92**, 161–171

179. Jardetzky, O. (1996): Protein dynamics and conformational transitions in allosteric proteins, Prog. Biophys. Molec. Biol. **65**, 171–219

180. Jaynes, E.T. (1978): Where do we stand on maximum entropy? In: *The Maximum Entropy Formalism*, ed. by R.D. Levine and M. Tribus (Cambridge, Massachusetts) pp. 211–314

181. Jäckle, J. (1986): Models of the glass transition, Rep. Prog. Phys. **49**, 171–232

182. Jones, R.A.L. (2004): *Soft Machines, Nanotechnology and Life* (Oxford University Press, Oxford)

183. Jordan, P., et al. (2001): Three-dimensional structure of cyanobacterial photosystem I at 2.5-Å resolution, Nature **411**, 909–917

184. Jülicher, F., Ajdari, A., Prost, J. (1997): Modelling molecular motors, Rev. Mod. Phys. **69**, 1269–1281

185. Jülicher, F., Bruinsma, R. (1998): Motion of RNA polymerase along DNA: A stochastic model, Biophys. J. **74**, 1169–1185

186. Kabsch, W., Mannherz, H.G., Suck, D. Pai, E.F., Holmes, K.C. (1990): Atomic structure of the actin:DNAse I complex, Nature **347**, 37–44

187. Kauffman, S.A. (1993): *The Origins of Order: Self-Organization and Selection in Evolution* (Oxford University Press, Oxford)

188. Keller, D., Bustamante, C. (2000): The mechanochemistry of molecular motors, Biophys. J. **78**, 541–556

189. Kern, D., Zuiderweg, E.R.P. (2003): The role of dynamics in allosteric regulation, Curr. Opin. Struct. Biol. **13**, 748–757

190. Kidera, A., Gō, N. (1990): Refinement of protein dynamic structure: Normal mode refinement, Proc. Natl. Acad. Sci. USA **87**, 3718–3722

191. Kigawa, T., et al. (2001): Selenomethionine incorporation into a protein by cell-free synthesis, J. Struct. Funct. Genom. **2**, 87–114

192. Kitamura, K., Tokunaga, M., Iwane, A.H., Yanagida, T. (1999): A single myosin head moves along an actin filament with regular steps of 5.3 nanometers, Nature **397**, 129–134

193. Kitao, A., Gō, N. (1999): Investigating protein dynamics in collective coordinate space, Curr. Opin. Struct. Biol. **9**, 164–169

194. Kitao, A., Hayward, S., Gō, N. (1998): Energy landscape of a native protein: Jumping-among-minima model, Proteins **33**, 496–517

195. Kittel, C. (1996): *Introduction to Solid State Physics* (Wiley, New York)

196. Knaff, D.B. (1996): Ferredoxin and ferredoxin-dependent enzymes. In: *Oxygenic Photosynthesis: The Light Reactions*, ed. by D.R. Ort and C.F. Yokun, Advances in Photophysics, Vol. 4 (Kluwer, Dordrecht) pp. 333–361

197. Kondepudi, D., Prigogine, I. (1998): *Modern Thermodynamics* (Wiley, Chichester)

198. Kongsaeree, P., Cerione, R.A., Clardy, J.C. The structure determination of Cdc42Hs and GDP complex, to be published

199. Konno, K., Ue, K., Khoroshev, M., Martinez, H., Ray, B., Morales, M.F. (2000): Consequences of placing an intramolecular crosslink in myosin S1, Proc. Natl. Acad. Sci. USA **97**, 1461–1466

200. Kováč, L. (1987): Overview: Bioenergetics between chemistry, genetics, and physics, Curr. Top. Bioenerg. **15**, 331–372

201. Kriegl, J.M., Nienhaus, G.U. (2004): Structural, dynamic, and energetic aspects of long-range electron transfer in photosynthetic reaction centers, Proc. Natl. Acad. Sci. USA **101**, 123–128

202. Krimm, S., Bandekar, J. (1986): Adv. Protein Chem. **38**, 181–360

203. Kull, F.J., Sablin, E.P., Lau, R., Fletterick, R.J., Vale, R.D. (1996): Crystal structure of the kinesin motor domain reveals a structural similarity to myosin, Nature **380**, 550–555

204. Kurisu, G., Zhang, H., Smith, J.L., Cramer, W.A. (2003): Structure of the cytochrome b_6f complex of oxygenic photosynthesis: Tuning the cavity, Science **302**, 1009–1014

205. Kurzyński, M. (1990): Chemical reactions from the point of view of statistical thermodynamics far from equilibrium, J. Chem. Phys. **93**, 6793–6799

206. Kurzyński, M. (1993): Enzymatic catalysis as a process controlled by protein conformational relaxation, FEBS Lett. **328**, 221–224

207. Kurzyński, M. (1994): A model of reversible reaction with slow intramolecular relaxation, J. Chem. Phys. **101**, 255–264

208. Kurzyński, M. (1997a): Diffusion on fractal lattices. A statistical model of chemical reactions involving proteins, Acta Phys. Polon. B **28**, 1853–1889

209. Kurzyński, M. (1997b): Protein machine model of enzymatic reactions gated by enzyme internal dynamics, Biophys. Chem. **65**, 1–28

210. Kurzyński, M. (1998): A synthetic picture of intramolecular dynamics of proteins. Towards a contemporary statistical theory of biochemical processes, Progr. Biophys. Molec. Biol. **69**, 23–82

211. Kurzyński, M., Palacz, K., Chełminiak, P. (1998): Time course of reactions controlled and gated by intramolecular dynamics of proteins: Predictions of the model of random walks on fractal lattices, Proc. Natl. Acad. Sci. USA **95**, 11685–11690

212. Kurzyński, M., Chełminiak, P. (2003): Mean first-passage time in stochastic theory of biochemical processes. Application to actomyosin molecular motor, J. Stat. Phys. **110**, 137–181

213. Kurzyński, M., Chełminiak, P. (2004): Stochastic action of actomyosin motor, Physica A **336**, 123–132

214. Kühlbrandt, W. (1994): Structure and function of the plant light-harvesting complex, LHC-II, Curr. Opin. Struct. Biol. **4**, 519–528

215. Lee, A.G. (2002): A calcium pump made visible, Curr. Opin. Struct. Biol. **12**, 547–554

216. Lee, A.L., Wand, A.J. (2001): Microscopic origins of entropy, heat capacity and the glass transition in proteins, Nature **411**, 501–504

217. Lee, C.L., Lin, C.T., Stell, G., Wang, J. (2003): Diffusion dynamics, moments, and distribution of first-passage time on the protein-folding energy landscape, with applications to single molecules, Phys. Rev. E **67**, 041905

218. Liebovitch, L.S., Fischbarg, J., Koniarek, J.P., Todorova, I., Wang, M. (1987): Fractal model of ion-channel kinetics, Biochim. Biophys. Acta **896**, 173–180

219. Lilley, D.M.J. (2003): The origins of RNA catalysis in ribozymes, TIBS **28**, 495–501

220. Linderstrom-Lang, K.U., Schellman, J.A. (1959): In: *The enzymes*, ed. by P.D. Boyer, H. Lardy and K. Myrback, 2nd edn., Vol. 1 (Academic, New York) p. 443

221. Lionne, C., Travers, F., Barman, T. (1996): Mechanochemical coupling in muscle: Attempts to measure simultaneously shortening and ATPase rates of myofibrils, Biophys. J. **70**, 887–895

222. Liu, X., Pollack, G.H. (2004): Stepwise sliding of single actin and myosin filaments, Biophys. J. **86**, 353–358

223. Lymn, R.W., Taylor, E.W. (1971): Mechanism of adenosine triphosphate hydrolysis by actomyosin, Biochemistry **10**, 4617–4624

224. Lyubarev, A.E., Kurganov, B.I. (1989): Supramolecular organization of tricarboxilic acid cycle enzymes, BioSystems **22**, 91–102

225. Ma, Y.Z., Taylor, E.W. (1994): Kinetic mechanism of myofibril ATPase, Biophys. J. **66**, 1542–1553

226. Mandelbrot, M.M. (1982): *The Fractal Geometry of Nature* (Freeman, San Francisco)

227. Mandelkov, E., Mandelkow, E.M. (1994): Microtubule structure, Curr. Opin. Struct. Biol. **4**, 171–179

228. Manz, J., Wöste, L., Eds (1995): *Femtosecond Chemistry*, Vols. 1 and 2, especially Chap. 18 by Elsässer, T.: Femtosecond intramolecular proton transfer in the condensed phase (VCH, Weinheim)

229. Marcus, R.A., Sutin, N. (1985): Electron transfer in chemistry and biology, Biochim. Biophys. Acta **811**, 265–322

230. Margittai, M., et al. (2003): Single-molecule fluorescence resonance energy transfer reveals a dynamic equilibrium between closed and open conformations of syntaxin 1, Proc. Natl. Acad. Sci. USA **100**, 15516–15521

231. Margulis, L. (1981): *Symbiosis in Cell Evolution. Life and Its Environment on the Early Earth* (Freeman, San Francisco)

232. Margulis, L. (1998): *Symbiotic Planet: A New Look at Evolution* (Sciencewriters, Amherst)

233. Marx, J. (2003): How cells step out, Science **302**, 214–216

234. Mattick, J.S. (2004): The hidden genetic program of complex organisms, Sci. Amer., October

235. May, V., Kühn, O. (2004): *Charge and Energy Transfer Dynamics in Molecular Systems*, 2nd edn. (Wiley-VCH, Berlin)

236. Mayr, E., Provine, W.E., Eds (1980): *The Evolutionary Synthesis, Perspectives on the Unification of Biology* (Harvard University Press, Cambridge)

237. McCammon, J.A., Harvey, S.C. (1987): *Dynamics of Proteins and Nucleic Acids* (Cambridge University Press, Cambridge)

238. McCammon, J.A., Gelin, B.R., Karplus, M. (1977): Dynamics of folded proteins, Nature **267**, 585–590

239. McClare, C.W.F. (1971): J. Theor. Biol. **30**, 1–34

240. McSween, Jr., H.Y. (1995): *Stars to Planets* (St. Martin's Griffin)

241. Mehta, A.D., Rief, M., Spudich, J.A., Smith, D.A., Simmons, R.M. (1999): Single molecule biomechanics with optical methods, Science **283**, 1689–1695

242. Mermall, V., Post, P.L., Mooseker, M.S. (1998): Unconventional myosins in cell movement, membrane traffic and signal transduction, Science **279**, 527–533

243. Michel, H. (1998): The mechanism of proton pumping by cytochrome c oxidase, Proc. Natl. Acad. Sci. USA **95**, 12819–12824

244. Millar, D.P. (1996): Time-resolved fluorescent spectroscopy, Curr. Opin. Struct. Biol. **6**, 637–642

245. Mirny, L., Shakhnovich, E. (2001): Protein folding theory: From lattice to all-atom models, Annu. Rev. Biomol. Struct. **30**, 361–396

246. Miyazawa, A., Fujiyoshi, Y., Unwin, N. (2003): Structure and gating mechanism of the acetylcholine receptor pore, Nature **424**, 949–955

247. Møller, J.V., Juul, B., le Maire, M. (1996): Structural organization, ion transport, and energy transduction of P-type ATPases, Biochem. Biophys. Acta **1286**, 1–51

248. Morals-Cabral, J.H., Zhou, Y., MacKinnon, R. (2001): Energetic optimization of ion conduction rate through the K^+ selectivity filter, Science **414**, 37–42

249. Montroll, E.W., West, B.J. (1987): On an enriched collection of stochastic processes. In: *Fluctuation Phenomena*, ed. by E.W. Montroll and J.L. Lebowitz (North-Holland, Amsterdam) pp. 61–206

250. Morange, M. (2000): *A History of Molecular Biology* (Harvard University Press, Cambridge)

251. Mukamel, S. (1995): *Principles of Nonlinear Optics and Spectroscopy* (Oxford University Press, New York)

252. Murata, K., Mitsuoka, K., Hirai, T., Walz, T., Agre, P., Heymann, J.B., Engel, A., Fujiyoshi, Y. (2002): Structural determinants of water permeation through aquaporin-1, Nature **407**, 599–605

253. Nailor, C.E., Eaton, J.T., Howells, A., Justin, N., Moss, D.S., Titball, R.W., Basak, A.K. (1998): Structure of the key toxin in gas gangrene, Nature Struct. Biol. **5**, 738–746

254. Nakayama, T., Yakubo, K., Orbach, R.L. (1994): Dynamical properties of fractal networks: Scaling, numerical simulations, and physical realizations, Rev. Mod. Phys. **66**, 381–443

255. Nelson, P. (2004): *Biological Physics. Energy, Information, Life* (Freeman, New York)

256. Neumann, J. von (1963): The general and logical theory of automata. In: *J. von Neumann, Collected Works*, Vol. 5, ed. by A.H. Tamb (MacMillan, New York) pp. 288–328

257. Neves, S.R., Ram, P.T., Iyengar, R. (2002): G protein pathways, Science **296**, 1636–1638

258. Newton, R.G. (1993): *What Makes Nature Tick?* (Harvard University Press, Cambridge)

259. Nicolis, G., Prigogine, I. (1977): *Self-Organization in Non-Equilibrium Systems* (Wiley, New York)

260. Nield, J., Rizkallah, P.J., Barber, J., Chayen, N.E. (2003): The 1.45 Å three-dimensional structure of C-phycocyanin from the thermophilic cyanobacterium *Synechococcus elongatus*, J. Struct. Biol. **141**, 149–155

261. Nienhaus, G.U., Müller, J.D., McMahon, B.H., Frauenfelder, H. (1997): Exploring the conformational energy landscape of proteins, Physica D **107**, 297–311

262. Nieselt-Struwe, K. (1997): Graphs in sequence spaces – A review of statistical geometry, Biophys. Chem. **66**, 111–131

263. Nishikawa, T., Gō, N.(1987): Normal modes of vibration in bovine pancreatic trypsin inhibitor and its mechanical property, Proteins **2**, 308–329

264. Nitao, L.K., Reisler, E. (2000): Actin and temperature effects on the cross-linking of the SH1–SH2 helix in myosin subfragment 1, Biophys. J. **78**, 3072–3080

265. Nitschke, W., Rutherford, A.W. (1991): Photosynthesis reaction centres: Variations on a common structural scheme, TIBS **16**, 241–245

266. Nogales, E., Wolf, S.G., Downing, K.H. (1998): Structure of the alpha beta tubulin dimer by electron crystallography, Nature **391**, 199–203

267. Northrup, S.H., Hynes, J.T. (1980): The stable states picture of chemical reactions, J. Chem. Phys. **73**, 2700–2714

268. Nymeyer, H., Garcia, A.E., Onuchic, J.N. (1998): Folding funnels and frustrations in off-lattice minimalist protein landscapes, Proc. Natl. Acad. Sci. USA **95**, 5921

269. Ochoa, G., Corey, M. (1995): *The Timeline Book of Science* (Stonesong, New York)

270. Onuchic, J.N., Luthey-Schulten, Z., Wolynes, P.G. (1997): Theory of protein folding: The energy landscape perspective, Annu. Rev. Phys. Chem. **48**, 545–600

271. Onuchic, J.N., Wolynes, P.G. (2004): Theory of protein folding, Curr. Opin. Struct. Biol. **14**, 70–75

272. Onuchic, J.N., Wolynes, P.G., Luthey-Schulten, Z., Socci, N.D. (1995): Toward an outline of the topography of a realistic protein-folding funnel, Proc. Natl. Acad. Sci. USA **92**, 3626–3630

273. Orgel, L.E. (1998): The origin of life. A review of facts and speculations, TIBS **23**, 491–495

274. Ormos, P., Ansari, A., Braunstein, D., Cowen, B.R., Frauenfelder, H., Hong, M.K., Iben, I.E.T., Sauke, T.B., Steinbach, P., and Young, R.D. (1990): Inhomogeneous broadening in spectral bands of carbonmonoxymyoglobin. The connection between spectral and functional heterogeneity, Biophys. J. **57**, 191–199

275. Ormos, P., Száraz, S., Cupane, A., Nienhaus, G.U. (1998): Structural factors controlling ligand binding to myoglobin: A kinetic hole-burning study, Proc. Natl. Acad. Sci. USA **95**, 6762–6767

276. Ostap, E.M., Barnett, V.A., Thomas, D.D. (1995): Resolution of three structural states of spin-labeled myosin in contracting muscle, Biophys. J. **69**, 177–188

277. Palmer, R.G. (1982): Broken ergodicity, Adv. Phys. **31**, 669–735

278. Pappa, H., et al. (1998): Crystal structure of the C2 domain from protein kinase C-delta, Structure **6**, 885–894

279. Parak, F.G. (2003a): Physical aspects of protein dynamics, Rep. Prog. Phys. **66**, 103–129

280. Parak, F.G. (2003b): Proteins in action: The physics of structural fluctuations and conformational changes, Curr. Opin. Struct. Biol. **13**, 552–557

281. Pate, E., Franks-Skiba, K., Cooke, R. (1998): Depletion of phosphate in active muscle fibers probes actomyosin states with the powerstroke, Biophys. J. **74**, 369–380

282. Pate, E., White, H., Cooke, R. (1993): Determination of the myosin step size from mechanical and kinetic data, Proc. Natl. Acad. Sci. USA **90**, 2451–2455

283. Pauling, L., Pauling, P. (1975): *Chemistry* (Freeman, San Francisco)

284. Penrose, O. (1979): Foundations of statistical mechanics, Rep. Prog. Phys. **42**, 1937–2006

285. Perelson, A.S., Weisbuch, G. (1997): Immunology for physicists, Rev. Mod. Phys. **69**, 1219–1267

286. Perutz, M.F., Mathews, F.S. (1966): J. Mol. Biol. **21**, 199–202

287. Peskin, C.S., Oster, G. (1995): Coordinated hydrolysis explains the mechanical behaviour of kinesin, Biophys. J. **68**, 202s–211s

288. Phillips, M., Fletterick, R.J. (1992): Proteases, Curr. Opin. Struct. Biol. **2**, 713

289. Piazzesi, G., Lombardi, V. (1995): A cross-bridge model that is able to explain mechanical and energetic properties of shortening muscle, Biophys. J. **68**, 1966–1979

290. Post, F., Doster, W., Karvounis, G., Settles, M. (1993): Structural relaxation and nonexponential kinetics of CO binding to horse myoglobin. Multiple flash photolysis experiments, Biophys. J. **64**, 1833–1842

291. Prigogine, I. (1980): *From Being to Becoming: Time and Complexity in the Physical Sciences* (Freeman, San Francisco)

292. Prigogine, I., Stengers, I. (1984): *Order Out of Chaos: Man's New Dialogue with Nature* (Random House, New York)

293. Prigogine, I., Stengers, I. (1997:) *The End of Certainty. Time, Chaos and the New Laws of Nature* (The Free Press, New York)

294. Privalov, P.L. (1989): Thermodynamic problems of protein structure, Annu. Rev. Biophys. Biophys. Chem. **18**, 47–69

295. Qian, H. (1997): A simple theory of motor protein kinetics and energetics I, Biophys. Chem. **67**, 263–267

296. Qian, H. (2000): A simple theory of motor protein kinetics and energetics II, Biophys. Chem. **83**, 35–43

297. Qin, F., Auerbach, A., Sachs, F. (1996): Estimating single-channel kinetic parameters from idealized patch-clamp data containing missed events, Biophys. J. **70**, 264–280

298. Quelle, D.E., Zindy, F., Ashmun, R.A., Sherr, C.J. (1995): Alternative reading frames of the INK4A tumor suppressor gene encode two unrelated proteins capable of inducing cell cycle arrest, Cell **83**, 993–1000

299. Ramakrishnan, V., White, S.W. (1998): Ribosomal protein structures. Insights into the architecture, machinery and evolution of ribosome, TIBS **23**, 208–212

300. Rammal, R., Toulouse, G., Virasoro, M.A. (1986): Ultrametricity in physics, Rev. Mod. Phys. **58**, 765–788

301. Rastogi, V.K., Girvin, M.E. (1999): Structural changes linked to proton translocation by subunit c of the ATP synthase, Nature **402**, 263–268

302. Rayment, I., Holden, H.M., Whittaker, M., Yohn, C.B., Lorenz, M., Holmes, K.C., Miligan, R.A. (1993): Structure of the actin-myosin complex and its implications for muscle contraction, Science **261**, 58–65

303. Rayment, I., Rypniewski, W.R., Schmidt-Base, K., Smith, R., Tomchick, D.R., Benning, M.M., Winkelmann, D.A., Wesenberg, G., Holden, H.M. (1993): Three-dimensional structure of myosin subfragment-1: A molecular motor, Science **261**, 50–58

304. Rejto, P.A., Freer, S.T. (1996): Protein conformational substates from X-ray crystallography, Prog. Biophys. Molec. Biol. **66**, 167–196

305. Ren, G., Reddy, V.S., Cheng, A., Melnyk, P., Mitra, A.K. (2001): Visualization of a water-selective pore by electron crystallography in vitreous ice, Proc. Natl. Acad. Sci. USA **98**, 1398–1403

306. Renger, T., May, V., Kühn, O. (201): Ultrafast excitation energy transfer dynamics in photosynthetic pigment–protein complexes, Phys. Repts. **343**, 137–254

307. Rock, H.S., et al. (2001): Myosin VI is a processive motor with a large step size, Proc. Natl. Acad. Sci. USA **98**, 13655–13659

308. Rojewska, D., Elber, R. (1990): Molecular dynamics study of secondary structure motions in proteins: Application to myohemerythrin, Proteins **7**, 265–279

309. Rottenberg, H. (1998): The generation of proton electrochemical potential gradient by cytochrome c oxidase, Biochem. Biophys. Acta **1364**, 1–16

310. Sablin, E.P., Fletterick, R.J. (2001): Nucleotide switches in molecular motors: Structural analysis of kinesins and myosins, Curr. Opin. Struct. Biol. **11**, 716–724

311. Sackmann B., Naher. E. (1995): *Single-Channel Recording*, 2nd edn. (Plenum, New York)

312. Sansom, M.P.S., et al. (1989): Markov, fractal, diffusion, and related models of ion channel gating. A comparison with experimental data from two ion channels, Biophys. J. **56**, 1229–1243

313. Sato, C., et al. (2001): Nature, **409**, 1047–1050

314. Schlitter, J. (1988): Chem. Phys. **120**, 187–197

315. Schluenzen, F., Tocilj, A., Zarivach, R., Harms, J., Gluehmann, M., Janell, D., Bashan, A., Bartels, H., Agmon, I., Franceschi, F., Yonath, A. (2000): Structure of functionally activated small ribosomal subunit at 3.3 angstrom resolution, Cell **102**, 615–623

316. Schopf, J.W. (1999): *Cradle of Life. The Discovery of Earth's Earliest Fossils* (Princeton University Press, Princeton)

317. Schreiber, G. (2002): Kinetic studies of protein–protein interactions, Curr. Opin. Struct. Biol. **12**, 41–47

318. Schrödinger, E. (1967): *What is Life?* (Cambridge University Press, Cambridge)

319. Schulten, K., Tesch, M. (1991): Coupling of protein motion to electron transfer: Molecular dynamics and stochastic quantum mechanics study of photosynthetic reaction centers, Chem. Phys. **158**, 421–446

320. Schultz, B.E., Chan, S.I. (2001): Structures and proton-pumping strategies of mitochondrial respiratory enzymes, Annu. Rev. Biophys. Biomol. Struct. **30**, 23–65

321. Schulz, G.E., Schirmer, R.H. (1979): *Principles of Protein Structure* (Springer, New York)

322. Selkov, E.E. (1968): Eur. J. Biochem. **4**, 79–89

323. Shah, A., et al. (1999): Photovoltaic technology, Science **285**, 692–698

324. Shaitan, K.V., Rubin, A.B. (1982): Conformational dynamics of proteins and simple molecular 'machines', Biophysics (USSR) **27**, 386–390

325. Shakhnovich, E.I., Gutin, A.M. (1989): Formation of unique structure in polypeptide chains. Theoretical investigation with the aid of a replica approach, Biophys. Chem. **34**, 187–199

326. Singer, A.L., Koretzky, G.A. (2002): Control of T cell function by positive and negative regulators, Science **296**, 1639–1640

327. Soler-Lopez, M., et al. (2004): Structure of an activated *Dictyostelium* STAT in its DNA-unbound form, Molecular Cell **13**, 791

328. Solomon, E.P., Berg, L.R., Martin, D.W. (2004): *Biology*, 7th edn. (Brooks/Cole)

329. Sørensen, T.L.-M., Møller, J.V., Nissen, P. (2004): Phosphoryl transfer and calcium ion occlusion in the calcium pump, Science **304**, 1672–1676

330. Spudich, J.A. (1994): How molecular motors work, Nature **372**, 515–518

331. Squire, J.M. (1997): Architecture and function in the muscle sarcomere, Curr. Opin. Struct. Biol. **7**, 247–257

332. Srere, P.A. (1987): Complexes of sequential metabolic enzymes, Annu. Rev. Biochem. **56**, 89–124

333. Stivers, J.T., at al. (1996): ^{15}N NMR relaxation studies of free and inhibitor-bound 4-oxalocrotonate tautomerase: Backbone dynamics and entropy changes of an enzyme upon inhibitor binding, Biochemistry **35**, 16036–16047

334. Stehle, R., Brenner, B. (2000): Cross-bridge attachment during high-speed active shortening of skinned fibers of the rabbit psoas muscle: Implications for cross-bridge action during maximum velocity of filament sliding, Biophys. J. **78**, 1458–1473

335. Stein, D.L., Ed. (1992) *Spin Glasses in Biology* (World Scientific, Singapore)

336. Steitz, T.A., Moore, P.B. (2003): RNA, the first macromolecular catalyst: The ribosome is a ribozyme, TIBS **28**, 411–418

337. Stock, D., et al. (2000): The rotary mechanism of ATP synthase, Curr. Opin. Struct. Biol. **10**, 672–679

338. Stryer, L., Berg, J.M., Tymoczko, J.M. (2002): *Biochemistry*, 5th edn. (Freeman, New York)

339. Sussman, J.L., Holbrook, S.R., Warrant, R.W., Church, G.M., Kim, S.H. (1978): Crystal structure of yeast phenylalanine transfer RNA. I. Crystallographic refinement, J. Mol. Biol. **123**, 607–630

340. Suzuki, S., Sung, H.C., Hayashi, M., Lin, S.H. (1995): Femtosecond processes and ultrafast biological electron transfer, Physica A **221**, 15–29

341. Suzuki, Y., Yasunaga, T., Ohkura, R., Wakabayashi, T., Sutoh, T. (1998): Swing of the lever arm of a myosin motor at the isomerization and phosphate-release step, Nature **396**, 380–383

342. Tanaka, H., et al. (2002): The motor domain determines the large step of myosin V, Nature **415**, 192–195

343. Tanford, C. (1980) *The Hydrophobic Effect: Formation of Micelles and Biological Membranes*, 2nd edn. (Wiley, New York)

344. Terada, T.P., Sasai, M., Yomo, T. (2002): Conformational change of the actomyosin complex drives the multiple stepping movement, Proc. Natl. Acad. Sci. USA **99**, 9202–9206

345. Thüne, T., Badger, J. (1995): Thermal diffusive X-ray scattering and its contribution to understanding protein dynamics, Prog. Biophys. Molec. Biol. **63**, 251–276

346. Titball, R.W. (1993): Bacterial phospholipases C, Microbiol. Rev. **57**, 347–366

347. Tournier, A.L., Smith, J. (2003): Principal components of the protein dynamical transition, Phys. Rev. Lett. **20**, 208106

348. Toyoshima, C., Nakasako, M., Nomura, H., Ogawa, H. (2000): Crystal structure of the calcium pump of sarcoplasmic reticulum at 2.6-Å resolution, Nature **405**, 647–655

349. Troyer, J.M., Cohen, F.E. (1995): Protein conformational landscapes: Energy minimization and clustering of a long molecular dynamics trajectory, Proteins **23**, 97–110

350. Tsallis, C. (1999): Nonextensive statistics: Theoretical, experimental and computational evidence and connections, Braz. J. Phys. **29**, 1–35

351. Tsallis, C. (2001): Nonextensive statistical mechanics and thermodynamics: Historical background and present status. In: *Nonextensive Statistical Mechanics and Its Applications*, ed. by S. Abe and Y. Okamoto, Lecture Notes in Physics, Vol. 560 (Springer, Berlin) pp. 3–98

352. Tsukihara, T., et al. (1996): Structure of bovine heart cytochrome c oxidase at the fully oxidated state, Science **272**, 1136–1142

353. Tuszyński, J.A., Brown, J.A., Hawrylak, P. (1998): Dielectric polarization, electrical conduction, information processing and quantum computation in microtubules. Are they plausible?, Phil. Trans. R. Soc. Lond. A **356**, 1897–1926

354. Tuszyński, J.A., Hameroff, S.R., Satarić, M., Trpišová, Nip, M.L.A. (1995): Ferroelectric behavior in microtubule dipole lattices: Implications for information processing, signaling, and assembly/disassembly, J. Theor. Biol. **174**, 371–380

355. Tuszynski, J.A., Kurzynski, M. (2003): *Molecular Biophysics* (CRC Press, Boca Raton)

356. Unwin, N. (2005): Refined structure of the nicotinic acetylcholine receptor at 4 Å resolution, J. Mol. Biol. **346**, 967

357. Vale, R.D., Milligan, R.A. (2000): The way things move: Looking under the hood of molecular motors, Science **288**, 88–95

358. Vale, R.D., Oosawa, F. (1990): Protein motors and Maxwell's demons: Does mechanochemical transduction involve a thermal ratchet?, Adv. Biophys. **26**, 97–134

359. van Andel, T.H. (1994) *New Views on an Old Planet. A History of Global Change* (Cambridge University Press, Cambridge)

360. Van de Peer, Y., De Rijk, P., Wuyts, J., Winkelmans, T., De Wachter, R. (2000): The European small subunit ribosomal RNA database, Nucleic Acids Res. **28**, 175–176

361. van Kampen, N.G. (2001): *Stochastic Processes in Physics and Chemistry*, rev. edn (North-Holland, Amsterdam)

362. Veigel, C., et al. (1999): The motor protein myosin-I produces its working stroke in two steps, Nature **398**, 530–533

363. Verkhiver, G.M., et al. (2002): Complexity and simplicity of ligand–macromolecule interactions: The energy landscape perspective, Curr. Opin. Struct. Biol. **12**, 197–203

364. Volkmann, N., Hanein, D. (2000): Actomyosin: Law and order in motility, Curr. Opin. Cell Biol. **12**, 26–34

365. Wales, D.J., Doye, J.P.K., Miller, M.A., Mortenson, P.N., Walsh, T.R. (2000): Energy Landscapes: From clusters to biomolecules, Adv. Chem. Phys. **115**, 1–111

366. Walker, M., Zhang, X-Z., Jiang, W., Trinick, J., White, H.D. (1999): Observation of transient disorder during myosin subfragment-1 binding to actin by stopped-flow fluorescence and millisecond time resolution cryomicroscopy: Evidence that the start of the crossbridge power stroke in muscle has variable geometry, Proc. Natl. Acad. Sci. USA **96**, 465–470

367. Wang, H.Y., Elston, T., Mogilner, A., Oster, G. (1998): Force generation in RNA polymerase, Biophys. J. **74**, 1186–1202

368. Wang, H.Y., Oster, G. (1998): Energy transduction in the F_1 motor of ATP synthase, Proc. Natl. Acad. Sci. USA **91**, 5740–5747

369. Wang, Z., Tang, J., Norris, J.R. (1992): The general treatment of dynamic solvent effects in electron transfer at high temperature, J. Chem. Phys. **97**, 7251–7256

370. Warshaw, D.M., et al. (1998): Myosin conformational states determined by single fluorophore polarization, Proc. Natl. Acad. Sci. USA **95**, 8034–8039

371. Weinberg, R.A. (1996): How cancer arises, Sci. Amer., September

372. Weinberg, S. (1980): *The First Three Minutes* (Bantam, New York)

373. Weiss, S. (1999): Fluorescence spectroscopy of single biomolecules, Science **283**, 1676–1683

374. Welch, G.R., Ed. (1985): *Organized Multienzyme Systems* (Academic Press, Orlando)

375. Westerhoff, H.V., van Dam, K. (1987): *Thermodynamics and Control of Biological Free-Energy Transduction* (Elsevier, Amsterdam)

376. Whittaker, M., Wilson-Kubalek, E.M., Smith, J.E., Faust, L., Milligan, R.A., Sweeney, H.L. (1995): A 34-Å movement of smooth muscle myosin on ADP release, Nature **378**, 748–751

377. Wider, G., Wüthrich, K. (1999): NMR spectroscopy of large molecules and multimolecular assemblies in solution, Curr. Opin. Cell Biol. **9**, 594–601

378. Widom, B. (1965): Reaction kinetics in stochastic models, Science **148**, 1555–1560

379. Widom, B. (1971): Molecular transitions and chemical reaction rates, J. Chem. Phys. **55**, 44–52

380. Williams, R.J.P. (1979): The conformational properties of proteins in solution, Biol. Rev. **54**, 389–437

381. Williams, R.J.P. (1993): Are enzymes mechanical devices?, TIBS **18**, 115–117

382. Wilson, R.J. (1996): *Introduction to Graph Theory*, 4th edn. (Addison Wesley Longman, London)

383. Winkler-Oswatitsch, R., Eigen. M. (1992): *Steps Towards Life: A Perspective on Evolution* (Oxford University Press, Oxford)

384. Woese, C. (1998): The universal ancestor, Proc. Natl. Acad. Sci. USA **95**, 6854–6859

385. Woledge, R.C., Curtin, N.A., Homsher, E. (1985): *Energetic Aspects of Muscle Contraction* (Academic, London)

386. Wüthrich, K. (1986): *NMR of Proteins and Nucleic Acids* (Wiley, New York)

387. Xiao, M., et al. (1998): Conformational changes between the active-site and regulatory light chain of myosin as determined by luminescence resonance energy transfer: The effect of nucleotides and actin, Proc. Natl. Acad. Sci. USA **95**, 15309–15314

388. Xie, X., et al. (1994): Structure of the regulatory domain of scallop myosin at 2.8-Å resolution, Nature **368**, 306–312

389. Xiong, J., et al. (2000): Molecular evidence for the early parallel transfer of genes, Science **289**, 1727

390. Xu, J., Root, D.D. (2000): Conformational selection during weak binding at the actin and myosin interference, Biophys. J. **79**, 1498–1510

391. Yakushevich, L.V. (1993): Nonlinear dynamics of biopolymers: Theoretical models, experimental data, Q. Revs. Biophys. **26**, 201–223

392. Yamamoto, T. (1960): Quantum statistical mechanical theory of the rate of exchange chemical reactions in the gas phase, J. Chem. Phys. **33**, 281–289

393. Yennawar, N.H., Yennawar, H.P., Farber, G.K. (1994): X-ray crystal structure of gamma-chymotrypsin in hexane, Biochemistry **33**, 7326–7336

394. Yildiz, A., Tomishige, M., Vale, R.D., Selvin, P.R. (2004): Kinesin walks hand-over-hand, Science **303**, 676–678

395. Yomosa, S. (1985): Solitary excitations in muscle proteins, Phys. Rev. A **32**, 1752–1758

396. Zhang, Z., et al. (1998): Electron transfer by domain movement in cytochrome bc_1, Nature **392**, 677–684

397. Zheng, J. H., et al. (1993): 2.2-angstrom refined crystal-structure of the catalytic subunit of cAMP-dependent protein-kinase complexed with ATP and a peptide inhibitor, Acta Crystallogr. D **49**, 362–365

398. Zhou, M., Morals-Cabral, J.H., Mann, S., MacKinnon, R. (2001): Potassium channel receptor site for the inactivation gate and quaternary amine inhibitors, Science **411**, 657–661

399. Zwanzig, R. (1988): Diffusion in a rough potential, Proc. Natl. Acad. Sci. USA **85**, 2029–2030

400. Zwanzig, R. (2001): *Nonequilibrium Statistical Mechanics* (Oxford University Press, New York)

Index

Printing: Krips bv, Meppel
Binding: Stürtz, Würzburg